PROJECT MANAGEMENT TOOLBOX

Second Edition

Russ J. Martinelli
Dragan Z. Milosevic

WILEY

Cover image: Wiley
Cover design: abstract background © yewkeo/iStockphoto

Published by John Wiley & Sons, Inc., Hoboken, New Jersey
Published simultaneously in Canada

Library of Congress Cataloging-in-Publication Data:

ISBN 978-1-118-97312-7 (hard back), 978-1-118-97321-9 (ePDF), 978-1-118-97320-2 (ePUB) and 978-1-119-17482-0 (oBook)

FSC www.fsc.org
MIX
Paper from responsible sources
FSC® C013604

CONTENTS

PREFACE

Much has changed since the publication of the first edition of this book, as the field of project management (PM) is continually evolving. Part of that evolution has involved a new approach for selecting project management tools from an ad-hoc "choose as you use" approach to a more systematic approach of creating a PM Toolbox that can be applied to many project situations. From this perspective, we feel fortunate to have been part of the recent project management evolution.

We also feel fortunate to have had the opportunity to work firsthand with many project managers and project office directors as they set about the task of creating their initial PM Toolboxes based on the teachings of this text. Our personal understanding of how project management is evolving and how it affects the needs for PM tools has been greatly enhanced. This new understanding became the basis for the changes introduced in this second edition.

The most significant changes in this edition are in four areas. First, we have focused the content of the book on the fundamental project management practice areas to create more depth in content. Next, we have maintained the traditional view of project management tools but have also provided a contemporary set of tools that reflect the changes in PM practices. Then, to strengthen an area that has created some of the most positive reader feedback, we have enhanced the various tips, tricks, and examples found throughout the book. Finally, we worked to create a stronger message concerning the importance of creating a PM Toolbox that enables stronger alignment between business strategy and project execution, between strategic goals and project deliverables, and between the work of senior leaders and project managers.

This book has established itself as both educational lecture material and an industry practice reference, which we hope to maintain with this second edition. Our heartfelt thanks to the existing and future readers of this book; we hope you find it both enjoyable and useful to read.

Of course, we would like to hear from you directly and get your feedback at www.programmanagement-academy.com. Supplemental materials and templates can be found on our web site as well.

ACKNOWLEDGMENTS

We would like to thank the many people who have helped in making this book a reality.

To our contributing authors, whose subject matter expertise is appreciated: Debra Lavell, Jim Waddell, Tim Rahschulte, Peerasit Patanakul, James Henrey, and Jeffrey Leach.

To the team at John Wiley & Sons, who continue to provide outstanding support and guidance. In particular, we want to thank our executive editor, Margaret Cummins, and our assistant editor, Amanda Shettleton. Your continued partnership and collaboration is greatly valued.

To our many colleagues and coworkers who have contributed to the concepts presented in this work in many ways.

To our families who provide the support and encouragement necessary to complete the writing process.

We are truly blessed to be associated with such a wonderful and supportive community of people!

PART

I

The PM Toolbox

1

INTRODUCTION TO THE PM TOOLBOX

Conventional wisdom holds that project management (PM) tools are enabling devices that assist a project manager in reaching an objective or, more specifically, a project deliverable or outcome. While this traditional role of PM tools is more than meaningful, we believe that there is greater opportunity to provide value to an organization and its project managers. In particular, each PM tool can be part of a set of tools that makes up a project manager's PM Toolbox.

The PM Toolbox, then, serves a higher purpose: (1) to increase efficiency of the project players, (2) to provide the right information to support problem-solving and decision-making processes, and (3) to help establish and maintain alignment among business strategy, project strategy, and project execution outcomes.

Project management tools support the practices, methods, and various processes used to effectively manage a project.[1] They are enabling devices for the primary players on a project: the project manager, the specialists who make up the project team, the executive leadership team, and the governance body.

PM tools include procedures, techniques, and job aids by which a project deliverable is produced or project information is created. Similarly, *A Guide to the Project Management Body of Knowledge* and other sources use the phrase "tools and techniques" in place of what we define as PM tools.[2]

PM tools may be either qualitative or quantative in nature. To illustrate, consider two examples: the team charter and Monte Carlo analysis. They differ in the type of information they process. The team charter provides a systematic procedure to process qualitative information about authorizing a team to implement a project. Monte Carlo analysis is a risk-planning tool that uses an algorithm to quantify risks. The heart of both the qualitative and quantitative groups of tools—and all PM tools belong to one of these groups—is in their systematic procedure.

Note that we don't talk about software tools here. True, many PM tools that we discuss in this book exist in a software format. However, our focus is not on tool formats. Rather, we concentrate on the substance of PM tools: the use of tools to manage projects more effectively and efficiently.

The design of a PM Toolbox should mirror the approach an organization takes for establishing standardized project management methodologies and processes. A highly standardized set of methods and processes will in turn require an equally high level of standardization of the PM Toolbox. Less standardization introduces more variability in PM Toolbox design and use, and therefore more possibility for inconsistent results.

In practice, as organizations strive to grow and mature, project execution efficiency and repeatability become increasingly important as the leaders of the organization look for consistency in achieving business results. This means that project managers must be armed with the right tools—those that support the business strategy, project strategy, and project management methodologies and processes. It also means that the same tools should be used across the gamut of projects with limited exceptions.

Standardization of a firm's PM Toolbox does not happen overnight. Rather, it is an evolutionary process. In a practical sense, PM Toolboxes will look quite ad hoc at first. The tendency is to begin building the PM Toolbox with existing tools due to a project manager's familiarity with them. So the early-stage PM Toolbox has more to do with familiarity of use than with standardization. As a firm begins to mature its project management practices, standardization of methodologies and processes begins to take hold. This is when the PM Toolbox also begins to become more standardized, as well as more aligned with the project strategy and the business strategy of the firm.

Construction of a PM Toolbox should be systematically driven, meaning that PM tools are a vital part of an organization's overall project execution mechanism. However, project execution must first be aligned to company strategy to be most effective. When this is the case, the PM Toolbox becomes strategically aligned as well, as illustrated in Figure 1.1.

As illustrated by the downward arrow, business strategy drives the project strategy, which in turn drives methods and processes, which influences the PM Toolbox design. For this downward flow to work, the PM Toolbox supports the project management methodology and processes implemented by an organization. The methodology and process in turn helps to implement the project strategy, which supports and is aligned to the business strategy of a company in its quest for growth (upward arrow).

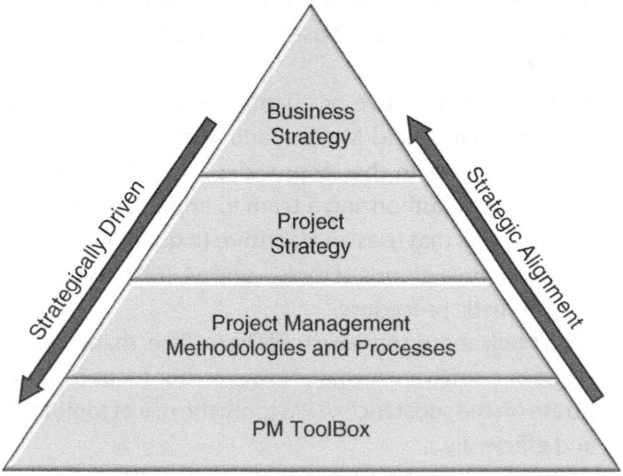

Figure 1.1: Strategically Aligned PM Toolbox

ENABLING BUSINESS AND PROJECT STRATEGY

Looking at how projects and the management of those projects support the business strategy of an organization is critical to understanding the strategic importance of the PM Toolbox. Since alignment between the PM Toolbox and business strategy is driven from the top of the pyramid (Figure 1.1), we start from there.

Historically, the strategic management and project management functions and processes of a company have been defined and performed as independent entities, each with its own purpose and set of activities.[3] Companies have come to realize, however, that the time, money, and human effort invested in refining and improving each of these independent functions and processes have not brought them closer to turning their ideas into positive business results. Increasingly, this fact is leading business leaders to the realization that strategy and project execution can no longer remain independent if they wish to repeatedly achieve their desired business benefits and business value. Rather, they must be integrated so that the formation of strategy and the execution of strategy are tightly aligned.

Use of the Porter model is a simple approach to demonstrate at a high level the alignment among business strategy, project strategy, and PM Toolbox design (Figure 1.2).[4]

The essence of business strategy lies in devising ways to create both short-term and long-term growth and sustainability for an enterprise. To equip themselves with the opportunity, companies rely on their organizational resources.[5] Visualize, for example, project management as an organizational resource. Useful for this visualization can be the framework of generic strategies, shown in Figure 1.2.[6]

		DIFFERENTATION	
		Low	High
COST	High		**Business Strategy:** Differentiation **Project Strategy:** Fast Cycle Time **Business Strategy:** Schedule Planning, Schedule Management, Risk Management
COST	Low	**Business Strategy:** Low-Cost **Project Strategy:** Cost Containment **Business Strategy:** Cost Planning, Cost Management	**Business Strategy:** Best - Cost **Project Strategy:** Cost and Quality **Business Strategy:** Cost Planning, Cost Management, Performance

Figure 1.2: Aligning Business Strategy, Project Strategy, and PM Toolbox

To understand the effect of business strategy, let's use Porter's model as an example to evaluate the strategies for three companies producing liquid crystal display (LCD) projectors.

The core of differentiation strategies (high differentiation/high cost quadrant in Figure 1.2) is an ability to offer customers something different from a company's competitors. This may include fast time to market (which we used as an example in Figure 1.2), high quality, innovative technology, special features, superior service, and so on. When striving for product superiority, LCD projector companies pursuing these strategies provide cutting-edge features that customers are willing to pay a premium price for.

Companies focusing on low-cost strategies aim at establishing a sustainable cost advantage over rivals (low-cost/low-differentiation quadrant). The intent is to use the low-cost advantage as a strategy to underprice rivals and take market share away from them. Another strategic option is to earn a higher profit by selling at the going market price. This is pursued with a good basic product that has few frills and continuous quest for cost reduction without giving up quality and essential features.

Best-cost companies combine upscale features with low cost (low-cost/high-differentiation quadrant). This should lead to superior value by meeting or exceeding customer expectations on product features and surpassing their expectations on price. At the same time, the aim is to become the low-cost provider of a product that has good-to-excellent features and use that cost advantage to underprice rivals with comparable features. Because such a company has the lowest cost compared with similarly positioned rivals, the strategy is called a best-cost strategy.

In Figure 1.2, the blank quadrant of high cost/low differentiation is not a viable option in the quest for short- or long-term business growth.

Now, let us use the model to see how the business strategy shapes project strategies. Examples of three companies—Sirius, Park, and Prima—will help us illustrate the point.

Sirius's business strategy is one of differentiation. The strategy uses innovation and time-to-market speed as competitive advantages. The business strategy is executed through product development projects, whose job it is to roll out new advanced LCD projector chips faster and faster. This is where the project strategy comes into play, focusing on overlap across project phase activities to shorten the project cycle time and the management of risk due to a number of new technologies. The project strategy emphasizes time and risk management.

Park's business strategy is quite different from that of Sirius. Instead of the differentiation and time-to-market emphasis that Sirius relentlessly pursues, Park has set out to become the low-cost leader in the industry. To develop the ability and become the leader in the industry, Park has had to employ a project strategy to continuously lower project and product cost goals. Part of that effort has been perfecting project cost planning and management methods for managing cost cutting within the projects. Nurturing these competencies supports Park's low-cost advantages.

The strategies of Sirius and Park exploit their schedule- and cost-focused project strategies, respectively. In contrast, Prima relies on a best-cost strategy. The goal is to have the best cost relative to competitors whose LCD projectors are of comparable quality. Accordingly, their project strategies emphasize high quality and low development cost. Project management methodologies and practices aim to accomplish cost and quality goals through excellent cost and performance management.

These examples provide a context from which we can construct a common base of understanding. First, companies select business strategies as a means of operating within the markets that they serve. Although each type of strategy has the same goal—to create and sustain business growth, ways to accomplish the goal differ. One company builds a strategy on the basis of differentiation, another on low cost, and still another on a best-cost approach.

Second, companies align their project strategies with their business strategy. Consequently, in the case of Sirius, Park, and Prima, each company's project strategy is focused differently: schedule focus (Sirius), cost focus (Park), and cost/quality focus (Prima).

Any of these approaches is, of course, acceptable. What is critically important, however, is that care should be taken to ensure that the projects and their associated project strategies align to and support the business strategies of the enterprise.

PROJECT MANAGEMENT METHODOLOGIES AND PROCESSES

As an organization grows and becomes more mature in its practices, the need for standardization of methodologies and processes invariably arises. This is due to increased need for repeatability and consistency of project outcomes.

But what does standardization really mean? If we seek a standardized sequence of project activities (that culminate in project deliverables and outcomes), then *standardized* means the degree of absence of variation in implementing such activities.[7] Let's use Figure 1.3 to explain this.

At one extreme, there may be a complete variation in the project management methods and processes. Literally, every time a process is performed, it is performed in a different way. Obviously, 100 percent variation means that standardization is equal to zero. This is often referred to as an ad-hoc approach. At the other extreme, methods and processes may be 100 percent standardized, meaning a process is performed in the same way every time. In this case, variation is zero percent. Between the two extremes lies a continuum of methodologies and processes with different ratios of standardization and variation.

Take, for example, process S on the *x*-axis of Figure 1.3, one of the many possible PM methods (e.g., the critical chain scheduling methodology). The degree of standardization and the degree of variation add up to 100 percent. If we go down the diagonal line to other methods, the degree of standardization will increase, and the degree of variation will decrease; but their sum will remain constant at 100 percent. Moving up the diagonal line will lead to a higher variation and lower standardization, still with the sum of 100 percent. Using plain language, the lower the variation, the higher the standardization; and the more varied the implementation of project activities, the less standardized they are.

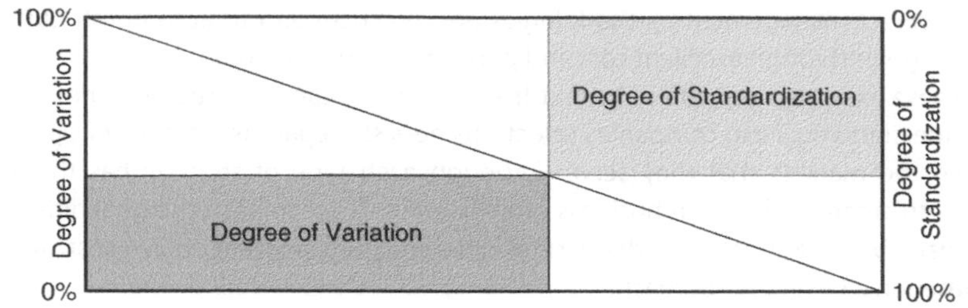

Figure 1.3: Continuum of PM Methods and Processes

This means that organizations have a host of options when developing their methodologies and processes—they can be more standardized or less standardized. The rationale behind standardization is to create a predictable process that prevents activities from differing completely from project to project, and from project manager to project manager. Put simply, standardization saves project players the trouble of reinventing a new method and process for each individual project.[8] As a result, the process is repeatable despite changes in customer expectations or management turnover. The higher the standardization, the higher the repeatability.

When establishing standardized methodologies and processes, organizations have a host of options to choose from. Some companies adopt one of the well-known project management methodologies such as the PMBoK, PRINCE2, or Agile Scrum. Many establish their own methodologies and processes based on how they normally perform their project work. Still others combine approaches by utilizing elements of the standard methodologies and then augmenting and customizing based on the culture of their organization.

The decision about how much to standardize project management methodologies and processes is a decision about the ratio of standardization and variation (popularly called *flexibility*). It is driven by business strategy and by the types of projects needed to realize the business strategy. Generally, projects of higher certainty will strive for higher levels of standardization and lower levels of flexibility. According to experts, the majority of projects in organizations belong to this group.[9] Projects that face high uncertainty require lower standardization and higher flexibility.

Selecting PM tools one at a time demands a substantial amount of resources and expertise. It is not reasonable to presume that each project manager—especially if he or she is less than experienced, as is the case with many—would have the resources and expertise to quickly, smoothly, and consistently select his or her own set of tools. Rather, such managers end up struggling to find the right PM tools and how to use them, introducing variability in results. In contrast, having a standardized PM Toolbox capable of supporting the methods and processes results in minimum variation (see Table 1.1).

Often, project managers assume that the PM Toolbox is of a one-size-fits-all nature. This, of course, is incorrect. The PM Toolbox can come in many sizes, shapes, and flavors. Logically, this is an issue related to the project management methodology and types of projects the methodology serves. Since the PM Toolbox is aligned with the PM methodology used, it is understandable that the level of standardization of the methodology

Table 1.1: One-Tool-at-a-Time versus the PM Toolbox Approach

	Impact on SPM Process	
Requirement	**One-Tool-at-a-Time**	**PM Toolbox**
Speed	Lower	Higher
Repeatability	Less repeatable	More repeatable
Concurrency	Less likely	More likely

impacts the standardization level of the PM Toolbox. For example, a methodology that is highly standardized will probably be supported by a highly standardized PM Toolbox.

Regardless of whether an organization's project management methods and processes are standardized, flexible, or semiflexible, a PM Toolbox needs to be designed so that it aligns with both the PM methods and processes employed as well as the strategy of the project and the business strategies driving the need for the project. To accomplish this, a process for selecting and adapting the PM Toolbox is needed.

CONSTRUCTING AND ADAPTING A PM TOOLBOX

PM tools serve two roles. First, in their conventional role, the tools are enabling devices for reaching a project deliverable. Second, in their new role, they serve as basic building blocks to construct the PM toolbox.

There are three major steps, each including several substeps, in constructing and adapting a PM Toolbox for specific projects or a project organization (Figure 1.4):

1. Secure strategic alignment
2. Customize the PM Toolbox
3. Improve continuously

As detailed in the previous sections, aligning the PM Toolbox with the organization's business strategy tells us in broad terms what categories of project management tools to select. This alignment drives the next step—customization of the PM Toolbox—by

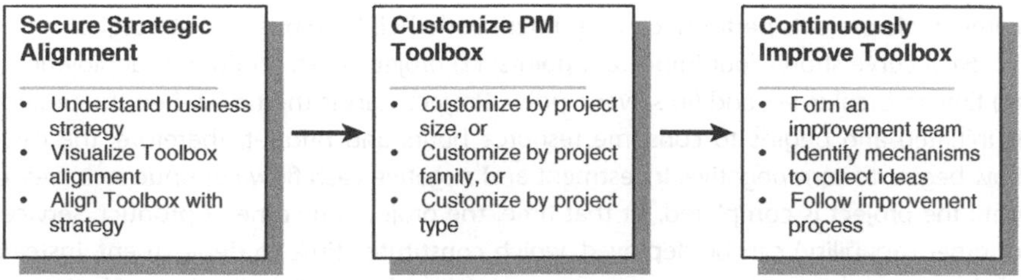

Figure 1.4: Steps for Constructing and Adapting a PM Toolbox

selecting specific tools to use on the projects. The deployment of the PM Toolbox in real-world projects will reveal its glitches and generate new learning, which leads to the third step—continuous improvement of the toolbox. Details about each step follow.

Secure Strategic Alignment

One of the primary purposes of the PM Toolbox is to enable the implementation of projects that affect the organization's strategic business goals. To make this purpose happen, the PM Toolbox needs to be in alignment with both the business and project strategies, as we discussed earlier in this chapter. To be successful in designing the Toolbox, therefore, project managers must have an understanding of the business strategy, at least knowing if their company follows a fundamental strategy of being a market leader, a market follower, a cost leader, or a customer service leader. However, many of them do not have this level of understanding. Why? Among many reasons for this is the fact that in many organizations, strategy formulation and implementation is viewed as the executive's domain. They are tasked with charting the business strategy for the enterprise. Project managers often are not in a position to access this knowledge or show little interest in gaining it. Project managers need to be tenacious by probing and digging to comprehend the strategic reasons for executing the projects they are in charge of, even if the strategy is not communicated to them.

This lack of strategic knowledge can create substantial obstacles for project managers and will limit the strategic alignment of their PM Toolbox. To remove the obstacles, project managers need to have conversations with top managers and convince them that business strategy is key to planning and implementing projects and that project managers need this knowledge in order to secure expected returns on their projects. Our mandate is simple: Gain an understanding of your organization's business strategy, or designing the toolbox will be like shooting an arrow into the fog—we don't know where the target is or whether we hit it.

Visualizing Alignment

Part of understanding how a toolbox should align to business strategy is the ability to clearly visualize the relationship. Earlier in the chapter, we laid the foundation for the alignment by using examples of three companies—Sirius, Park, and Prima—to illustrate how the PM Toolbox can be focused to support business strategies.

To visualize this alignment, in Figure 1.5 we show what we conveniently call investment curves—a more precise term is the net present value curves—for three comparable projects performed in alignment with their base business strategies.

Each curve shows four important points: (1) project start, (2) time to deployment, (3) time to breakeven, and (4) salvage point. Project start is the time when the project is initiated and begins to consume resource hours and budget; therefore, the cash flow begins to turn negative. Investment and negative cash flow continue to increase until the project is completed. At that time, the project outcome (a product, service, or other capability) can be deployed, which constitutes time to deployment. Instead of time to deployment, some project managers prefer the term *project cycle time* or, simply, *project completion*. Note that *negative* cash flow usually reaches its peak at the

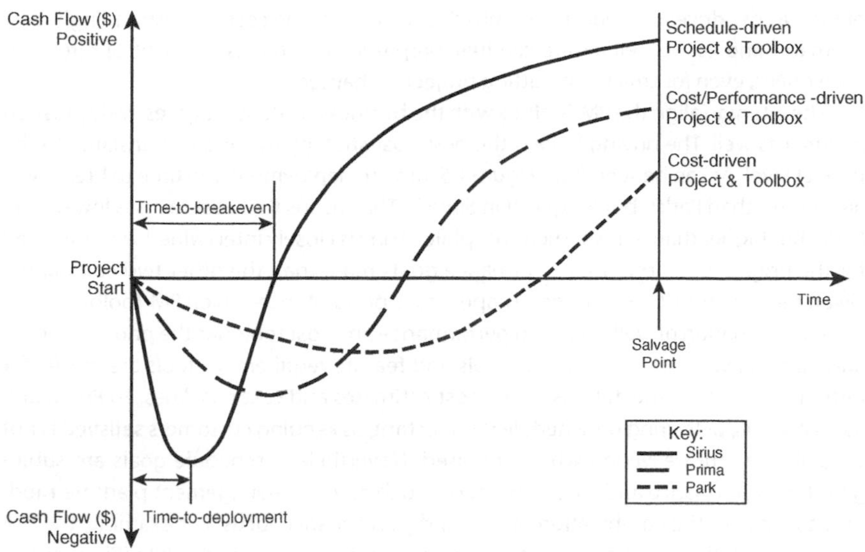

Figure 1.5: Visualizing a Strategy-Driven Toolbox

time-to-deployment point. After that, the use of the project output begins to generate returns (revenue, cost savings, efficiency gains), and the curve begins to turn upward.

Hopefully, the upward trend will continue until at least the time-to-breakeven point is reached. This is the point where all investments in the project are equal to returns generated by the use of the project output. Beyond that point, the cash flow turns positive and typically continues to do so until the project output is salvaged.

We use the curves to explain the nature of the PM Toolbox's alignment with the business strategy for each of the three companies discussed earlier. Consider, for example, Sirius. A primary element of Sirius's differention strategy is project cycle time speed. Figure 1.5 illustrates that point: Time-to-deployment and time-to-breakeven points are reached much sooner than for the other two companies. For this to be possible, Sirius needs a timeline-driven toolbox in which the central role and priority belong to tools that can help enable fast cycle times. These may include tools such as the Gantt chart, time-scaled arrow diagram, critical path diagrams, milestone charts, and so on (Chapter 6). This does not mean that other components of the typical PM Toolbox such as cost, risk, and stakeholder tools are ignored. Quite to the contrary—they are important and have their role in the toolbox as well, but they are subjugated to timeline-driven tools.

The case is different for the toolbox for Park, a company that concentrates on cost leadership. Logically, then, most projects within Park are cost driven, searching to minimize project cost whenever possible. This logic is apparent in Figure 1.5. The Park curve shows less negative cash flow than those of Sirius and Prima. It is the intended goal and realized outcome of project actions. To accomplish the project strategy, Park is willing to take the longest time to reach time to deployment and time to breakeven. Crucial in this effort is a cost-driven toolbox that emphasizes cost, cost, and cost. Correspondingly, cost estimates and cost baselines are carefully prepared, as is the assessment of return on investment, even for small cost-cutting projects (Chapter 5).

The intent to align the PM Toolbox with the business strategy is aggressively pursued in Prima as well. The driving force is the best-cost strategy that is also translated to the project level. As can be seen from Figure 1.5, time to deployment and time to breakeven are shorter than Park's, but longer than Sirius's. This means that cost focus is lower than Park's but higher than Sirius's. Such cost philosophy is closely intertwined with the need for the project to emphasize performance goals more than the other two companies. Given this situation, how does one shape a cost-performance-driven PM Toolbox?

A combination of well-balanced performance and cost tools has the priority. Formal and informal voice of the customer tools and feature requirement tools are crucial for hitting customers' expectations, as are cost estimates and cost baselines. To Prima and its customers, delivering on schedule is important, as keeping customers satisfied is not possible without delivering when promised. Nevertheless, schedule goals are subjugated to performance and cost. Other tools, such as a risk management plan, are modified to support the combination of cost and performance focus. For example, the risk management plan may be focused on lowering cost rather than schedule (Chapter 14).

As can be gleaned from our discussion, the nature of alignment of the toolbox is reflected in the balance of two issues. First, many of the tools show up in all three toolboxes. The second issue concerns the situational approach: adapting tools to account for the characteristics of the three toolboxes (see Table 1.2)

Customize the PM Toolbox

There are multiple options for customizing a strategically aligned PM Toolbox. Three options are perhaps the most viable:

1. Customization by project size
2. Customization by project family
3. Customization by project type

The options are three different ways to select and adapt the toolbox. Each option has the purpose of showing which specific project management tools to select and adapt for the PM Toolbox. For this to be possible, each option is based on the particular methodology used, which has a large influence on the choice of tools.

An in-depth knowledge of individual tools is a prerequisite to each of the options because you need to understand how each tool can support a project deliverable. We will describe the customization options in turn and offer guidelines for selecting one of them for implementation.

Customization by Project Size

Some organizations use project size as the key variable when customizing a PM Toolbox. Their logic is that larger projects are more complex than smaller ones, or that size drives differences in project management methodology complexity. The reasoning here is that as the project size increases, so does the number of activities and resulting project deliverables associated with a project, as well as the number of interactions among

Table 1.2: Characteristics of Strategically Aligned Toolboxes

	Company's Core Business Strategy		
	Differentiation	Low-Cost	Best-Cost
	Nature of PM Toolbox		
Characteristics of the PM Toolbox	**Schedule Driven**	**Cost Driven**	**Performance-Cost Driven**
Central role and priority belong to schedule tools	✓		
Management attention is on schedule performance	✓		
PM spends majority of time managing to schedule	✓		
Schedule tools are primary basis for decisions	✓		
Other tools adapted to support schedule tools	✓		
Central role and priority belong to cost tools		✓	
Management attention is on cost performance		✓	
Project manager spends majority of time managing cost		✓	
Cost tools are primary basis for decisions		✓	
Other tools adapted to support cost tools		✓	
Central role and priority belongs to cost-performance tools			✓
Management attention on performance and cost			✓
PM spends majority of time managing performance requirements and cost			✓
Performance and tools are primary basis for decisions			✓
Other tools adapted to support performance tools			✓

them. Worst of all, this number of interactions grows by compounding, rather than linearly.[10] Such increased complexity, then, has its penalty—larger projects require more work to coordinate the increased number of interactions.

Since different project sizes require different processes and tools, we first need a way to classify projects by size and then customize their toolboxes. For size classification we draw on the experience of some companies. In Table 1.3, we present three examples. All companies use three classes of project size: small, medium, and large. The units used to measure project size are dollars or person-hour budgets. On the basis of the size, the companies determined the managerial complexity of its project classes and processes. The complexity further dictated the PM Toolbox makeup, a simplified example of which is illustrated in Table 1.4. For the sake of simplicity, only the toolbox is shown, leaving out the project deliverables.

Table 1.3: Examples of Project Classification per Size in Three Companies

Project and Company Type	Project Size		
	Small	**Medium**	**Large**
Product development projects in a $1 billion/year high-technology manufacturer	$1–2m	$2–10m	> $10m
Infrastructure technology projects in a $300 million/year food processing company	< $50k	$50–150k	> $150k
Software development projects in a $40 million/year customer relationship management software company	300–400 person-hours	1,000–3,000 person-hours	>3,000 person-hours

Table 1.4: Examples of PM Toolbox Customization by Project Size

	Project Phases			
Project Size	**Initiation**	**Planning**	**Execution**	**Closure**
Small	Project charter	Scope statement	Progress report	Final report
		WBS		
		Responsibility matrix		
		Milestone chart		
Medium	Project charter	Scope statement	Progress report	Final report
	Skill inventory	WBS or PWBS	Change process	Change log
		Responsibility matrix	Change log	Postmortem report
		Cost estimate	Gantt chart	
		Gantt chart	Cost burn down	
		Risk plan	Risk register	
Large	Project charter	Scope statement	Progress report	Final report
	Stakeholder matrix	WBS and PWBS	Project indicators	Postmortem report
	Stakeholder strategy	Responsibility matrix	Change process and log	Closure checklist
		Cost estimate	Time-scaled arrow diagram	
		Time-scaled arrow diagram	Slip chart	
		P-I matrix	EVM	
			Risk register	

EVM = earned value management; P-I = probability-impact; PWBS = program work breakdown structure; WBS = work breakdown structure.

As Table 1.4 indicates, some of the tools in the toolboxes for projects of different size are the same, while others are different. For example, all use the summary status report (Chapter 12) because all projects need to report on their performance. Since managerial complexity of the three project classes and their processes call for different tools, some of the tools differ. A P-I matrix (Chapter 14), for example, is needed only in large projects. To be successful, the process team designing the toolbox should carefully balance the standard tools with those that account for the specific size of the project.

Experience of these companies offers several guidelines for customizing the PM Toolbox by project size:

- Identify a small number of project classes and their methodologies.
- Define each class by the size parameter.
- Match the project size with the proper toolbox, each tool supporting a specific project deliverable.

Note that while customization by project size offers advantages of simplicity, it also carries a risk of being generic, disregarding other situational variables. To some, these other variables may be of vital importance, as will be pointed out in the next section on customization by project family.

Customization by Project Family

When the PM Toolbox is strategically aligned, you can opt to customize it by family types within an industry. Many companies choose such options in a belief that project families in their industry are sufficiently unique to merit an industry-specific project family methodology and toolbox.[11]

As a group of organizations that compete directly with each other, an industry is characterized by the nature of its environment and business risk. For example, companies in the high-technology industry face an environment of dynamic technology change. Because of this, their portfolio abounds with fast time-to-market projects driven by the desire of their customers to continuously buy the latest and greatest technological products and services. Combined, the business environment and risk profile create similar challenges in families of projects. For example, a family of new product development projects in high-tech industries face similar challenges. So do facilities management projects, manufacturing projects, marketing projects, and information technology projects within the same industry.

Often, project families are defined by the novelty of the capabilities the projects produce. Generally, the more novel the capability, the more complex the projects.[12] This is because increasing novelty (newness or uniqueness) in projects leads to more uncertainty, elevating the need for more flexibility in the processes and the supporting toolbox. For example, as novelty grows:

- The more evolving the scope statement and WBS become.
- The project time line becomes more fluid.
- The cost estimates follow the fluidity of the schedules and scope.
- More risks need to be identified and managed.

Table 1.5: Customizing the Toolbox by Project Family

Project Family (Novelty)	Project Phases			
	Initiation	**Planning**	**Execution**	**Closure**
Derivative projects	Project charter	Milestone chart	Progress report	Final report
	Financial scoring model	Requirements baseline		
		WBS		
Incremental projects	Project charter	Scope statement	Progress report	Final report
	Financial scoring model	WBS or PWBS	Change log	Change log
	Stakeholder map	Requirements baseline	Gantt chart	Retrospective
		Cost estimate	Cost burn down	
		Gantt chart	Risk register	
		Risk plan		
Breakthrough projects	Project charter	Scope statement	Progress report	Final report
	Voting Models	WBS or PWBS	Project indicators	Postmortem report
	Stakeholder map	Requirements baseline	Change process and log	Closure checklist
	Stakeholder strategy matrix	Responsibility matrix	Milestone chart	
		Cost estimate	Slip chart	
		Milestone chart	EVM	
		P-I matrix	Risk register	

EVM = earned value management; P-I = probability-impact; PWBS = program work breakdown structure; WBS = work breakdown structure.

A simple example reflecting these trends in adapting the toolbox for the three classes of project families is illustrated in Table 1.5.

As the table shows, the toolboxes of the three classes of projects are similar in some and different in other aspects. For example, all use schedules and progress reports. Still, the schedules differ in that simple projects rely on a simple milestone chart, while complex projects use a rolling wave type of the time-scaled arrow diagram. Obviously, the variation in the novelty of the project is the source of the differences.

Customization by Project Type

While the previous two approaches to PM Toolbox customization rely on one dimension each—project complexity and project family as defined by novelty, respectively—customization by project type uses both dimensions.[13]

To make it more pragmatic, we will simplify the model, while maintaining its comprehensive nature. Each of the two dimensions includes two levels: (1) novelty of the capability under development (low, high) and (2) project complexity (low, high). This helps to create a two-by-two matrix that features four types of projects: routine, administrative, technical, and unique (see Figure 1.6).

A routine project is one having a low level of capability novelty (less than half of the features are new) and low complexity (few cross-project interdependencies). Due to the low levels of novelty and complexity, the project scope can normally be frozen before project execution begins or early in the execution stage. Scope also remains fairly stable, with few scope changes. With scope remaining stable, project scheduling, cost management, and performance management are also quite static.

Typically, routine projects are performed within a single organization or organizational function (e.g., infrastructure technology). Examples include the following:

- Continuous improvement project in a department.
- Upgrading an existing software application or existing product.
- Adding a swimming pool to an existing hotel.
- Developing a derivative model in a washing machine product line.
- Expanding an established manufacturing line.

Administrative projects are similar to routine projects in terms of novelty. Business goals and scope are normally well defined, stable, and detailed. The added complexity requires the coordination of multiple organizational functions and the mapping of the many functional interdependencies, but the lack of capability novelty allows for standard scheduling techniques. The same added complexity generally means larger project size, with higher financial exposure, justifying the need for detailed bottom-up cost estimates reconciled with financial targets contained in the project business case. Risk is primarily related to the increased number of interactions between the function's project teams; therefore, additional risk planning and analysis is required.

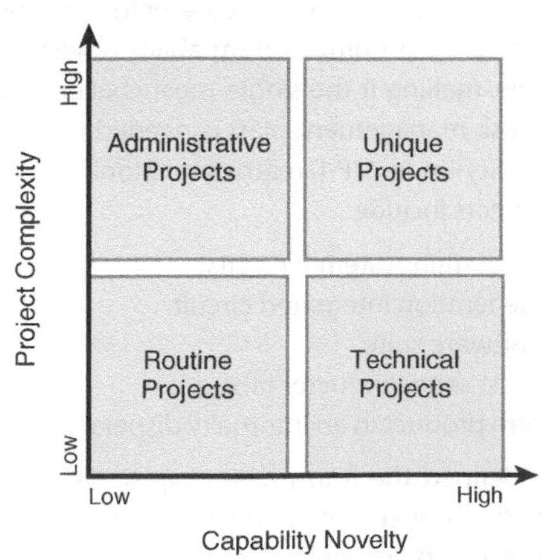

Figure 1.6: Four Project Types

Some examples of administrative projects are as follows:

- Corporate-wide organizational restructuring.
- Deploying a standard information system for a geographically dispersed organization.
- Building a traditional manufacturing plant.
- Developing a new automobile model.
- Upgrading an enterprise computer system.

Technical projects consist of more than 50 percent of new technologies or features at the time of project initiation. This creates a higher degree of uncertainty that requires project flexibility. The goals, scope, and work breakdown structure (WBS) are simple due to the low level of complexity, but they may take longer to fully define. The rolling wave or similar approach can be used, meaning that only the schedule for the following 60 to 90 days can be planned in detail, while the remainder of the project schedule is represented only by milestones. Similarly, cost estimates are fluid as well. A detailed cost estimate for the next 60 to 90 days can be detailed, while cost estimates for the remainder of the project are at the summary or rough order of magnitude level. The increased technical novelty results in increased technical risk and the need for a more rigorous risk management implementation and tools. Here are some examples:

- Reengineering a new product development process in an organization.
- Developing a new software program.
- Adding a line with the latest manufacturing technology to a semiconductor fab.
- Developing a new model of a computer game.

For unique projects, business goals, detailed scope definition, and WBS development takes time to evolve as a result of many new features and cross-project interdependencies. The evolving nature of scope leads to the need for fluid schedules. Project mapping and rolling-wave scheduling processes can be used to contend with the fluidity. Similarly, cost estimates for milestones are more detailed in the near term and more summary level for the longer term. A high level of project complexity exists due to multiple organizational functions required to execute unique projects, requiring integration tools such as the project map. Combined capability novelty and project complexity push risks to the extreme, making it the single most challenging element to manage. In response, a rigorous risk management plan is needed, as well as a combination of tools such as the probability-impact (P-I) matrix and Monte Carlo analysis (Chapter 14). Example technology projects include:

- Building a new light rail train system for a city.
- Developing a new-generation integrated circuit.
- Developing a new software suite.
- Constructing that latest semiconductor fab.
- Developing a platform product in an internally dispersed corporation.

Now that we have defined the four project types, we can move on to the next step: Describe how the two dimensions impact the construction of the PM Toolbox. Taken overall, the growing technical novelty in a project generates more uncertainty, which consequently requires more flexibility in the tools chosen. In Figure 1.7 we show

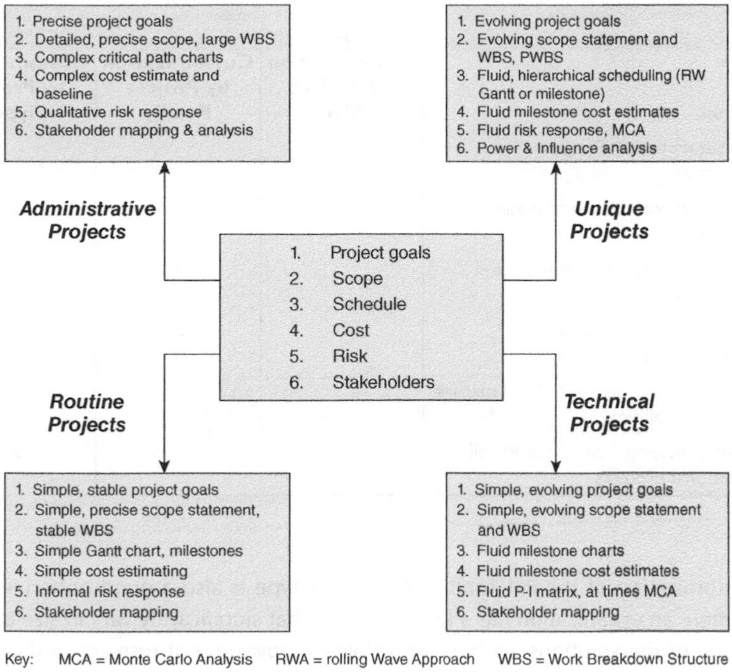

Figure 1.7: Customizing the PM Toolbox by Project Type

examples of several tools that have to be adapted to account for different processes driven by different project types.

A summary comparison of the tools for the four project types reveals that they use very similar types of tools. For example, all use the WBS. Still, when the same type of tool is used, there are differences in their structure and how they are used. Consider, for instance, Gantt and milestone charts. Both are used in the routine and unique projects, but terms of use are significantly different. This is the situational approach—as the nature of the PM processes changes, so does the PM Toolbox.

Which Customization Option to Choose?

We offer three options for the customization of the PM Toolbox. Each has its advantages, disadvantages, and risks, and fits some situations better than others. To assist with the selection, refer to Table 1.6. Customization by project size is a good option when an organization has projects of varying size and needs a simple start toward more mature forms of customization. In addition, projects of varying size characterized by mature processes lend themselves well to this customization option. In an organization that has a stream of projects that feature both mature and novel capabilities but project size is not an issue, customization by project family may be the best option. This is also a good option to go for when projects are dominated by a strong industry or professional culture.

Table 1.6: Project Situations and PM Toolbox Customization

Situation	Customization by Project Size	Customization by Project Family	Customization by Project Type
Simplest start to PM Toolbox customization	✓		
Projects of varying size with mature capabilities	✓		
Projects with both mature and novel capabilities, size not an issue		✓	
Projects with strong industry or professional culture		✓	
Projects of varying size with both mature and novel capabilities			✓
Need for a unifying framework for all organizational projects			✓

Customization of the PM Toolbox by project type is also a good option in situations where an organization has a lot of projects that significantly vary in size but also in novelty of the solutions, such as a portfolio of government research and procurement projects. Organizations searching for a unifying framework that can provide the customization for all types of projects—from facilities to product development to manufacturing process to customer service to information systems—may find customization by project type an appropriate choice.

Continuously Improve the PM Toolbox

Once the Toolbox has been customized, it will be more effective if it is continuously improved. Without such improvement, the Toolbox will gradually lose its effectiveness and its ability to support the project management methods and tools employed and the business strategy of the organization.[14] Avoiding such a predicament and instead sustaining an effective toolbox can be achieved through the following steps:

1. Form a PM Toolbox improvement team.
2. Identify mechanisms for collecting improvement ideas.
3. Follow an improvement process.

Form an Improvement Team

The toolbox improvement team is usually part of the process team responsible for designing and managing project processes. This team has the total responsibility for simplifying, improving, and managing the implementation of the PM Toolbox. Each team member owns a piece of the toolbox, and, overall, the responsibility should be distributed as evenly as possible across the team. When forming a team, it is important to understand that management enforces, while the team operates and owns the toolbox. Since it is mostly project managers that must use the toolbox, we recommend that the majority of the toolbox improvement team come from the PM ranks.

Identify Mechanisms for Collecting Improvement Ideas

Ideally, there should be a continuous stream of suggestions and ideas to improve the customized toolbox. To secure such a stream, you can require that project teams address PM Toolbox improvement suggestions as part of the retrospective or postmortem reviews (Chapter 13). If the reviews find a need to change the toolbox, the team should submit a change request. Change requests may come at any time from anyone involved in projects. Note that requests are not the only way to collect the toolbox improvement ideas. A survey, brown bag information-gathering sessions, or focus groups may also be viable options to collect improvement ideas.

Follow an Improvement Process

A toolbox improvement process should define steps for acting on change requests, including an escalation process for brokering requests that are turned down. Quickly collecting and responding to PM Toolbox change requests is of vital importance. Also significant are requests to deviate from various tools that are included in a toolbox, usually the standardized tools. Deviations from standardized tools help to ensure that a toolbox remains flexible. Since most deviation requests are submitted while a project is in progress, it is important to respond as soon as possible. At a later time, the requests can be evaluated to determine if the toolbox should be permanently modified to include the requests.

Effectively constructing and adapting a PM Toolbox is predicated on the user's knowledge of individual PM tools. To help increase our readers' knowledge, the chapters that follow will detail a multitude of useful tools that can be chosen for inclusion in your own PM Toolbox.

References

1. Martinelli, Russ, James Waddell, and Tim Rahschulte. *Program Management for Improved Business Results*, 2nd ed (Hoboken, NJ: John Wiley & Sons, 2014).
2. Project Management Institute. *A Guide to the Project Management Body of Knowledge*, 5th ed (Drexell Hill, PA: Project Management Institute).
3. Martinelli, Waddell, and Rahschulte, 2014.
4. Spencer, J. C. *Business Strategy: Managing Uncertainty, Opportunity, & Enterprise* (London, England: Oxford Press, 2014).
5. Pearce, John A. II, and Richard B. Robinson Jr. *Strategic Management: Formulation, Implementation, and Control*, 12th ed (New York, NY: McGraw-Hill, 2010).
6. Porter, Michael E., W. Chan Kim, and Renee Mauborgne. *HBR's 10 Must Reads on Strategy* (Boston, MA: Harvard Business Review Press, 2011).
7. Stevenson, W. J. *Production and Operations Management* (Boston, MA: Irwin, 1993).
8. Kerzner, H. *Applied Project Management* (New York, NY: John Wiley & Sons, 2000).

9. Hammer, M., and J. Champy. *Reengineering the Corporation* (New York, NY: Harper Business, 1993).
10. Kahn, Kenneth B. *The PDMA Handbook of New Product Development* (Hoboken, NJ: John Wiley & Sons, 2012).
11. Pinto, Jeffrey K., and Jeffrey G. Covin. "Critical Factors in Project Implementation: A Comparison of Construction and R&D Projects." ScienceDirect website: http://www.sciencedirect.com/science/article/pii/0166497289900400. Accessed April 2014.
12. Tatikonda, M. V., and R. S. Rosenthal. "Technology Novelty, Project Complexity, and Product Development Project Execution Success: A Deeper Look at Uncertainty in Product Innovation." *IEEE Transitions on Engineering Management* 47 (1): 74–87, 2009.
13. Shenhar, A. J. "One Size Does Not Fit All Projects: Exploring Classical Contingency Domains." *Management Science* 47 (3): 394–414, 2001.
14. Boutros, Tristan, and Tim Purdie. *The Process Improvement Handbook: A Blueprint for Managing Change and Increasing Organizational Performance* (New York, NY: McGraw-Hill, 2013).

PART

II

Project Initiation Tools

2

PROJECT SELECTION

We have yet to witness an organization that had an overabundance of resources to execute on their full list of ideas and projects. Quite the opposite, in fact. Commonly, resources become overcommitted, and an organization needs to make choices among possible projects.

As is usually the case, organizations have many more product, service, and infrastructure solutions or transformational change ideas than available resources to execute them. As a result, an organization must find a way to broker competing demands for its limited resources. Project selection techniques are used to identify and prioritize projects that best support attainment of the business goals of an enterprise. Projects are ranked and prioritized based on a set of criteria that represents business value to the organization. Senior management can then allocate available resources to the highest-value and most strategically significant projects. This is sometimes referred to as *resource capacity planning*.

Additionally, the mix of projects within a portfolio is in a state of constant flux, as an organization reevaluates its selections on an ongoing basis to respond to changes in the business environment and other needs. Project selection tools allow an organization to select projects for initiation and termination as conditions change.

This chapter is intended to help with the selection of projects based on the highest value of the project's return to an organization. We describe a number of tools that can be added to a PM Toolbox that help to better evaluate the value a project offers, the benefits it can deliver, and how well it aligns to business strategy. There are, of course, many tools for selecting projects. In this chapter, we present the tools that we find most ubiquitously used across a variety of industries and organizations, beginning with the benefits map.

THE BENEFITS MAP

Effective project selection is about evaluating the potential value a project idea can return to an organization, and then making deterministic choices about which of the project ideas will be funded and resourced. We talked about the importance of taking a strategy-driven approach to managing a project and to constructing a PM toolbox. This is because organizations are in business or mission for the long haul. The outcome of their projects must contribute to the long-term viability of the enterprise by delivering both short-term and long-term value. Projects that contribute to the achievement of the long-term goals of the organization provide the greatest value.[1]

But what is value? The British European Standard (BSI) describes value as the relationship between the satisfaction of need and the resources used to achieve that satisfaction.[2] They use the diagram shown in Figure 2.1 to demonstrate this definition of value.[3]

In the United States as well as other parts of the world, value with respect to business and project management has come to be associated with the generation and delivery of anticipated business benefits. In turn, value management is often described in terms of benefits management. The value diagram can therefore be updated as shown in Figure 2.2 to demonstrate this more specific view of value.

So what is all this leading to? First, project managers must be able to describe the value proposition of their projects in terms of the business benefits the project will create for the organization. Second, the management team of the organization must also sift through the value promise of each of its project opportunities in order to prevent an over commitment of its limited resources.

Projects should be viewed as investments that a management team makes in hopes of gaining an acceptable return. For many organizations, the objective measure of value is in financial terms. Other organizations make a qualitative measure of value by assessing the business benefits delivered in correlation to the strategic goals. In either approach, business benefits are the outcomes of a project that provide value to the organization in return for the investments made in the project. We begin by looking at the qualitative assessment of business benefits value.

Enabling Benefits Management

Benefits management is about realizing the business results desired from the investment in a project. It is about management of the business goals that are driving the need for a project and the achievement of the business results intended. A critical part of benefits management is the development of a benefits management strategy. Creating a

$$\text{Value} = \frac{\text{Satisfaction of Need}}{\text{Use of Resources}}$$

Figure 2.1: BSI Description of Value

$$\text{Value} = \frac{\text{Business Benefits Achieved}}{\text{Resources Expended}}$$

Figure 2.2: Business Benefits as a Measure of Value

benefits management strategy involves identifying the specific business results that a project must deliver to directly support the strategic goals of the enterprise.

It is the benefits management strategy that forms the foundation of the project business case (Chapter 3). The benefits map is an effective tool that documents the expected benefits that are to be realized from the investment. It specifically charts the path from the organization's business strategy to the distinct benefits that are to be derived from the output of a project.[4] The information provided by the benefits map supports the cost/benefit analysis for a project and should become part of the project business case.

The benefits map can be depicted in visual form to provide a useful means to demonstrate alignment of project deliverables to the business success factors for a project, and to the expected business goals as displayed in Figure 2.3.

As demonstrated in the figure, the benefits map provides traceability between project outcomes and deliverables to the benefits intended. It is essential to first establish the overall vision and scope for a project, then communicate how the project contributes to the objectives, and finally track the execution of the project to final delivery of strategic business goals.

This tool is used to assist in the characterization of how specific project objectives are met. However, benefits maps can become complex and confusing due to the *one-to-many* relationships between project deliverables and outcomes to the objectives. The critical component in building an effective benefits map is to ensure each project deliverable or outcome is mapped to an objective, and every objective to the business success factors.

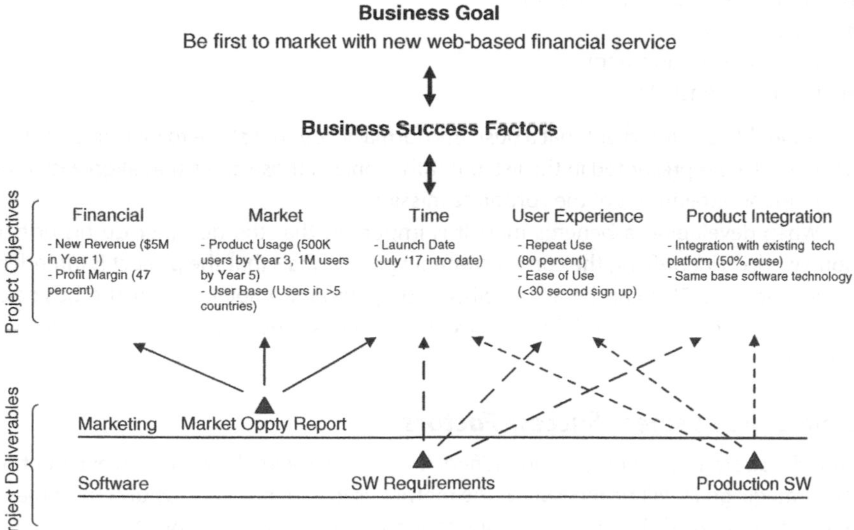

Figure 2.3: Example Project Benefits Map

Developing a Benefits Map

The development of a benefits map is not a simple exercise because it requires knowledge of both business strategy and detailed project execution. Because of this, however, it is an extremely effective tool for aligning the *setting* of strategy and the *implementation* of strategy. These two sides of strategy were discussed in Chapter 1. The major steps in creating a benefits map are described next.

Identify the Strategic Business Goals

The first activity involved in developing a benefits map is to define the strategic business goals that are underwriting the need for the project. Strategic business goals define *what* the company wants to achieve within a specific period of time, and are normally defined at two levels of an organization: corporate strategic goals and business or operating unit strategic goals.[5]

The purpose of corporate-level strategic goals is to align the various business units toward a common purpose and direction, while business unit strategic goals serve to focus the functional departments and work efforts of the people within a business unit.[6] Every enterprise is unique, and therefore every enterprise will have its own set of strategic goals. It is common, however, to find strategic goals centered in a number of areas, including:[7]

- Profitability
- Competitive position
- Employee relations
- Product or solution leadership
- Productivity
- Employee development
- Public responsibility

It should be pointed out that a business normally does not strive to identify strategic goals in all areas presented in the list, but rather only in those areas that align with and fully support attainment of the corporate mission.

When developing a benefits map, it is important that the designer do his or her homework on identifying the correct strategic goal or goals that the project is intended to help achieve. This can be accomplished only through knowledge of the business aspects of an enterprise and through a series of discussions with the senior leaders of the organization.

Define the Business Success Factors

With the strategic goal or goals identified, it is time to define how the achievement of the strategic goals will be measured. The metrics used to define successful achievement of the strategic goals become the business success factors for the project.

The business success factors transform the business results derived from strategic goals into a specific statement of success that guides the organization as to how to plan and execute their work. We define business success factors as *the set of quantifiable measures that describe the successful achievement of a project's business results.* It is the business success factors that bind the activities of strategy setting to those of strategy implementation.

It is important to minimize the number of business success factors to the critical few—three to six is an ideal number. During the early stages of a project, the factors are used to align the project sponsor, executive stakeholders, and the project team on what project success will likely mean. They form the foundation of the project business case, and for the broader project team they establish the end state they will be working to achieve.

Identify Project Outcomes

Through the use of the project work breakdown structure (Chapter 5) the primary outcomes for the project are identified. These are normally defined at Level 2 or Level 3 of the breakdown structure, before the detailed tasks associated with the creation of the outcomes are documented.

Once project outcomes are identified, it normally works well to categorize the outcomes by ownership. In other words, group all outcomes by each project sub team that will be responsible for creating and delivering the outcome.

As shown in Figure 2.3, it works well to document the outcomes in what we refer to as a series of team *swim lanes*. Even though the swim lanes look linear in nature, there is not a time element to the benefits map, so the outcomes for each team do not have to be documented in time sequence order. In fact, it is normally best not to, but rather to try to align them with the business success factors that they support.

Perform the Mapping

This step simply involves graphically demonstrating, through the use of interconnecting lines, the relationship between project outcomes and project objectives, and between the project objectives and the business success factors. The use of color-coding of lines associated with the particular project outcomes is especially helpful for larger projects.

Validating the Results

The final step in creating a benefits map is to validate that each project outcome directly supports the achievement of a project objective, and that the project objective (through the delivery of the project outcome) directly supports the achievement of the business success factor it is associated with.

When completed, a project manager has a visual mapping of how a project outcome supports the strategic goals of the firm via a direct link between project objectives and business success factors.

Using a Benefits Map

Along with the project work breakdown structure, the benefits map is a useful tool for establishing the overall scope of a project and for demonstrating the alignment between project outcomes and deliverables to the business success factors. The benefits map can also be used to communicate to top management, the project team, and other stakeholders how the strategy of the organization and project are melded together, and how each business benefit will be realized.

The benefits map is intended to be used throughout the life of the project to analyze consequences caused by adjustments and changes as they occur to the original project strategy and scope. The first use of the benefits map normally occurs as part of the business case development process, where a high-level mapping of benefits to project objectives to strategic intent is established. Further detail is then added during detailed planning when the full comprehension of scope and traceability of project outcomes to business benefits is necessary.

In organizations with established governance policies and a set of related processes, the benefits map is often used as one of their key business-level monitoring and tracking devices for the project.

Variations

At times, business benefits are described in terms of solutions to problems that have been identified. If this is the case, a variation of the benefits map, often called the *objectives tree,* is better suited to assess project value. The goals defined in the objectives tree can be used to refine the project objectives.[8]

An objectives tree provides a visual representation that allows you to quickly and completely articulate the scope of the problem you are attacking. It is used primarily in the earliest stages of project definition.

As illustrated in Figure 2.4, the objectives—which are the means to solving a problem—are decomposed into lower levels of detail and specificity.

When using the objectives tree to solve a problem, the core problem is reworded into an objective that describes a desired end state. For example, suppose we are the new product development team in charge of defining, designing, and producing our

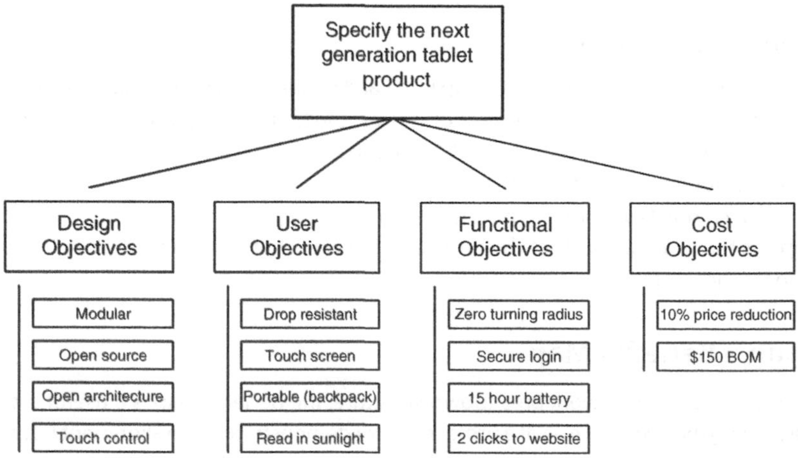

Figure 2.4: Example Objectives Tree

company's tablet products. A primary object we would likely have at some point would be to *specify the next tablet that will be designed and manufactured*. This primary objective would then be decomposed into sub objectives that define solutions to the primary. Decompose to at least three levels and then work your way back up the tree to validate that the sub objectives are sufficient to achieve the objectives at the next-highest level.

The resulting objectives tree can be used as the basis for defining the core requirements for the project, as discussed in Chapter 4.

Benefits

There are several advantages to be gained by both top management of an organization and the project manager through the use of the benefits map. It helps to create better clarification and understanding of the project strategy and scope, and establishes direct alignment between objectives, outcomes, and the business benefits to be realized.

The benefits map also provides a systematic process to assess business benefits as part of the project's cost-benefit analysis, which is a critical element of the business case of a project and in project selection.

Finally, it enables focused tracking and monitoring of progress toward realization of the benefits as part of the project reporting process and establishes an effective means for evaluating success of a project from a benefits realization perspective.

ECONOMIC METHODS

When quantitative methods are used to define project value to guide the project selection process, they are normally economically focused. We discuss the three most common economic methods for describing project value: payback time, net present value (NPV), and internal rate of return (IRR).

Payback Time

Payback time is simply the length of time from when the project is officially initiated until the cumulative cash flow (or cost savings) becomes positive. At that point all the funds invested in the project have been recovered. As can be seen from the three project cash flow scenarios in Table 2.1, the cash flow for Project 1 turns positive in six years, for Project 2 in five years, and for Project 3 in eight years.

Payback time is a very conservative criterion and provides more protection against future uncertainties than do either of the other economic methods (NPV and IRR). However, it is insensitive to project size, since a project with massive investment requirements may still have short payback time. Moreover, it takes no account of future economic potential once payback is reached.

Net Present Value

Net present value takes into account the time value of money, in that a dollar a year from now is worth less than a dollar today due to inflation. NPV discounts both future costs

Table 2.1: Project Payback Time Examples

	Project 1			Project 2			Project 3		
Year	**Cost**	**Revenue**	**Cum Cash**	**Cost**	**Revenue**	**Cum Cash**	**Cost**	**Revenue**	**Cum Cash**
1	20		−20	30		−30	20		−20
2	50		−70	80		−110	50		−70
3	80		−150	105	147	−68	78		−148
4	120		−270	110	154	−24	80		−228
5	200	200	−170	125	175	**26**	150		−378
6	450	630	**10**	150	210	86	375	525	−228
7	500	775	260	160	220	146	525	735	−18
8	500	775	535	170	223	199	600	840	**222**
9	450	700	785	175	230	254	800	1,120	542
10	450	650	985	170	228	312	750	1,200	992

and revenues by the interest rate, according to the formula:

$$\text{NPV}(i, N) = \sum_{t=0}^{N} \frac{R_t}{(1+i)^t}$$

In this formula, R_t represents the net cash flows (cash inflow – cash outflow, at time t), i is the discount rate or the rate the company pays for borrowed money, expressed as a decimal fraction, and N is the total number of periods (years, months, etc).

Spreadsheet programs can be used to calculate NPV directly. You only need to enter the discount rate and the values (or a vector of cells) into the NPV function. The result is computed and displayed in the cell holding the function. Using the NPV function, the NPV for three projects at a discount rate of 5, 10 and 15 percent are shown in Table 2.2.

It can be seen that the more the future is discounted (i.e., higher the discount rate), the less the NPV of the project. When comparing the three projects for project selection, Project 3 delivers greater value than either Project 1 or Project 2 at all discount rates. That is, the higher the NPV, the greater the economic value of a project.

Table 2.2: Project Net Present Value Examples

Project 1	Project 2	Project 3
NPV at	**NPV at**	**NPV at**
5% discount rate = $5,283	5% discount rate = $2,320	5% discount rate = $6,400
10% discount rate = $2,841	10% discount rate = $1,254	10% discount rate = $3,275
15% discount rate = $1,563	15% discount rate = $688	15% discount rate = $1,679

NPV takes into account the magnitude of the project and the discount rate. It is not a particularly conservative measure, however, since it incorporates estimated future revenues that may not actually materialize.

Internal Rate of Return

Internal rate of return is simply the discount rate at which NPV for the cash flow is zero. There is no closed-form formula for it. IRR must be computed iteratively by "honing in" on the exact discount rate that produces an NPV of zero. Most spreadsheets have an IRR function that allows the user to obtain an IRR. You need to only enter the list of values (or a vector of cells) and a guess value of IRR. The function then carries out the iterative calculation. For the projects shown in Table 2.2, the IRR values are as follows:

Project 1 IRR = 42 percent

Project 2 IRR = 40 percent

Project 3 IRR = 36 percent

By the IRR criterion, Project 1 is superior in economic value to the other two projects. While IRR discounts future values, it takes no account of the size of the project. Project 3 promises a significantly greater total return than Project 1, but those returns are farther in the future and follow a longer course of investment before cash flow turns positive. Hence, the IRR for Project 2 is lower than for Project 1.

Using Economic Methods

Financially driven quantitative measures of value require data about revenues (or cost savings) and costs expected to result from a project. They are thus appropriate primarily for capital projects and projects intended to improve existing capabilities (products, services, or infrastructure). In such cases, they allow a direct comparison of such projects with alternative capital investments.

Care should be taken to ensure that the same economic method is used for all project valuation calculations to provide a direct "apples-to-apples" comparison of projects. Also ensure that you know your organization's standard discount rate before performing the calculations. For a comparison of the three economic methods, see the example titled "Choosing an Economic Method."

Choosing an Economic Method

As we have demonstrated, the three methods can give differing results on the same set of projects. Which method should be chosen depends on the considerations important to the decision maker. NPV is best used for projects with

(continued)

large payoffs but gives little protection against future uncertainties. IRR likewise gives little protection against future uncertainties and may tend to give preference to a project with modest total payoff but high return on a modest investment. Payback time is a very conservative method, giving more protection than the other two methods against future uncertainty, but it does not factor in the size of project payoff nor of the discounted value of future costs and revenues. The consideration that is most important will govern the choice of methods. Moreover, since these methods treat projects as capital investments, it would be appropriate to use whichever method the company uses for evaluating its other capital investments, thus allowing a direct comparison between project investments and other investments.

Benefits of Economic Methods

Quantitative measures are readily understandable. They enable the decision maker and project managers to communicate more readily about financial considerations of projects. They also make it easier to compare projects with other opportunities that are vying for capital investment.

Also, once the necessary data are obtained, they are easy to compute. They also make sensitivity testing easy by adjusting discount rates, time, and other input data. Alternate scenarios about future costs and returns can be compared to test various assumptions being made about project value against future uncertainty.

SCORING MODELS

A scoring model is little more than a list of criteria the decision maker wishes to take into account when selecting projects from the list of candidate projects. Projects are then rated by decision makers on each criterion, typically on a numerical scale with anchor phrases. Finally, multiplying these scores by weightings and aggregating them across all criteria will produce a score that represents the merit of the project. Higher scores designate projects of higher value.

Developing a Scoring Model

Like other methods, scoring models follow the basic steps of the project selection process: 1) Create the menu or list of candidate projects, 2) Develop the relevant project selection criteria, 3) Rate projects against the selection criteria, and 4) Choose the projects to invest funds and resources. To be fully functional and meaningful, the models need the following inputs: (1) a menu of candidate projects, (2) strategic goals, (3) project proposal or business case, and (4) historical information.

Since the purpose of the models is to help maximize the value of the selected projects for the company, understanding which of the company's strategic goals a project supports is a key point. While these goals are described in the strategic and tactical plans of the organization, project proposals offer specifics on projects. To make better decisions, decision makers should also rely on the historical information about both the results of past project selection decisions and past project performance. When these inputs are available, you can move to choose relevant project selection criteria.

Identify Relevant Scoring Criteria

A key factor in developing a successful scoring model is identifying an appropriate set of scoring criteria that will reflect the strategic, financial, technical, and behavioral situation of the company. The challenge often seems to be in overcoming the temptation to develop a detailed and, therefore, cumbersome list of criteria that becomes unmanageable. Narrowing down to the critical few criteria that really matter is rather difficult, but necessary. Consider, for example, the criteria listed in the example titled "Criteria to Be Considered in Project Selection." To be used effectively, most of the criteria need to be described in more specific terms. This further elevates the challenge of sticking with the critical few. An effective approach used by some companies is in conceiving the list and refining over time with the intent to reduce the number of scoring criteria to five or less.

Criteria to Be Considered in Project Selection

Scoring criteria that are relevant in project selection depend on the types of projects and their situation. For example, the following criteria are typically considered in choosing research and development projects. This list is intended to be suggestive rather than comprehensive:

Cost to develop	Total expected revenue
Probability of technical success	Probability of market success
Market size	Market share
Alignment to strategic goals	Strength of competition
Availability of required staff	Degree of organizational commitment
Regulatory alignment	Alignment to company policy

Although many of these criteria may be used in selecting different types of projects, the important thing is to include in the analysis whatever criteria are relevant to your project situation.

Constructing the Model

To construct a scoring model, you must understand and resolve several issues:

1. The form of model you want to use.
2. Categories of scoring criteria you want to use.
3. Value and importance of the criteria.
4. Measurement of the criteria.

First, we will deal with formation of the model. A *generic* scoring model would have the following form:

$$\text{Score} = \frac{A(bB + cC + dD)(1 + eE)}{fF(1 + gG)}$$

The symbols *A, B, C, D, E, F,* and *G* represent the criteria to be included in the score for the project. The *value* of each criterion for a given project is substituted in the formula. The symbols *b, c, d, e, f,* and *g* represent the *weights* assigned to each criterion. In the model, the criteria in the numerator are *benefits,* while the criteria in the denominator are costs or other *disbenefits*. The *values* of the criteria are project specific and are normally provided by the project team.

This model uses three categories of criteria:

1. *Overriding criteria (e.g., A)*. These are factors of such great importance that if they go to zero, the entire score should be zero. For instance, factors to be included in the model might be measures of performance such as efficiency or total output. A performance measure of zero should disqualify a project completely, regardless of any other merit.
2. *Tradable criteria (B, C, D, F)*. These are factors that can be traded against one another; a decrease in one is acceptable if accompanied by a sufficient increase in another. For instance, a designer may be willing to trade between reliability and maintainability, so long as "cost of ownership" remains constant. In this case, the weights would reflect the relative costs of increasing reliability and making maintenance easier. Cost F is shown as a single criterion that is relevant to all projects. Typically, this would include monetary costs of the project. This might be disaggregated into cost categories such as wages, materials, facilities, and shipping, if there is the possibility of a trade-off among these cost categories. If no such trade-off exists, the costs should simply be summed and treated as a single factor.
3. *Optional criteria*. These are factors that may not be relevant to all projects: If they are present, they should affect the score; but if they are absent, they should not affect the score. Note that either costs or benefits may involve optional factors. For instance, *E* in the formula represents a benefit that may not be a consideration with all projects. It should be counted in the score only if it is relevant to a project. For example, this might be a rating of *ease of consumer use*, which would not be relevant to a project aimed at an industrial purpose. *G* in the formula represents an "optional" cost that might not be relevant to all projects. Typically, this type of cost is one in which the availability of some resource is a more important consideration than its monetary costs. For instance, there may be limits on the availability of a testing device, specialized computers, or a scarce skill such as a programmer. In such a case, the hours or other measure of use should be included separately from monetary cost and should apply only to those projects requiring that resource.

The second issue focuses on value and importance of criteria. Once the form of the model is selected, the designers of the model need to distinguish between the *value* of a criterion and the *weight* or *importance* of that criterion. In the preceding formula, *B, C,* and *D* are the values of their respective factors for a specific project, while *b, c,* and *d* are the weights assigned to those factors, reflecting the importance assigned to them by the decision maker. In the case of the tradable factors, the ratio *b/c* represents the trade-off relationship between factors *B* and *C*. If *B* is decreased by one unit, *C* must be increased by at least the amount *b/c* for the sum of the tradable factors to remain constant or increase. That is, the decision maker is willing to trade one factor for another according to the ratios of their weights, so long as the total sum remains constant or increases.

Finally, the third issue to resolve is one of measurement of criteria. Some criteria are objectively measurable, such as costs and revenues. Others, such as probability of success or strategic importance, must be obtained judgmentally. Scoring models can readily include both objective and judgmental criteria. It is helpful if the judgmental criteria are estimated with a scale and descriptor phrases to obtain consistency in estimating the magnitude of the factor for each project. The estimates should be made on some convenient scale, such as 1 to 10. (see "An *Example* of Subjective Measure of Criteria.") A similar scale should be devised to aid in making estimates of each of the criteria to be obtained judgmentally, as was done in the example in Table 2.3.

An Example of Subjective Measure of Criteria

10 All skills are in ample supply.
9 All skills are available with no excess.
8 All technical skills are available.
7 Most professional skills are available.
6 Some technical skill retraining is required.
5 Some professional skill retraining is required.
4 Extensive technical skill retraining is required.
3 Extensive professional skill retraining is required.
2 Most technical skills must be hired.
1 Most technical and professional skills must be hired.
0 All technical and professional skills must be hired.

Most scoring models are more complex than a simple sum of criteria. Suppose the factors we wish to include in the score are probability of success, payoff, and cost. Suppose further that we are willing to trade payoff and probability of success (e.g., we are willing to accept a project with higher risk if the payoff is high enough), and we think payoff is twice as important as either probability of success or cost. Probability of success and payoff are benefits, where cost is a disbenefit. Then the scoring model will be as follows:

$$Score = \frac{P_{Success} + 2 * Payoff}{Cost}$$

The designer of the scoring model is free to include whatever factors are considered important and to assign weights to reflects relative importance.

Table 2.3: Rating a New Product Development Project with a Scoring Model

Criteria/Factor	Item	Score Out of 10 (Points)	Average Criterion/Factor Score (Points)
	Criteria/Factors (Scored 0–10)		
Strategic positioning	Degree of project's alignment with business unit strategy (strategic significance)	8	8.0
Product/Competitive advantage	Unique product functionalities	8	8.0
	Provides better customer benefits	9	
	Meets customer value measures better	7	
Market appeal	Market size	8	7.0
	Market share	8	
	Market growth	6	
	Degree of competition	6	
Alignment with core competencies	Market alignment	8	7.0
	Technological alignment	7	
	Manufacturing alignment	6	
Technical merit	Technical gap	9	8.0
	Technical complexity	6	
	Technical probability of success	9	
Financial merit	Expected net present value	9	8.0
	Expected internal rate of return	9	
	Payback time	7	
Total Project Score			**130 out of a possible 170 points (77%)**

Scoring Projects

When the criteria for the model have been selected, the form of the scoring model chosen, weights established, and measurement scales defined, you are ready to rank the candidate projects. Note that while the decision maker(s) must obtain the criteria and their weights from management, this is a one-time activity. The project-specific data for individual projects will in most cases come from those proposing the project. They will provide either objective data (e.g., costs, staff hours, machine use) or ratings based on the scales the decision maker has established. In some cases, project-specific data may be obtained from sources other than the project originators. For instance, data on probability of market success or payoff might be obtained from marketing rather than from research and development (R&D). While this data must be obtained for each project being ranked, the criteria and their weights will remain fixed until management decides they must be revised.

In most cases, the project data will be in units that vary in magnitude: probabilities to the right of the decimal, monetary costs to the left of the decimal, scale rankings in integers, and so forth. It is necessary to convert all the factors to a common range of values. Assuming that the project-specific values are approximately normally distributed, the result should be standardized values ranging from about –3 to about +3. These must now be restored to positive values. If any of the original values in a column was zero, the standardized value for that factor should also be zero. Add to every value the absolute value of the most negative number in the column resulting from the subtraction and division process. This will result in standardized values ranging from zero to approximately 6. If none of the original values was zero, add to every number 1 plus the absolute value of the most negative number. This will give values ranging from 1 to approximately 7. These standardized values should then be substituted in the model. Each project then receives a score based on weights supplied by management and project data supplied by the project originators.

Using a Scoring Model

Table 2.4 shows the results of the model applied to standardized scores. The rows have been reordered in decreasing magnitude of the score. If the standardized values are available in a spreadsheet, the process of computing project scores is trivial. Likewise, sorting the projects in order of scores is readily accomplished using a spreadsheet. In this

Table 2.4: Ranking of Projects Using the Scoring Model

Project	Cost ($k)	P (Sxs)	Payoff ($m)	Score
4	1.89	2.67	3.35	4.96
6	2.12	3.38	2.78	4.22
3	1.00	2.13	1.00	4.13
5	2.17	3.33	2.78	4.10
8	2.51	3.88	2.37	3.43
1	1.45	2.13	1.42	3.42
12	3.58	3.56	4.22	3.35
11	3.70	3.61	3.34	2.78
2	1.62	1.00	1.58	2.57
7	2.49	3.67	1.34	2.56
14	4.44	3.78	3.54	2.45
16	6.39	3.88	5.74	2.4
9	3.11	3.56	1.91	2.38
10	4.13	3.65	1.91	2.38
13	4.11	3.98	2.53	2.20
15	5.43	3.86	2.88	1.77

example, Project 4 is the highest-ranked project. The other projects fall in order of their score. The next step would be to approve projects starting from the top of the list and working down, until the budget and/or resources are exhausted. Note that the difference between Projects 8 and 1 comes only in the third significant figure. Since most of the original data were good only to one or two significant figures, this difference should not be taken seriously.

While scoring models can be used for any type of project, they are especially useful in the earlier phases of a project life cycle, when major project selection decisions are made. Take, for example, new product development projects. In earlier project phases, market payoff is distant or even inappropriate as a measure of merit. In such projects, considerations such as technical merit—a frequent criterion in scoring models—may be of greater significance than economic payoff. Selection of other types of projects, large and small, widely relies on scoring models as well. The final score is typically used for two purposes:

1. *Go/kill decisions*. These are located at certain points within the project management process, often at the ends of project phases. Their purpose is to decide which new projects to initiate and which of the existing ones to continue or terminate.
2. *Project prioritization*. This is where resources are allocated to the new projects with a "go" decision, and the total list of new and existing projects, which already have resources assigned, are prioritized.

Although the principles behind the scoring models are relatively simple, developing an effective scoring model can be an arduous endeavor.

Benefits

The value of the scoring model is that it can be tailored to fit the decision situation, taking into account multiple goals and criteria, both objective and judgmental, which are deemed important for the decision.[9] This prevents putting a heavy emphasis on financial criteria that tend not to be reliable early in the project life. With such an approach, decision makers are forced to scrutinize each project on the same set of criteria, focusing rigorously on critical issues but recognizing that some criteria are more important than others (by means of weights).

Scoring models are also conceptually simple. They trim down the complex selection decision to a handy number of specific questions and yield a single score, a helpful input into a project selection effort. This is perhaps a major reason for scoring models' wide popularity.

Finally, scoring models produce results. Several studies showed that they yield good decisions. For example, Procter & Gamble claims that their scoring models provide an 85 percent predictive ability.[10]

VOTING MODELS

For some organizations, numeric scoring models can become complex and cumbersome. We have found that when scoring models fail within an organization, they do so for a couple of primary reasons. First, teams get embroiled in trying to design the perfect scoring model. An enormous amount of time and energy can be lost in debating the right scoring criteria, the definition of numeric value for each criterion (e.g., how to describe a scoring value of 1 versus 2 versus 3, etc.), and which criteria are more important than the others so weights can be assigned.

Second, numeric scoring models quite often fail to provide good scoring separation between projects. One executive from a financial services organization describes this outcome as "the scores tend to munge to the middle" to form a bell curve effect. When this happens, it gets very difficult to evaluate which projects are the best investment choices.

For these reasons, some organizations adopt an approach that provides the necessary structure and information to make project investment choices but relies on the experience and judgment of a cross-section of informed stakeholders. Voting models are effective in facilitating this judgmental approach by providing a technique to tap the *collective knowledge* of a group of experts with diverse perspectives. Voting models take advantage of these diverse perspectives to create a clear understanding of the highest-priority projects, increasing the organizational knowledge about the value proposition of each project, creating broad buy-in of project priorities across the organization, and clearly separating the wheat from the chaff.

Developing a Voting Model

The process for developing a voting model is very similar in nature to developing a numeric scoring model, but simplified. The primary difference is in how project value is evaluated and scored. Simplification is achieved through the limitations imposed on the scoring structure. Project value for any one criterion is limited to three values (1-2-3, H-M-L, etc.), and no weighting of criteria is needed. The major steps in developing a voting model are described next.

Identify the Stakeholders

Critical to a successful outcome in using voting models is the assembly of the right set of stakeholders who will have a vote on project value and prioritization. The intent is to assemble a set of stakeholders who have a vested interest in the prioritization outcome and represent a good cross-section of functional perspectives. It is advisable that *no more* than 12 to 15 stakeholders participate in a voting event.

Develop Value Propositions for Each Project

For each of the candidate projects, a brief value proposition should be prepared for use in the voting event. Limit the amount of content to that which can be presented to the stakeholders within a maximum of three to five minutes. If the potential value of a project cannot be adequately communicated within a three- to five-minute time limit, additional vetting of the project is needed and should be held until a later date so it won't be mistakenly viewed as being of lesser value than other candidate projects that have a mature value proposition prepared.

Create the Prioritization Criteria and Value Anchors

For voting models, the prioritization criteria have to be limited to the *critical few* criteria. The number of criteria *must* be limited to a maximum of three to five. For each criterion, a description of the criterion and the value anchors must be prepared prior to the voting event. Table 2.5 shows an example of criteria and value descriptions.

Create a Voting Template

Voting models are best used in a facilitated work session where all critical stakeholders can be assembled and provided the opportunity to discuss and debate the value of each of the candidate projects. To facilitate the collection of discussion outcomes and information, it is best to create a work template for use in the voting event. Figure 2.5 illustrates an example voting model template that can be used in either physical or electronic format. Some of the most productive prioritization sessions we have witnessed have been those that use a physical voting template that is large enough to hang on a conference room wall.

Using the Voting Model

Since voting models rely on the expert judgment of a cross-section of stakeholders, it is best if a face-to-face work session is used to ensure that proper collaboration and communication takes place. It is within this context that the following steps are recommended for using a voting model.

Step 1: Validate the voting criteria. The initial set of criteria and voting anchor descriptions developed prior to the work session are presented to the stakeholders in this initial step. The intent is to level set the stakeholders on the criteria and how the criteria will be evaluated, and gain buy-in from the cross-section of stakeholders. If necessary, the criteria and voting descriptions can be modified during this step.

Table 2.5: Example Criteria Description and Value Anchors

Criteria	A	B	C
Monetary value	Clear path to money, high ROI	Clear path to money, low ROI	No clear path to money
	> $100m ROI (3 yr)	$50–100m ROI (3 yr)	<$50m ROI
Strategic value	Severe competitive threat	Moderate competitive threat	Low competitive threat
	Must do for leadership or time-critical neutralizer	Moderate for neutralization activities	No direct map to strategy
Market pull	Capability requested by customers	Valid interest by customers	No pull or interest from customers
Complexity and risk	Low complexity, low risk	Moderate complexity and risk	High complexity, high risk
Effort and cost	Low effort, cost < $3m	Low effort, cost $3–9m	Large effort, cost > $9m

ROI = return on investment.

Voting Model Work Template							
			Prioritization Criteria				
Candidate Projects	Project Priority	Wisdom of Crowds Votes	Criterion 1	Criterion 2	Criterion 3	Criterion 4	Criterion 5

Figure 2.5: Example Voting Model Template

Step 2: Review the list of candidate projects. This step involves reviewing all candidate projects that the stakeholders will be evaluating during the prioritization process. To expedite time, the projects can be prepopulated in the voting template, as shown in Figure 2.6. The intent is to ensure that all projects are listed and also make an initial determination if any of the projects should be eliminated from consideration or if others should be added to the list.

Voting Model Work Template							
			Prioritization Criteria				
Candidate Projects	Project Priority	Wisdom of Crowds Votes	Criterion 1	Criterion 2	Criterion 3	Criterion 4	Criterion 5
Project 1							
Project 2							
Project 3							
Project 4							
Project 5							
Project 6							
Project 7							

Figure 2.6: Candidate Projects Prepopulated in Work Template

Step 3: Project value proposition and initial voting. Taking each candidate project in the list in order, the project representative describes the project value proposition in three to five minutes. Stakeholders will then have an opportunity to ask clarifying questions and debate the value proposition. It is in this step that an expert facilitator is needed to gauge the discussion and debate for focus, value, and time constraints.

At the end of each discussion, the stakeholders are asked to vote on the project for each criterion by a show of hands (or electronic vote if blind voting is desired). The project receives a vote of H, M, L (or 1, 2, 3) for each criterion, as shown in Figure 2.7. This process is repeated for each candidate project on the list.

Through this step in the process, two significant things are accomplished. First, an initial assessment of candidate project value is debated and scored. Second, the stakeholder's knowledge about the intent and potential value of each candidate project is increased. It is common at this point to want to make a final judgment about project priority. However, doing so creates a fatal flaw that many numeric scoring models fall into. Little separation will still exist between the highest-value projects and lowest-value projects. To create more separation, one final step is needed.

Step 4: The "wisdom of crowds" vote. Tapping into a technique that James Surowiecki termed the *wisdom of crowds* provides the opportunity to let your new *wise* set of cross-organizational stakeholders determine the highest-value projects.[11] In this step, each stakeholder is given a set number of votes (e.g., ten) that he or she can place against the candidate projects. However, each stakeholder has a maximum limit on the number of votes (e.g., three) that he or she can place against any one project. This constraint prevents a stakeholder from placing all votes on one particular project. Once again, this voting can take place in the open using the voting model template or electronically if blind voting is preferred.

Voting Model Work Template							
			Prioritization Criteria				
Candidate Projects	Project Priority	Wisdom of Crowds Votes	Criterion 1	Criterion 2	Criterion 3	Criterion 4	Criterion 5
Project 1			M	H	H	L	L
Project 2			L	H	H	M	M
Project 3			L	M	H	L	L
Project 4			L	M	M	H	H
Project 5			L	M	L	H	L
Project 6			H	M	M	M	L
Project 7			L	H	H	M	L

Figure 2.7: Initial Value Voting Assessment Results

Voting Model Work Template							
			Prioritization Criteria				
Candidate Projects	Project Priority	Wisdom of Crowds Votes	Criterion 1	Criterion 2	Criterion 3	Criterion 4	Criterion 5
Project 1	25	H	M	H	H	L	L
Project 2	19	H	L	H	H	M	M
Project 3	17	H	L	M	H	L	L
Project 4	12	H	L	M	M	H	H
Project 5	11	H	L	M	L	H	L
Project 6	9	M	H	M	M	M	L
Project 7	8	M	L	H	H	M	L
Project 8	8	M	L	H	H	M	M
Project 9	5	L	M	H	H	L	L
Project 10	3	L	L	M	M	H	L

Figure 2.8: Ranking of Projects into High, Medium, and Low Priority

A final tally of votes is conducted, and the top candidate projects are identified. Figure 2.8 shows an example of a fully populated voting model with three bands of prioritized projects based on the results of the voting. The top band (projects 1-5) contains the highest-value projects, which need to be funded and initiated. The bottom band (projects 9-10) contains the lowest-value projects, which need to be eliminated or recast to provide more value. The middle band of projects needs additional scrutiny but can be funded and initiated if resources are available after all top-level projects are initiated.

Benefits

Voting models provide an alternative project selection approach for organizations that get themselves bogged down in the minutia of numeric scoring models. They often provide an accelerated path from project opportunity identification to full project ranking, especially if a large number of candidate projects are involved.

As shown in Figure 2.8, voting models can provide clear separation between projects that the selected group of stakeholders deem of highest value and those that are deemed of low value. With those of highest and lowest value identified, the decision maker has fewer candidate projects requiring further scrutiny and debate.

Finally, since a cross-section of organizational stakeholders are involved in the determination of project ranking via the voting model technique, a greater level of buy-in of the final project ranking decision tends to occur (see "A Rock Star Votes").

A Rock Star Votes

Prioritizing a set of technology development projects can be a challenging task. This is especially true in an organization like Intel, where a significant number of projects are vying for funding and resources, and each project has an influential and respected technologist championing its value proposition for the company.

Being the person responsible for the final decision about which new technology development projects will be funded and which won't requires a track record of success, such as the track record of a gentleman named Ajay. Ajay was a co-inventor of the universal serial bus (USB) and was featured in one of the company's "Intel Rock Stars" television commercials.

However, as Ajay learned, being an expert in your field and possessing a track record of success does not give you a decision-making mandate. Making project ranking and selection decisions is relatively easy; the hard part comes in gaining organizational buy-in for the decisions you have made. In Ajay's case, two problems consistently emerged once a mandate decision was made:

1. People continued to work on technology projects that were not officially selected and funded.
2. The product planners who made the decisions on which technologies to include in a project sometimes didn't agree with Ajay's decision, making it difficult for the project to transition from technology development to project development.

Ajay turned to the voting model as a potential solution to the problems. He began by inviting both technology experts and product planners to the project prioritization discussions. Through the voting model technique, the product planners have an equal say in which technology projects were most valuable, but from a product and user perspectives (everyone has the same number of votes).

Two other behaviors were also critical. First, Ajay remained quiet during the questioning and debate on project value; second, he made sure he voted last in order to not influence the voting of the other members in the session.

At the end of the session, he had a prioritized project list that was not too far from how he would have personally ranked the projects. But most importantly, he gained alignment with the product planners on which project technologies would eventually make their way into Intel products. Some of them we are now using with our personal computers.

PAIRWISE RANKING

When a small number of candidate projects are involved in the selection process, the pairwise ranking technique is an effective tool for identifying priority order of the projects. A number of decisions have to be made when comparing candidate projects, so it is necessary that the decision method be identified up front (i.e., consensus, autocratic, consultative). Additionally, it must be determined whether a numeric or judgmental approach to comparing one project against another will be used.

Developing a Pairwise Ranking Tool

Development of a pairwise ranking tool is a very simple exercise consisting to two primary steps. First, a comparison matrix must be constructed that represents the number of candidate projects to be compared.[12] Second, a set of criteria must be identified and documented.

Construct a Comparison Matrix

The construction of the comparison matrix is a simple exercise that is dependent on the number of candidate projects that will be ranked. Figure 2.9 shows a pairwise comparison matrix for six candidate projects.

Care should be taken to ensure that the numbering schema is accomplished correctly. The numbers representing each of the six projects should be laid out in *ascending* order from top to bottom in both the vertical and diagonal axis.

Document the Comparison Criteria

A set of criteria from which each project pair will be compared must be identified and documented to ensure consistency in comparison. These criteria will serve as the basis for comparing each of the candidate projects against one another. Try to limit the criteria to three items, five as an absolute maximum. Anything greater than five criteria adds unnecessary complexity and makes the tool and comparison exercise more complicated than it needs to be. Pairwise ranking criteria may include some of the following:

- Net present value
- Return on investment
- Payback period
- Level of strategic alignment
- Cost savings
- Market share increase
- Affordability
- Usability

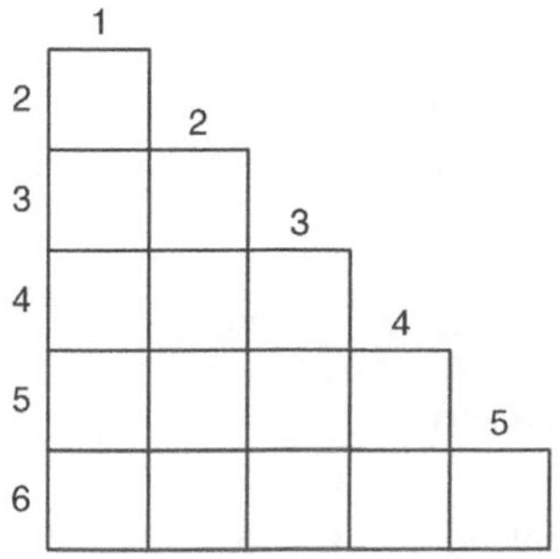

Figure 2.9: Pairwise Ranking Comparison Matrix

Using the Pairwise Ranking Matrix

With a comparison matrix constructed, the comparison criteria identified, the decision method agreed upon, and the right set of stakeholders assembled to compare the candidate projects, the following steps are recommended for using pairwise ranking.

Step 1: Rank each project pair. Sytematically work through the candidate projects, comparing one against another, but following the structure created by the comparison matrix. For each pair of projects, the stakeholders will determine which of the two candidate projects is preferred based on the criteria identified. The number of the preferred project is then inserted into the comparison matrix. This project-to-project comparison is repeated until the comparison matrix is completely filled, as demonstrated in Figure 2.10.

Step 2: Tally comparison results. Using a simple scorecard (Table 2.6), tally the number of times each candidate project was preferred during the project-to-project comparison exercise. This is accomplished by viewing the inputs in the fully populated comparison matrix.

Step 3: Rank the candidate projects. Using the tally information from the previous step, rank the candidate projects based on the number of times they are preferred. Table 2.7 illustrates the priority ranking of the six projects used in this example.

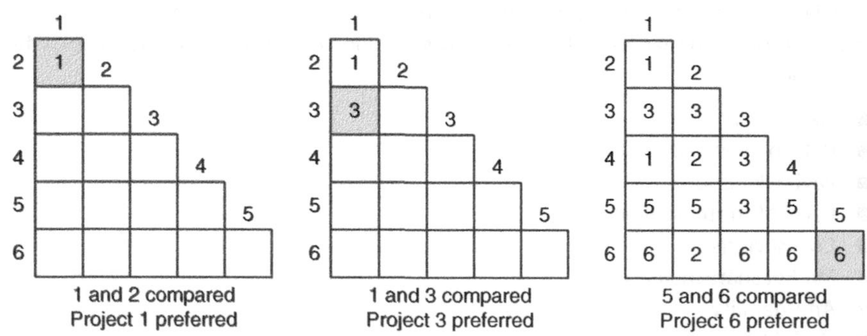

Figure 2.10: Project-to-Project Comparison

Table 2.6: Project Preference Tally

Candidate projects	1	2	3	4	5	6
Times preferred	2	2	4	0	3	4

Table 2.7: Candidate Project Ranking

Candidate projects	1	2	3	4	5	6
Times preferred	2	2	4	0	3	4
Project ranking	**4**	**5**	**2**	**6**	**3**	**1**

In cases where there is a tie between two candidate projects, refer to the comparison matrix created in Step 1 and find the box where the two projects were compared against one another. The project that was preferred between the two receives the higher ranking. In the preceding example, Project 3 and Project 6 were each preferred four times. By referring back to Figure 2.10, one can see that when the two projects were compared against one another, Project 6 was preferred. Therefore, Project 6 receives the higher ranking.

Benefits

Pairwise ranking helps to make a complicated process (prioritizing a set of candidate projects) much more simple by eliminating the tendency to try to sort through a list of projects holistically. Through the project-to-project comparison methodology, only two projects at a time have to be evaluated.

The tool itself forces a structured process and makes the project comparison results immediately visible to the stakeholders and decision makers. The tablature nature of the final ranking results is also more visually effective than other project scoring and ranking tools.

THE ALIGNMENT MATRIX

The project alignment matrix is used to establish the degree to which a project is aligned with the organization's business strategy (see Figure 2.11). This is, of course, also

Example Business Strategies	Project 1	Project 2	Project ...	Project n
Provide clearly differentiated products from our competitors	P	F		P
Consistently deliver performance increase in high-speed devices	F	N		P
Be the first to market with new products	F	F		F
Supports the common platform architecture	F	F		F
Continuously reduce manufacturing cost	N	F		P

Legend: F = Fully Supports P = Partially Supports N = Does Not Support

Figure 2.11: Example Alignment Matrix

balanced against the specific customer needs that the firm is attempting to meet. The alignment assessment aids the project manager and top managers in understanding how well a project supports the strategic goals of the firm. With this in place, each project concept can be evaluated on the basis of cost, benefit, risk, and strategic importance.

Developing the Alignment Matrix

The first column of the matrix contains the list of the organization's strategic business goals that serve as criteria to align projects with the organization's strategy. Then, in each of the remaining columns, the degree of alignment of individual projects with each goal is assessed using a qualitative scale. As an outcome, a qualitative goal-by-goal alignment evaluation for each project is generated that may be used for different strategic purposes.

The assessment of a project's alignment with an organization's business strategy calls for information that typically comes from three inputs:

1. Approved business strategy.
2. The portfolio of projects.
3. Project business case (or preliminary business case information).

The approved business strategy provides a list of the organization's business goals that the strategy is striving to accomplish. To assess the degree of alignment of individual projects to the strategic goals, a list of current and future projects is needed. This information is typically found in project portfolio documents. Finally, to understand how well each project is aligned with the strategic goals, the preliminary project business case is needed.

Identify the Organization's Strategic Goals

Strategic goals are defined by the organization's senior management, sometimes formally in strategic plans, at other times informally. In either case, the goals should be used for the alignment matrix assessment. Since each organization is a unique entity, the list of strategic goals found in the alignment matrix will be unique as well. Additionally, as the strategic goals of an organization are updated and modified, the list of goals in the alignment matrix needs to be updated accordingly.

Identify the Projects

There are two steps to this action. First, the names of the new and existing projects are entered into the columns of the alignment matrix. As stated earlier, the list of the projects is normally part of the portfolio of projects documentation. If a formal portfolio of projects does not exist, an active project roster will suffice. Second, the project strategy and goals should be developed and documented to secure the information needed to assess alignment of projects to strategy.

Define the Alignment Scale

Scales vary, of course, and the choice of scale to use is organization specific. We believe that a simple, qualitative scale is completely adequate and provides the value we want from this matrix. An example of a simple qualitative scale is the three-level scale shown in Figure 2.11. The scale includes *Fully Supports* for the highest degree of alignment, *Partially Supports* for the medium-level alignment, and *Does Not Support* for no alignment of the project with a specific goal.

Assess the Degree of Alignment

Now it is time to assess each project's alignment with each organizational business goal using the adopted scale of assessment. Typically, decision makers who do the assessment should use a collaborative work session format, where information from multiple perspectives is exchanged, and the assessment decisions are shared and consensual.

Variations

An information technology (IT) organization for a major financial institution uses a variation of the alignment matrix, which they call the strategy alignment map (Figure 2.12) to gain alignment between the IT strategy and business strategy. The map is used to ensure that their IT projects complement the needs of the business. Their IT Project Management Office director, Melida Ramos, explains

> The alignment chart provides a strategic mapping of the business goals, the business values, and the IT projects. As part of the strategic changes which occurred last year, we were tasked by the company to design the alignment process, part of which was accomplished by the alignment map which helped us visualize the alignment between business goals, business value, and the projects.

A bubble on the intersection of Strategic Business Goal 4 and Business Value 3 means that they are aligned. Further, a bubble indicates that Business Value 3 intersects with Project 3, meaning they are aligned. In summary, Project 3 delivers Business Value 3,

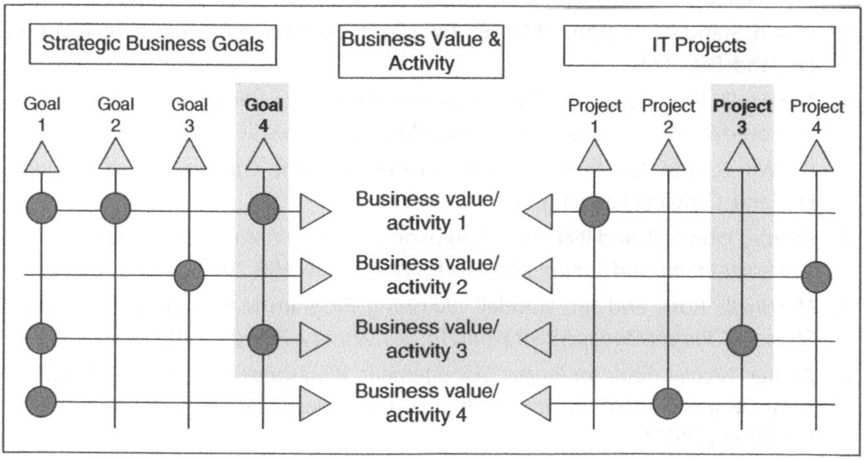

Figure 2.12: Strategy Alignment Map

which helps to achieve Strategic Business Goal 4. The alignment is about IT projects contributing to business benefits by helping to achieve the strategic business goals.

Benefits

Typically, the alignment matrix is prepared for project portfolio management reviews when projects are assessed with respect to alignment with the strategies, goals, and objectives of an organization. Based on the alignment information obtained by means of the alignment matrix, the preliminary selection and risk balancing of projects may be changed and the final project portfolio adopted, providing improved alignment of projects with the strategic goals of the organization.

The alignment matrix enables an organization to refine the selection and risk balancing of the preliminary portfolio of projects by pointing to projects that are best aligned with the organization's business goals.[13] Based on that, one can eliminate some preliminary selected projects that are not strategically aligned; one can also add new projects that are better aligned. That is the matrix's value, which is strategically precious given how difficult it is to select the most valuable projects that are risk balanced and are also aligned with the strategic goals of the organization.

Additionally, it requires that senior managers of an organization develop and document the organization's strategic goals. The alignment matrix aids the project manager in understanding how well his or her project supports the strategic objectives of the firm, therefore helping him or her in creating the project strategy.

References

1. Venkataraman, Ray R., and Jeffrey K. Pinto. *Cost and Value Management in Projects* (Hoboken, NJ: John Wiley & Sons, 2008).
2. Thiry, Michael. *A Framework for Value Management Practice* (Newton Square, PA: Project Management Institute, 2013).
3. British Standard Institute (BSI). *BS EN 12973:2000: Value Management* (London, England: BSI, 2000).
4. Martinelli, Russ, James Waddell, and Tim Rahschulte. *Program Management for Improved Business Results*, 2nd ed. (Hoboken, NJ: John Wiley & Sons, 2014).
5. De Wit, B., and Ron Meyer. *Strategy: Process, Content, Context*, 4th ed. (London, England: Cengage Learnings, 2010).
6. Pearce, John A. II, and Richard B. Robinson Jr. *Strategic Management: Formulation, Implementation, and Control* (New York, NY: McGraw-Hill, 2000).
7. Martinelli, Russ, and Jim Waddell. "Aligning Program Management to Business Strategy." *Project Management World Today,* January–February 2009.
8. Andler, Nicolai. *Tools for Project Management, Workshops and Consulting: A Must-Have Compendium of Essential Tools and Techniques* (Erlangen, Germany: Publicis Publishing, 2011).

9. Morris, P. W. G., and A. Jamieson. "Moving from Corporate Strategy to Project Strategy." *Project Management Journal* 36 (4): 5–18, 2005.
10. American Productivity and Quality Center. *New Product Development: Embracing an Adaptable Process*, Final Report, APQC, 2010.
11. Surowiecki, James. *The Wisdom of Crowds* (Harpswell, ME: Anchor Publishing, 2005).
12. Andler, 2011.
13. Pearce, John A. II, and Richard B. Robinson Jr. *Strategic Management: Planning for Domestic and Global Competition* (New York, NY: McGraw-Hill, 2014).

3
PROJECT INITIATION

The primary outcome of the project selection process is, of course, an approval decision to initiate a project. We must remember that while going through the selection process, projects are normally little more than a set of ideas, goals, and assumptions. If the ideas and goals stand up to the scrutiny of the project selection criteria and process, then a set of actions must take place to reformulate the ideas into a project construct in which the ability to execute upon the ideas can be evaluated. Project initiation is aimed at establishing a firm foundation to validate that the business case underwriting a project is achievable and that a successful project outcome is possible.[1]

Due diligence is needed during the project initiation process to ensure that financial, capital, and human resources are committed to a project that has its objectives adequately defined, level of complexity understood, roles and responsibilities defined, and the business case validated. All parties involved must be clear on what a project is intended to achieve, why it is needed, and how the outcome will be accomplished.[2]

A number of tools that we describe in this chapter are available to assist in the project initiation process. Consider adding the following tools to your PM Toolbox. We begin with the checklist questions for project initiation.

CHECKLIST QUESTIONS FOR PROJECT INITIATION

Management teams in successful and innovative companies fully understand that some of the greatest opportunities reside in the *fuzzy front end* of a project.[3] The ability to accurately forecast future customer, user, and market needs and then integrate those needs with leading-edge capabilities is critical for companies to survive in their respective industries.[4] This work is never simple, and high levels of ambiguity have served as a test in frustration and a lesson in patience for many.[5]

However, the presence of ambiguity is a characteristic of many projects, particularly in the early stages of project initiation and initial planning. The ability to efficiently and effectively converge on the business need, the value proposition, the roles and responsibilities, and a clear set of business objectives during project initiation is crucial. A set of checklist questions for project initiation is a very good tool for establishing an approach for managing the project initiation process and for ensuring that the necessary information is collected.

Developing Checklist Questions for Project Initiation

Taking the time to develop a standard set of questions for an organization is good practice, as it drives consistency in project initiation content and outcomes. The questions contained within the checklist can be developed by first understanding the various work activities and project artifacts that result from the activities (such as the project business case). Questions will center on what work has to be completed and what outcomes have to be created.

Then, additional questions can be developed by understanding what information the project sponsor and primary stakeholders require in order to make the critical decisions involved during project initiation. The checklist questions for project initiation will be different for every organization because every organization requires its own unique set of information and outcomes to support its project initiation activities.

Table 3.1 illustrates a sample set of questions that can be used as a reference for developing your own checklist questions for your PM Toolbox.[6]

Table 3.1: Sample Checklist Questions for Project Initiation

Status	Checklist Questions
☑	Do we understand the primary goal of the project?
☑	What are the business benefits derived from the project?
☑	What strategic goals does this project support and enable?
☑	What do we want to achieve and avoid?
☑	What are the project objectives: business, financial, organizational, social, etc.?
☑	Is there agreement and alignment on the objectives (in particular with the project sponsor)?
☑	Do we understand the problem(s) we intend to solve?
☑	Is the problem clearly defined, differentiated, and documented?
☑	Do we understand the solution requirements?
☑	Do we understand the priorities of the solution requirements?
☑	Is the solution concept convincing and realistic?
☑	Are we ready to make a decision about the proposed solution concept?
☑	What are our base assumptions?
☑	Have we verified all our assumptions with our stakeholders?
☑	What are our project constraints?
☑	What are the boundaries and limitations?
☑	Do we have primary roles and responsibilities documented?
☑	Do we understand the project risks?
☑	Do we have mitigation plans for the high-impact, high-likelihood risks?
☑	What will serve as evidence of project success or failure?
☑	Are we ready to make a decision to transition to project planning?
☑	Can we justify the decision internally and externally?

The status column can be used to indicate whether sufficient information has been discovered to answer the question. Some project managers add a third column to the checklist labeled "artifact" or "documentation" to indicate where the answers to the checklist questions are documented. For example, they may be documented in the project charter, responsibility matrix, or project business case.

Using the Checklist Questions

Once a project has been selected and a project manager is assigned ownership of it, he or she becomes immersed in a series of activities focused on learning as much as possible about the intent of the project as well as the environment in which the project will exist. The more ambiguous a project, the more there is to learn. The old saying, "We don't know what we don't know," applies when beginning the project initiation process.

Use of the checklist questions for project initiation can be a means of breaking through the ambiguity of a project and beginning to achieve clarity. It should therefore be developed very early in the project initiation process.

Obviously, not all answers to the questions contained within the checklist will come quickly. Depending on the complexity associated with the project, it may take a number of weeks or, in some cases, months to discover the right information needed to satisfy the requirements of project initiation. Use the checklist throughout the initiation process as a guide and focusing mechanism to peel away the layers of ambiguity to discover the core information needed to successfully initiate a project and prepare the project team to move into planning activities.

THE GOALS GRID

We have all read or heard about the reports that describe the staggering rate of project failures. It is true that most of the reports are now written by consultants who want to sell us services to improve our success rate. However, there is merit in the findings even though the situation may not be as dire as reported. One of the factors that is consistently identified as a contributing factor is the failure to define or document the criteria for project success.

Establishing project goals is a task that is many times delegated to the project manager. Saying that a project manager must define what success looks like for his or her project is easy to say but hard to do in many instances, because project success is usually outside of the project manager's purview. This is really a task for the senior sponsor and leadership team of an organization. The project manager may participate in the documentation of the project goals, but the senior leaders need to define what constitutes project success; because, ultimately, they will be the ones judging final success or failure.

For a project manager to document the project success factors, therefore, he or she must engage the project sponsor and primary stakeholders in conversations about what they believe constitutes project success. The outcome of these conversations then needs to be documented as a set of project goals, and then verified.

The goals grid is an effective tool for helping the project manager facilitate the discussions with the project sponsor and stakeholders, and then documenting the project goals. It was originally developed to help senior managers establish a set of strategic goals but can be effectively repurposed to help a project manager establish project-level goals.

The goals grid is a simple 2 × 2 matrix constructed to help answer two basic questions: (1) Do you want something?, and (2) Do you have it? By applying Yes and No answers to each of these questions, four categories of project goals emerge that are centered on the following questions:

1. What do we want to achieve?
2. What do we want to preserve?
3. What do we want to avoid?
4. What do we want to eliminate?

These four key questions help us to derive our project goals and achieve goal clarity by prompting us to think about our goals in a structured manner and from four different perspectives.

Developing a Goals Grid

Creating a goals grid is a very simple exercise. As illustrated in Figure 3.1, the goals grid is a 2 × 2 matrix.[7] Along the vertical axis is the first primary question to be answered, "Do you want it?"; along the horizontal axis lies the second primary question to be answered, "Do you have it?"

The upper left quadrant is then set up to explore the goals you want to *achieve*. Goals in this quadrant are identified from the perspective of *yes, I want something, and no, I don't have it.*

The upper right quadrant of the grid is set up to explore the goals that you want to *preserve*. Goals in this quadrant are identified from the perspective of *yes, I want something, and yes, I already have it.*

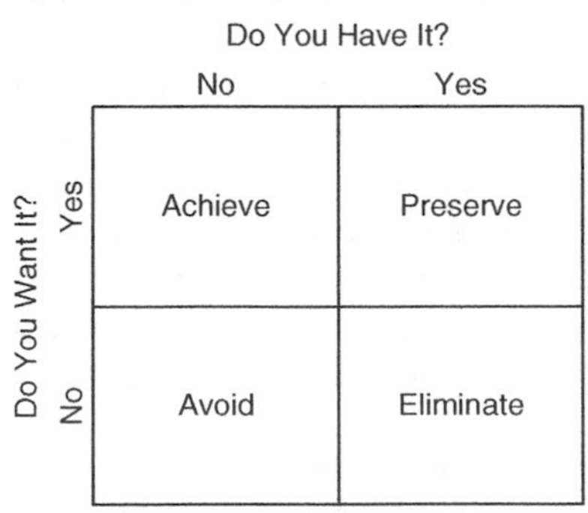

Figure 3.1: The Goals Grid Structure

The lower left quadrant is set up to explore things that should be *avoided*. Things that should be avoided are identified from the perspective of *no, I don't want something, and no, I don't already have it*.

Finally, the lower right quadrant of the grid is set up to explore project goals associated with *eliminating* things that you feel you should shed going forward. Goals in this quadrant are identified from the perspective of *yes, I have it, and no, I don't want it*.

With the basic structure of the goals grid in place, a project manager is now ready to use the tool to facilitate the necessary conversations with his or her sponsor and primary stakeholders in an effort to identify and document the goals of the project.

Using the Goals Grid

There are four primary steps in the use of the goals grid: (1) collecting input from stakeholders on what they believe are the goals for the project; (2) synthesizing the goals into the *critical few;* (3) validating that the goals align to the organization's values, priorities, and strategies; and (4) using the goals grid as the project compass.

- *Step 1: Collect input*. This step begins with the identification of the people from whom you will solicit feedback concerning the project goals. The project sponsor should always be involved, as well as those individuals who have a direct stake in the success or failure of the project. This group of individuals are your primary stakeholders (Chapter 15).

  Solicitation of feedback on project goals can be conducted in either a work session where the project manager facilitates the discussion around the questions associated with each quadrant of the goals grid, or a series of one-on-one discussions with each of the stakeholders. Either approach will result in the desired result, but it is normally a matter of personal preference on the part of the project manager as well as the cultural norms of the organization.
- *Step 2: Synthesize the goals identified*. It is normal that after the collection of project goals is complete, more goals are documented than could ever be realized. When this happens, the goals have to be synthesized to the critical few.

  Begin by identifying the duplicate goals within each quadrant and eliminate all but one that captures the essence of success the best. Next, identify goals that are in direct conflict with one another and work directly with the senior sponsor to broker the conflicting goals appropriately. Finally, within each quadrant, combine the goals into groups of related themes and work to create a single, common goal that encompasses the definition of success for each grouping.
- *Step 3: Validate the results*. With the goals pared down and synthesized into a manageable and meaningful set, validate with your sponsor and primary stakeholders that the set of goals contained within the goals grid adequately defines success for the project undergoing the initiation process.

  The goals grid is very effective in helping the project stakeholders and project manager think about goals in terms of the categories making up the grid, but it does not by itself ensure that the goals are consistent with the organization's values, priorities, and strategic objectives. Therefore, validation of the goals contained within the goals grid should also focus on alignment of the goals, and ultimately the project, to three factors: organizational values, priorities, and strategic objectives.

- *Step 4: Use the goals grid as your compass.* Throughout the course of a project, a project manager is faced with hundreds of decisions that have to be made. The goals grid should be used throughout project planning and execution as a guide to ensure that decisions remain in alignment with the goals of the project. This will help prevent the series of decision outcomes from gradually steering a project off course and away from project success.

Benefits

From a usage perspective, the greatest benefit is that the goals grid is simple to create, easy to use, and flexible, and guides the right conversations needed to adequately identify project goals.

From a functional perspective, the goals grid offers benefits to the project manager as well. Because of its structure, it guides those using it though a logical progression of thought for goal identification. Instead of only thinking about what it is you'd like to achieve, it also guides discussion about goals focused on what you would like to preserve, avoid, and eliminate. This gives the project manager a richer set of goals for the project.

Through the validation step described previously, the goals grid provides a check and balance of the alignment of the goals to the values, priorities, and strategic objectives of the organization. This provides verification that the project is on track to the intent of the investment decision that was made during the project selection process (Chapter 2).

Finally, use of the goals grid provides an effective means for developing group consensus among the project stakeholders on how project success is defined. It is better to establish consensus at the early stages of a project than to witness a debate about whether a project is successful at the end of the project cycle. This consensus building also creates broad buy-in that the project is of value to the organization.

THE RESPONSIBILITY MATRIX

By nature, project initiation is fraught with ambiguity caused by the sheer fact that the amount of information that is unknown is normally significantly greater than the amount of information that is known. Therefore, establishing clarity is a primary focus for the project manager during initiation of a project.

An important aspect of establishing early clarity is figuring out who needs to be involved with the project, what they will be responsible for, and how they will participate. A responsibility matrix is an effective tool for establishing this *who, what,* and *how* relationship between the project elements and project players.

Completing a high-level responsibility matrix during project initiation will give the project manager insight into how responsibility will be shared across the various aspects of the project without having to go into task-level detail, which is yet to be established. Doing so removes at least one element of project ambiguity.

Developing a Responsibility Matrix

Using a high-level responsibility matrix during the early stages of a project requires one to take a top-down approach to formulating the project. Much like the design of a residential or commercial building, top-down project formulation begins with creating the project architecture.

Create the Project Architecture

The term *architecture* refers to the conceptual structure and logical organization of a system. It includes the elements of the system and the relationships among them.[8] A *project architecture* is therefore the conceptual structure and logical organization of a project.[9]

By way of example, Figure 3.2 illustrates a project architecture for an information technology (IT) project focused on transitioning a firm's workforce work platforms from desktop systems to laptop systems. It is a high-level conceptual design showing the major components that comprise the project.

The initial responsibility matrix will include the people who are responsible for the successful development of each component contained in the project architecture. While this is a good beginning, it is not entirely sufficient. To derive more utility from a responsibility matrix, an initial step of identifying the primary outcomes and deliverables for each of the project components should also be completed. Figure 3.3 illustrates a possible initial responsibility matrix for the workforce mobilization project described previously.

Identify Project Players

With the project architecture defined and the primary outcomes for each of the project components identified, the next step involves determining who within the organization (or partner organizations) are responsible for the various outcomes. It is common during this early stage of a project that some project players are not yet identified by

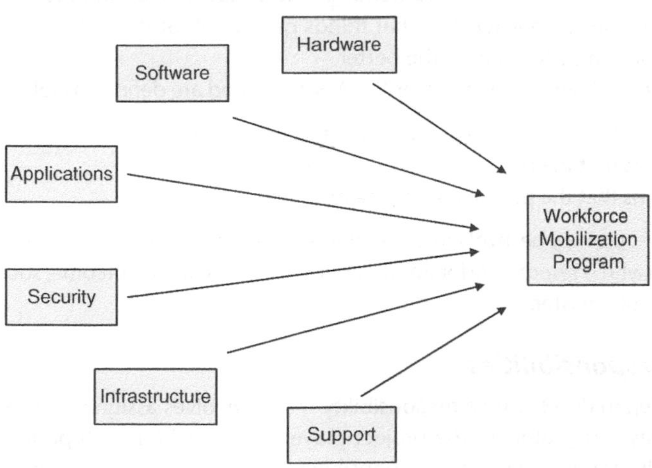

Figure 3.2: Workforce Mobilization Project Architecture

	David P.	System Designer	Marianne V.	Aaron M.	Bridget B.	Victor P.	Gina C.	Siva G.
HW Solution								
HW Definition								
HW Requirements								
SW Platform								
SW Architicture								
SW Requirements								
Applications								
System Security								
Infrastructure Definition								

Figure 3.3: Initial Responsibility Matrix Structure

name. When this is the case, use a role designation, such as "scrum master," instead of a person's name.

At this point, the basic structure of the responsibility matrix is created (see Figure 3.3). As shown, each project element and nested outcomes are listed in the left-hand column of the matrix. The project players identified (or roles) are in turn listed along the top row.

Select Responsibility Designations

In order for the matrix to be effective, it must accurately reflect people's expectations and responsibilities. This is accomplished by using different participation types that designate the appropriate level of responsibility for project outcomes.

Many variations of participation designators exist; we particularly like the "A-R-C" method because of its simplicity, something that is important during the early stages of a project. Our philosophy remains that things get complicated in a hurry on a project; the more you can do to simplify, the better.

The responsibility designators for the A-R-C method are defined as follows:

- "A" means that the person *approves* the project outcome.
- "R" means that the person *reviews* the outcome.
- "C" means that the person *creates* the outcome.

Many people may be involved in creating a project outcome, but usually there is only one person who is *responsible* for approving the creation of the outcome, such as a team leader or scrum master.

Assign Responsibilities

The final step in developing a responsibility matrix involves assigning the appropriate responsibility designators to the project players identified for each project outcome. Figure 3.4 illustrates a completed responsibility matrix for the project defined earlier.

	David P.	System Designer	Marianne V.	Aaron M.	Bridget B.	Victor P.	Gina C.	Siva G.
HW Solution		A	C					
HW Definition		R	C					
HW Requirements	A	R	C					
SW Platform		R			A		C	
SW Architecture		R			A		C	
SW Requirements						A	C	R
Applications					C	R	C	A
System Security	A		C	R				
Infrastructure Definition	A	R	C					C

Figure 3.4: Completed Project Responsibility Matrix

It is common during this exercise to realize that some project players have been mistakenly forgotten. When this happens, add the person's name or role on the project and assign them the appropriate level of responsibility.

Using the Responsibility Matrix

As stated previously, the responsibility matrix is used during the early stages of a project to document and communicate the types of involvement associated with the various key players on the project. It should be completed before a commitment to resources, budget, and timeline has been established. This means that it is best used to establish high-level responsibility and ownership of the project components and key project outcomes. Care should be taken to avoid going into the task-level detail at this juncture of the project even though it is a natural tendency for many project managers.

Ultimately, the project manager is responsible for the various outcomes and outputs associated with a project, as well as the successful achievement of the success factors. However, along the way, responsibility for the satisfactory completion of work is a shared responsibility. Use the responsibility matrix to explicitly demonstrate how responsibility will be delegated and shared on a project.

Once the responsibility matrix is developed as described in the previous section, it must be validated with the project sponsor, those listed as having responsibility, and other key stakeholders. This step is important in order to reach broad buy-in across the organization. Equally important, if multiple organizations are involved in a project, validation is necessary between partner organizations to avoid confusion and misunderstanding about who is responsible for what.

It is important to treat the responsibility matrix as a living project artifact. Project players and responsibilities will change during the course of a project; so, too, must the responsibility matrix. Update the matrix whenever major responsibilities for project outcomes shift, and then publish the changes to the key stakeholders to communicate the change.

It is good practice to incorporate the responsibility matrix into the materials used to present project status. If no changes in responsibility have occurred since the last status review, the matrix can be part of the informational backup material. If changes *have* occurred, the responsibility matrix should be included in the main project status review materials.

Benefits

Using a responsibility matrix during project initiation serves to accelerate the pace at which clarity about who is responsible for what becomes established. It brings together two critical activities that are part of the project initiation process: definition of the project architecture and formation of the core project team. Left unmanaged, ambiguity at the front end of the project cycle can be the greatest contributor to lost cycle time on a project. The responsibility matrix gives the project manager a tool for managing and reducing this inherent ambiguity.

As stated earlier, a project manager has to delegate responsibility of tasks and outcomes to other players associated with the project. Along with the delegation of responsibility must come empowerment to make the decisions necessary to ensure that the project outcomes are created successfully. By documenting delegated responsibility through the responsibility matrix, the necessary delegation and empowerment is established and communicated to primary stakeholders of the project and top management within an organization.

By documenting project responsibility, it removes possible implied assumption on the part of both the project players charged with sharing in the project responsibilities and the stakeholders and replaces it with explicit responsibility direction, again removing a factor that causes ambiguity in the early stages of projects—implicit assumptions. This paves the way to gain broad buy-in across an organization for responsibility and decision authority and final accountability for project outcomes. This is also the case *between* organizations when multiple companies are involved in a collaborative project agreement.

THE COMPLEXITY ASSESSMENT

Complexity is a characteristic of many projects. Contributing factors include the following: Designs have become more complex as features and integrated capabilities increase; the process to develop and manufacture solutions requires more partners, suppliers, and others throughout the value chain; the ability to integrate multiple technologies with end-user wants requires not only accuracy regarding requirements delivery, but also speed and agility to change; and the current global, highly distributed business environment requires work to occur in multiple sites across multiple time zones. Therefore, the ability to characterize and profile the degree of complexity associated with a project has become essential for both executive leaders and their project managers.

If the complexity assessment tool is part of a PM Toolbox, the information gained from its use helps to balance the portfolio of projects from a complexity perspective. For a project manager, it aids in the determination of the skill set and experience level required of the project team; guides the implementation of key project processes such as change management, risk management, and contingency reserve determination; and helps the project manager adapt his or her management style relative to the level of complexity of the project.

Developing the Complexity Assessment

The structure of the project complexity assessment features several parts. The tool includes various dimensions (first part) as defined by a business. Each dimension of complexity is assessed on an anchor scale (second part), and when the complexity scores of each dimension are connected, a line called the complexity profile (third part) is obtained. The complexity profile is a graphical representation of a project's multifaceted complexity. An example of a project complexity assessment is illustrated in Figure 3.5.

	Low Complexity	←			→	High Complexity
Complexity Dimension		1	2	3	4	
Business Climate	Stable			X		Uncertain
Market Novelty	Derivative			X		Breakthrough
Financial Risk	Low		X			High
Project Objectives	Clear		X			Vague
Requirements	Clear		X			Vague
Organization	Hierarchical			X		Matrix
Technology	Low-Tech		X			Very Hi-Tech
Speed to Market	Normal		X			Blitz
Geography	Local				X	Global
Team Members	Experienced	X				Inexperienced

Figure 3.5: Example Project Complexity Assessment

Determining the Complexity Dimensions

Every industry has unique characteristics, every business within an industry is unique, and every project within a business is unique. This means that a firm has to customize the complexity assessment tool for its use.[10] Project managers often start this work by determining the complexity dimensions that are specific to the organization. For example, technical complexity may be directly related to the technical aspect of the product, service, or other capability under development, or from the knowledge and capability of the existing resources of the firm. Structural complexity also has a number of subfactors that involve the organizational elements of a firm. Business complexity involves the business environment in which the firm operates.

Examples of variations in dimensions with respect to technical, structural, and business complexity are shown in Table 3.2.

Define the Complexity Scale

With the dimensions of complexity identified that are appropriate for a particular organization and project, the next step in developing a complexity assessment tool is to define how each dimension of complexity will be measured. We do this by choosing a scale for each dimension.

Table 3.2: Example Technical, Structural, and Business Complexity Dimensions

Technical Complexity	
Low Complexity	**High Complexity**
Feature upgrade to an existing product	New product architecture and platform design
Development of a single module of a system	Development of a full system
Use of existing and developed technologies	Use of new and undeveloped technologies
Structural Complexity	
Low Complexity	**High Complexity**
Team is co-located	Team is a geographically distributed
Mature processes and practices	Ad hoc processes and practices
High performing team	Low level of team cohesion
Single-site development	Multisite development
Single-geography development	Multigeography development
Single-cultural team	Multicultural team
Single-company development	Multicompany development
Business Complexity	
Low Complexity	**High Complexity**
Selling into traditional and mature markets	Selling into new and emerging markets
Receptive customers and/or stakeholders	Unreceptive customers and/or stakeholders
Flexible time-to-money requirements	Aggressive time-to-money requirements
Existing end-user usage models	New end-user usage models

Alternatives for the scales abound; we chose a scale in the preceding example where the complexity for each dimension is measured on a simple four-level scale (1 being the lowest complexity, 4 being the highest complexity). The important thing to establish is not so much the scale, but rather the anchor statements for each level of the scale. Anchor statements help build consistency when assessing the complexity for each dimension. Without anchor statements, each assessor may evaluate the levels differently, leading to inconsistent complexity evaluations. Well-defined anchor statements help to ensure that all assessors approach the scale for each complexity dimension from a consistent frame of reference.

Using the Complexity Assessment

As stated earlier, each organization should create a customized version of the assessment tool that is specific to the complexity dimensions they are dealing with. Once the complexity dimensions are identified, each dimension is then assessed based on the scale established. For example, in Figure 3.5, speed to market is assessed as a Level 2 complexity (fast and competitive). Once all complexity dimensions are assessed, connect the obtained scores for each dimension to produce the complexity profile, which helps to visually depict the overall project complexity. The profile in Figure 3.5, for instance, indicates that the program is of medium complexity, with all dimensions at Levels 2 and 3, except team members who are experienced (the least complex) and a globally distributed team (the most complex).

Typically, the project complexity assessment tool is prepared very early in the project cycle. However, this tool should be utilized dynamically and updated periodically in high-velocity environments where the project scope and business climate may frequently change. It is advantageous if the senior management team of an organization who manages the portfolio of projects uses this tool to inform them of the overall level of complexity for each project.

By using this tool, the senior management team and project manager can quickly get a feel for the level of complexity of each project. However, care should be taken to prevent the inclusion of too many complexity dimensions. In this case, the simpler the tool structure, the more effective its use will be.

Benefits

The project complexity assessment tool's value is multifold. First, knowing the project complexity helps balance the portfolio of projects with an appropriate mix of low-, medium-, and high-complexity projects. Further, the complexity assessment aids in the planning process, indicating how to adapt one's management style to the level of complexity of the project.

The project complexity assessment tool also helps the top management team determine the level of skill and experience needed, and thus aids in the selection of the project manager and the key leadership positions on the team to successfully define and execute the project. Additionally, the tool may influence how much contingency buffer to build into the project budget and schedule—the more complex the project, the higher the risk, and the bigger the buffer. Finally, the tool can help identify the categories of risk and the level of robustness you will need in your risk management plan.

THE PROJECT BUSINESS CASE

The project business case, sometimes called the project proposal, is a start-up document used by the project manager and top management to assess the feasibility of a project from multiple business perspectives. It demonstrates how the project will contribute to business results and how the project aligns with the strategy of the organization.[11]

The project business case establishes the project vision by describing a business opportunity in terms of alignment to strategy, market or customer needs, technology capability, and economic feasibility. It also provides a balanced view of business opportunity versus business risk. The project business case is used for the following purposes:

- To gain agreement on project scope and business success criteria.
- To obtain approval of funding and resource allocation for project planning and implementation.
- To evaluate a project against others in the portfolio of projects.
- To obtain approval to proceed from the initiation stage to the planning stage of the project cycle.

The project business case is a *must-have* tool for every project manager's PM Toolbox.

Developing the Project Business Case

The business case for a project must be correctly based on the knowledge available at the time it is created, and also be unbiased and clear. This requires quality information about the following:

- The business environment
- Customer requirements
- The business strategy
- Business success criteria

The business environment and customer requirements information are necessary to build the foundation of the business case. Understanding the needs of the customer as well as the state of the environment within which the business operates is needed to position and differentiate the project outcome being proposed. The business strategy specifies the strategic goals that the organization is striving to achieve and that the project is charged with enabling. The business success criteria consist of high-level directives from senior management to gauge initial feasibility of a project to meet the business needs. Table 3.3 can be used as a guide for developing a project business case. It suggests a minimum set of information to include in the business case.

Table 3.3: Minimum Elements of a Project Business Case

Business Case Element	Description
Project purpose	A succinct statement of the anticipated business benefits driving the need for the investment in the project
Value proposition	A succinct statement characterizing the value to be delivered (quantified when possible)
Business success factors	The set of quantifiable measures that describe business success for the project
Detailed cost analysis	The investment cost of the project
Critical assumptions	The events and circumstances that are expected to occur for successful realization of the project objectives
Project timeline	Critical project milestones and timing expectations on the part of key stakeholders
Risk analysis	A thorough analysis of the risks that may prevent realization of the business benefits of the project

Describe the Business Opportunity

There are two steps to this action: first, a description of the benefits that the project will fulfill is provided, with a focus on the *business* benefits; second, a description of value that the project brings to the organization from a business perspective should be described.

The business benefit is a statement that summarizes what the company should expect to gain from its investment in the project. It provides answers to the following questions: What business benefits will be gained from investment in this project? What organizational strategies does this project help to achieve? What opportunity in the market is this project going to capitalize on?[12]

The purpose of projects is to serve as basic building blocks for the execution of an organization's strategy. This should be the basis of the value proposition statement for all projects. The premise is that if projects are aligned with the firm's strategy, they will better support the goals of that strategy. Since one tangible way to express the strategy is to define its specific, measurable, attainable, relevant, and time-based goals, one can use the stated goals to assess how well an individual project supports them. A description of how the project aligns to and supports achievement of one or more strategic goals of the organization should be included in the value proposition statement of the project business case.

Additionally, a description of how the project outcome fulfills documented customer and market needs is included. Finally, a description of any new technologies that will be included in the project outcome being developed is provided in this step.

Define the Business Success Criteria

Identification of the business success criteria for a project should be accomplished during project initiation and documented in the project business case. The business success criteria ensure that the product, service, or infrastructure capability under development supports key business goals such as profitability, time to benefit, productivity gains, and technology advancement.[13]

Perform the Cost-Benefit Analysis

The heart of the project business case is the feasibility assessment that results from the cost versus benefit analysis of the project. The cost-benefit analysis should identify both tangible and intangible benefits, with the benefits expressed in quantifiable terms such as dollars gained or saved, hours saved, and gross margin increase. The cost-benefit analysis should answer the following questions:

- How much will this project cost to implement?
- How much will this project contribute to the company bottom line?
- Is the project outcome, in terms of achievement of specific business objectives, worth investing in?

List Critical Assumptions

Much of the work performed during project initiation is focused on trying to predict what will happen in the future. In order to do this, a series of assumptions about the future have to be made, where assumptions are events and circumstances that are expected to occur for successful realization of the project objectives.

Each project player and each project stakeholder has a set of assumptions in his or her mind that they use to guide their vision and perspective of the future relative to the project being initiated. By explicitly stating the primary assumptions on which the project business case is built, the project manager establishes a common vision of how it is assumed that the future will unfold. Discussing and debating this critical set of assumptions is as important as analyzing the cost-benefit portion of the business case.

Analyze Project Risk

In this final step, all potential risk events that may affect the business success of the project are identified. At this stage of the project, it is a high-level look at the known risks. The risk events are then analyzed, and a plan to minimize the impact and probability of occurrence for the high-level risks is developed. The risk analysis should answer the following questions:

- What is the probability of success for this project?
- What will be done to maximize the probability of success?
- How will the known risks be avoided or mitigated?
- Does the level of risk prevent continued investment in the project?

Using the Project Business Case

Presentation of the business case to top management stakeholders is normally used to drive the final investment and funding decision near the end of the initation stage of the project cycle and provides primary content for the project charter.

Although the project business case is first developed during project initiation, it should be viewed and used as a living document throughout the project cycle and as a primary guide for project tracking and governance. The project business case, along with the market or customer requirements document, forms the foundation on which the project plan is developed. The business case is updated as needed and reviewed as part of the implementation plan approval.

During project execution, the business case needs to once again be reviewed prior to releasing funds for large expenditures such as factory tooling. Finally, the business case information is used during the project retrospective to evaluate whether the project was successful in achieving the business goals intended.

Benefits

The benefits of creating a good business case for a project are many. The primary benefits are fourfold. First, the business case answers the critical question, "How will this project help our company meet its business and strategic goals?" In this sense, it helps top managers of a firm make sound decisions when considering investment options.

Second, the project business case establishes alignment between strategic goals and project execution outcomes based on multiple business perspectives.

Third, consistent use of the business case for all projects helps to make the portfolio management process more effective by enabling the evaluation of projects within the portfolio on a consistent set of criteria.

Finally, it establishes the vision, or future state, to effectively plan, execute, and deliver the output of the project.

THE PROJECT CHARTER

With every project, a project manager needs an effective way to define what work has to be accomplished and communicate how the project is going to achieve its objectives.[14] The project charter is a tool that formally authorizes a project and serves as the contract between the project manager and the organization (see Table 3.4).[15]

Typically ratified by a manager external to the project, it equips the project manager with the authority to deploy organizational resources on the project. This is especially important in environments where project managers have no direct authority over project team members and other resources but bear the responsibility for delivery of the project outcome. In such a situation, for the charter to be effective, the issuing manager has to be on a level that has control over the resources.

Developing a Project Charter

When comparing the type of information described in the charter and scope statement, you will note a lot of similarities. Both contain the same elements—project purpose, goals, and milestones, for example. Where these elements differ is their level of detail. More precisely, because it is an authorization tool, not a planning tool, the project

Table 3.4: Example Project Charter

Project Name:	**ISU Alumni Website**
Project Manager:	**Jen Cosgrove**
Project Sponsor:	**Dan Seales**

Project Mission

This project will provide the local chapter of the ISU Alumni Association a web site that will be used as a resource to enable continued social networking, information exchange, and information repository for alumni association members.

Project Goals

The university is looking for new ways to help alumni stay connected to the university post-graduation. The web site to be developed will (1) create the means to establish a strong alumni social network; (2) provide a portal for the university to communicate activities, information, and needs; and (3) establish a repository of academic research information for the alumni to access and contribute to. The project will be completed prior to the Fall 2017 academic semester, and cost no more than $60,000 to implement.

Project Scope

The scope of the project will include the design, development, test, and go-live activities necessary to create an operational web site. The web site will include four major capabilities: (1) HOME page for alumni information and navigation, (2) SOCIAL NETWORKING page, (3) ACADEMIC RESEARCH repository, and (4) ALUMNI ACTIVITIES page.

Dependencies, Risks, and Assumptions

Dependencies	Risks	Assumptions
■ Budget approval ■ Availability of IT resources	■ First use of open source software ■ Schedule is aggressive	■ Additional security software is not needed ■ Open source software can be leveraged

Major Milestones:

Usage study completion
Web site design completion
Prototype development completion
Development completion
Operational test completion
Go-live launch completion
30-day retrospective completion

Sponsor and Project Team

Project Sponsor:	Dan Seales
Project Manager:	Jen Cosgrove
User Exp. Designer:	Lynda Carmody
Web Site Developer:	Ajit Chattergee
Web Site Developer:	Fariba Rezzanie
Quality Assurance:	Will Torday

Budget and Completion Date

Budget:	$60,000
Completion Date:	August 1, 2017

Approvals

Project Sponsor: ______________________________

Project Manager: ______________________________

Finance Manager: ______________________________

charter tends to include fewer details, giving a mandate to the project team to proceed with a detailed planning cycle, part of which includes developing a detailed scope statement. Naturally, then, the scope statement has more details about these elements than the charter.

Collect Information Inputs

Issuing a project charter is a major decision because it commits resources to support organizational goals. For that reason, organizations tend to invest in generating information that will help make educated charter decisions. Crucial pieces of such information include the following:

- Project goals
- Project requirements
- Project business case
- Project selection information

Projects are vehicles for the delivery of organizational needs; therefore, the project goals cannot be overstated. Understanding which of the goals a project supports is therefore of crucial importance. As discussed earlier in this chapter, the information contained in the goals grid is sufficient to adequately provide content for the project charter. Additionally, for a project to be successful, the project requirements need to be documented, understood, and responded to. Also, to properly assess the viability of a project, you need to develop a feasibility study that should be completed during the project selection process.

Define the Project Mission

Precision and clarity are two key words in the charter's definition of what the project should accomplish.[16] Whether the charter is for a small process improvement or a multibillion-dollar semiconductor fab, a few words can usually do the trick. The statement may identify major tasks such as design, prototyping, and programming, or it can be as simple and directive as "develop a new product platform."

To express the accomplishment expected of the project, we use the term *project mission*. Project mission has an aura of significance, which may be why it is often used. Alternative terms, such as *project purpose* or *assignment*, have less gravitas, but may nonetheless be appropriate. The selection of the term is often dictated by organizational jargon.

Contained within the project mission is an understanding of what drives the implementation of the project. Is its purpose to increase customer satisfaction? Or is it to enter a new market, increase market share, develop new competencies, and so on? On a strategic level, there may be several different reasons for the existence of the project. The point is that we should know it and spell it out.

Once we are out of the strategic territory and in the tactical world of small projects, many project teams struggle with what exactly is their project's purpose. They assume it is simply to create the project output. However, it is not. Like any other project, your small project exists to accomplish some tactical gains supporting your organization's strategic business goals. For example, when you are buying and installing a piece of new factory equipment, the business purpose is not to buy and install the new equipment. Rather, the purpose may be to increase the efficiency of your operation and lower your manufacturing costs. Similarly, a project developing a standardized PM process is probably aiming at improving consistency and repeatability of performance. Certainly, its business purpose is not to have the process itself.

Define the Project Goals

By their nature, "project mission" and "business purpose" are broadly defined. To provide more specific guidelines to the project team, the charter needs to identify specific project goals (see the example titled "Stretch Goals or Not?"). At a minimum, these usually include schedule, cost, and quality targets. Your schedule target is your desired project completion date. Remember to focus not only on what you want to achieve, but also on what you want to preserve, avoid, and eliminate, as described in the goals grid. You may have important project goals associated with each perspective.

Stretch Goals or Not?

How attainable should the project goals in the charter be? Is it okay to write charters using a stretch goal? Empirical evidence suggests that those who set stretch goals—that is, outline goals that are typically quite difficult to attain—outperform those with routine goals, which are typically easy to attain. If you are purely driven by performance, the choice is clear: go for stretch goals.

At Google, many project managers deliberately use stretch goals in their project charters because corporate culture drives this behavior. What happens when they do not attain their stretch goals? According to one project manager, "No project sponsors use this practice to call people out. The idea is to always strive for more and do your best. If you do so, you won't be penalized if a stretch goal is missed."

Is this the case in all companies? According to a project manager for a business-to-business software company, "If you try stretch goals and miss, it is likely to be held against you in your next performance evaluation. That's why everybody goes for routine goals in our company." The point, then, is that the use of project stretch goals is dependent on company culture and likely related to the level of desire for industry leadership.

State the Scope

The project charter should include a summary-level scope statement in order to establish an agreement between the project team and the project sponsor by clarifying, identifying, and relating the work of the project to the sponsor's business goals.

A well-written scope statement is crucial to guide a project manager's decisions during the course of the project cycle. The more information you can document in the initiation stage of a project, the more adaptable you will be when having to deal with critical decisions that appear during the project.

List Dependencies, Risks, and Assumptions

Begin this section of the project charter by listing the major dependencies that will need to be in place for the project to proceed to planning and implementation activities. Examples may be the approval of the project budget by the selection committee or the availability of resources. It is of high importance that these types of dependencies be documented, as normal organizational behavior is to approve a project charter with the expectation that work will begin immediately.

If there are major dependencies that have to occur *before* project activities begin, it is best to include them in the charter to increase understanding and probability of a successful project start.

Likewise, it is advisable to list any known high-impact or high-probability risks that may affect the success of the project. At this stage in a project, there is a lot of excitement about the possibilities and opportunities associated with the project. But take some time to think negatively about the project in order to identify what may go wrong. This will be the project manager's first opportunity to ask for project sponsor assistance to overcome or eliminate high-level risk events.

Finally, list the critical assumptions associated with the project. During project initiation, all activities and planned outcomes are predicated on a set of assumptions of how the *future* will play out. Since the future rarely goes as predicted, some of the assumptions made at this early stage will be incorrect. This means that all activities and outcomes associated with the incorrect assumptions may need to be adjusted in the future. By listing the major assumptions in the charter, they become visible and can be tracked and validated as the project progresses.

Specify Major Project Milestones

Major milestones include completion of certain tasks or deliverables by a specified date and are typically requested by those issuing the charter. The key word here is *major;* you should limit the number of major milestones to those that are absolutely vital. Specifying three to five is a dominant practice. In other words, given the charter's purpose and the related level of detail, developing a long list of milestones is unnecessary, especially at this early stage of project initiation.

Identify Project Sponsor and Team

One of the purposes of issuing a charter is to formally announce the names of the primary stakeholders associated with the project, including the team members. However, it is not important that all team members be immediately identified. The expectation here is that functional managers will nominate those members after the charter is issued.

In some organizations, the use of project sponsors is a regular practice for major projects. Sponsors provide guidance for the project team, making sure that the functional managers fulfill their resource commitments to projects and serve as a communication link with customers.[17] Typically, the sponsor is a senior manager who has authority over budget and resources. In the case of less strategy-driven projects, the role of a sponsor may go to a middle-level manager. Whatever the level of the project sponsor, issuing the charter is a convenient way to visibly announce the name of the sponsor.

Include Project Budget and Completion Date

Explicitly document the project budget allocated, as well as the expected completion date for the project. Cost and schedule are normally the two primary project constraints, and as such, need to be elevated to the signature and approval level of the project charter.

Project Approvals

As an authorization document to officially start expending organizational resources, the project charter must include the names, titles, and signatures of the individuals who will sign off on the project. At a minimum, signatures should be required for the project sponsor, the project manager, and a manager from the finance or accounting department of the organization.

Refer to the Supporting Detail

What is immediately evident in a project charter is the decision about bringing a project to life, stated in a laconic manner. What is not evident is the process that led to the decision. The decision was a result of the process of project selection that was based on information developed in strategic and tactical plans, the high-level project requirements, the project proposal, and project selection methods. To make this visible and give credibility to the charter, refer to these documents in the charter.

Using a Project Charter

The project charter has been used in large projects since the beginning of formal project management. Because large projects engage substantial organizational resources originating in different functional groups, this approach is quite logical. For the same reason—resources that derive from various functional groups—the charter is popular with small, cross-functional projects as well. However, for other small projects that are not cross-functional, issuing a project charter is an infrequent practice—unless functional department members are not collocated, a growing phenomenon in our virtual world. For an example of the use of the project charter in different corporate situations, see "The Need for the Charter" on page 77.

Even though a large majority of organizations now make use of the project charter, not all companies create the charter at the same point in the project cycle.[18] Some charters are created as part of the project selection process, immediately after the feasibility study is complete. When used at this point in the project cycle, the project charter contains the results of the feasibility study as well as the underlying assumptions and constraints.

Most organizations create the charter after the project has been selected and a project manager has been assigned. In this case, the project charter contains some or all of the information described previously.

Still other organizations create the project charter after detailed planning is complete and the project plan is approved. When used at this point in the project cycle, the charter contains a summarized description of the detailed project plan.

Even though the project charter is an agreement between the project manager and executive sponsor, all functional groups or departments in the organization that will be supporting the project need to be informed correctly and promptly about the start of the project.[19] For that reason, they need to put them on the distribution of the charter.

Variations

The practice of project chartering exhibits many variations and nuances, including its name, content, pattern of use, and formality. For instance, some organizations call it the "project authorization notice," others the "project birth certificate." In all cases, the charter is meant to bring a project into existence.

As for the content, some organizations use charters that include specifics about budget and schedule for major milestones, as shown in the project charter example in Table 3.4. Others, especially for smaller projects, find it sufficient to announce the purpose of the project, the start of the project, the team composition, and the executive sponsor.

Benefits

Projects often require organizational arrangements that span functional boundaries. In such cross-functional designs, functional managers "own" resources, and the project charter is a practical way of communicating the need for the project and for making a request for resources. This practically defines specific resources, the amount and time of their use in the project, and who is responsible for providing them. Aside from this act of organizational legitimacy, the charter also helps a project get visibility by announcing its start and purpose, leaving the ball in the project manager's court.

The primary benefit of the project charter, however, is that it serves as the agreement or contract between the project manager and the executive sponsor to work as a team to execute the project to the intent of the project business case and to utilize the organization's resources to maximum business benefit.

The Need for the Charter

Do you need a charter for all projects? Consider that it normally takes a leading truck manufacturer located in the United States months of work to issue a charter for a new truck development project. With millions of dollars involved, the company develops multiple scenarios of scope, cost, and timeline, and evaluates them carefully before launching the effort. The launch begins with

(continued)

the issuance of a detailed charter, where the sponsor typically is a corporate vice president.

In contrast, a major information technology upgrade project within the same company typically starts with a sentence-long charter, e-mailed to the functional managers providing resources. No sponsor is identified, and not much of the charter is completed prior to project planning. The rule for these projects is that any major project consuming resources (over $10k) must issue a charter. Charters are not used for projects below $10k, usually performed within a functional group. The reason? It is considered an unnecessary step.

This is a good example to review when deciding whether you need a charter for all projects. The need for the charter should be matched with the size, complexity, and degree of cross-functional involvement on the project.

References

1. TSO. *Managing Successful Projects with PRINCE2* (London, England: TSO, 2012).
2. Cleland, David I., and Lewis R. Ireland. *Project Management: Strategic Design and Implementation* (New York, NY: McGraw-Hill, 2002).
3. Koen, P. A., G. M. Ajamian, et al. "Fuzzy Front End: Effective Methods, Tools, and Techniques." In P. Belliveau, A. Griffin, and S. Somermeyer, eds., *The PDMA ToolBook for New Product Development* (New York, NY: John Wiley & Sons, 2002): 5–35.
4. Smith, Preston G., and Donald G. Rinertsen. *Developing Products in Half the Time: New Rules, New Tools* (Hoboken, NJ: John Wiley & Sons, 1998).
5. Cooper, Robert G. *Winning at New Products: Accelerating the Process from Idea to Launch*, 3rd ed. (Cambridge, MA: Perseus Books, 2001).
6. Andler, N. *Tools for Project Management, Workshops, and Consulting* (Erlangen, Germany: Publicis Publishing, 2011).
7. Nichols, F. "The Goals Grid: A Tool for Setting and Clarifying Goals and Objectives;" 2012. http://www.nickols.us/goals_grid.htm
8. *New Oxford American Dictionary*, 3rd ed. (New York, NY: Oxford University Press Publishing, 2010).
9. Martinelli, Russ, James Waddell, and Tim Rahschulte. *Program Management for Improved Business Results*, 2nd ed. (Hoboken, NJ: John Wiley & Sons, 2014).
10. Edmonds B. "What Is Complexity?" In F. Heylighen & D. Aerts, eds., *The Evolution of Complexity* (Dordrecht, Netherlands: Kluwer, 2006).
11. Cohen, Dennis J., and Robert J. Graham. *The Project Manager's MBA: How to Translate Project Decisions into Business Success* (San Francisco, CA: Jossey-Bass, 2001).
12. Ibid.
13. Wysocki, Robert K. *Effective Project Management: Traditional, Agile, Extreme* (Hoboken, NJ: John Wiley & Sons, 2013).

14. Kerzner, Harold K. *Project Management: Best Practice: Achieving Global Excellence*, 2nd ed. (Hoboken, NJ: John Wiley & Sons, 2010).
15. Project Management Institute. *A Guide to the Project Management Body of Knowledge*, 5th ed. (Drexell Hill, PA: Project Management Institute, 2013).
16. Katzenbach, J. R., and D. K. Smith. *The Wisdom of Teams* (reprint) (New York, NY: HarperCollins, 2006).
17. Kerzner, Harold K. *Project Management: A Systems Approach to Planning, Scheduling, and Controlling* (Hoboken, NJ: John Wiley & Sons, 2009).
18. Kerzner, 2010.
19. Pyzdek, Thomas. *The Six Sigma Project Planner: A Step-by-Step Guide to Leading a Six Sigma Project through DMAIC* (New York, NY: McGraw-Hill, 2003).

PART III

Project Planning Tools

4

PROJECT REQUIREMENTS

Contributed by

Debra S. Lavell

James M. Waddell

One of the most important aspects of good project planning and execution is a comprehensive, clear, and valid set of requirements. When it comes to requirements, it is well documented that many of the contributing factors to failed projects are such things as lack of stakeholder input or involvement, inadequate understanding of the voice of the customer, and improper documentation and validation of the requirements for the intended solution.

The Project Management Institute (PMI) defines requirements management as:

> The discipline of planning, monitoring, analyzing, communicating and controlling requirements. It is a continuous process throughout a project. It involves communication among project team members and stakeholders and adjustments to requirements changes throughout the course of the project.[1]

A broader term, *requirements engineering*, focuses on using a systematic and repeatable process to ensure that solution requirements are discovered, documented, and maintained throughout the project cycle. For this chapter, we will focus on four major activities within requirements engineering within the context of project planning and the tools used to support those activities. They are:

1. *Requirements elicitation*: Gathering requirements from stakeholders and ensuring the voice of the customer is captured.
2. *Requirements analysis*: Assessing, negotiating, and ensuring requirements are correct.
3. *Requirements specification*: Documenting requirements.
4. *Requirements verification*: Assessing requirements for quality.

The purpose of requirements engineering is to ensure that projects fully meet the requirements intended by the external or internal customers who are the users of the solution. Most of us have experienced a situation where a key customer's needs and wants were improperly addressed, leading to a failure or, at the very least, project outcomes adversely impacted.

To begin, we must first ask what is a requirement? A requirement states what a solution must do (functional requirement) and how well it must do what it does (quality or nonfunctional requirement). In its most simplistic form, a requirement is anything that drives design choices. Done well, requirements establish a clear, common, and coherent understanding of what a solution must accomplish in order to meet customer expectations. The bottom line is that requirements are the foundation on which projects are built.

In an in-depth study conducted in 2014 by PMI, in which they elicited responses from 2,066 project managers, program managers, and business analysts, it was determined that 47 percent of unsuccessful projects failed to meet their original goals and business objectives due to poor requirements management.[2] These failures were identified as being caused by inadequate resources, insufficient skills development, informal processes and practices, and lack of understanding and support for the importance of requirements management by top management. By contrast, enterprises performing successful projects recognize the importance of the requirements engineering discipline and are much more likely to adopt formal processes for requirements gathering, writing, and management for their projects.

The major finding was that the more time spent on requirements definition and understanding up front, the more predictable project costs become, the higher the probability for achieving successful solutions from the customer's and user's perspectives, and the greater the chance for achieving the financial and business objectives for the project. Requirements are the foundation on which systems are built.[3]

Developing and managing requirements is hard work! There are no simple shortcuts or magic solutions.[4] This chapter is intended to assist the reader in understanding the process of project requirements engineering and to identify specific tools that are used in support of these processes. There are many requirements tools available through various sources. In this chapter, we have identified the tools and techniques that provide the most utility to the project manager.

The tools presented are designed to work holistically together, as a set of requirements tools that will guide you through planning the very first interaction with a stakeholder; assist you with the art of writing clear, concise, coherent, and measurable requirements; and provide you with a template to effectively verify that you have the requirements developed to the highest quality. These tools should be a part of every project manager's PM Toolbox.

THE ELICITATION PLAN

When in the initial stages of a project, there are several ways to approach collecting requirements from stakeholders. Many teams, unfortunately, spend an insufficient amount of time collecting requirements and jump directly into writing requirements. This behavior leads to key requirements being missed, incomplete information needed to make decisions, and multiple scope changes later in the project cycle.

The most important tool for gathering project requirements is an elicitation plan. This plan ensures that the best methods are used to gather requirements from the stakeholders and allows the team to document the methods they will use to ensure that they cover all possible sources.[5] The plan need not be long and complicated, but the detail level must be consistent with the risk and complexity of the project. This can be determined by the type of project, number of stakeholder groups, experience of the team, and so forth.

Why spend time to document a plan to gather requirements? Many teams make the mistake of diving into requirements discussions with a small set of stakeholders who they think have the information they need, and with whom they have an established relationship. Jumping in without a comprehensive and holistic view of who has a material interest in the project can lead to missing stakeholders and therefore missing requirements. The opposite is also true. It is impossible to interview everyone, so identifying key contributors and then narrowing down the list to the "right" stakeholders who can provide a well-rounded understanding of the requirements can save the team precious time. Interviews, focus groups, and surveys are better conducted if the right stakeholders are thoughtfully identified and approached.

Figure 4.1 provides a template for a simplified elicitation plan. If appropriate, based on the size of the project, or if the team has experience with a similar project, feel

PROJECT ELICITATION PLAN

Project Name: Highlands **Rev #:** 1.0 **Date:** 22 April 2016

REQUIREMENTS AUTHORS & KEY CONTRIBUTORS

Primary Authors: Simon B., Phillip C., Christine H.

Key Contributors: Carry M., Nesli S., Carl W.

INTRODUCTION

Problem Statement:

ELICITATION STRATEGY & PROCESS

Strategy and Process: *Group interviews with three key customers, followed by series of prototype reviews, and a direct observation of end users at five sites*

LIST OF STAKEHOLDERS

Stakeholder's Name	Current Role	Elicitation Technique	Desired Outcome
Jim Johnson	Medical Provider	Group Interview	Feedback on key features

SCHEDULE & RESOURCES

Item	Estimated Schedule	Resources	Range of Uncertainty
Group Interviews	5 weeks	Jim, Sury, Pat	+2 weeks

ASSUMPTIONS & RISKS

Risk	Magnitude of Risk	Likelihood	Mitigation Plan
Stakeholder Schedules	High	High	Add 2 week buffer

Figure 4.1: Simple Elicitation Plan Template

free to add or delete topics that are pertinent to your project. Again, the elicitation plan can be as simple or as broad as you like based on the needs of the team and the value from a deeper discussion. Be mindful that you are focused only on *gathering* project requirements. This is not an overall project plan; this is a specific plan to *elicit* requirements.

Developing an Elicitation Plan

The elicitation plan can be created using simple word processing or spreadsheet software. Begin by documenting the project name, version number, and date; list the primary requirements authors(s) and the key contributors to the elicitation plan. Next, spend adequate time documenting the following five key items.

What Is the Problem to Be Solved?

Document a clear understanding of the problem or opportunity the project will be addressing. This is generally high level because an elicitation plan is normally developed early in the project cycle, when the scope and assumptions are still ambiguous. This is usually translated into the project's scope and charter when combined with information about the user environment and business context.

What Are the Strategies for Gathering Requirements?

Identify the high-level objectives of the effort and the key strategies the team will use to gather requirements. Most requirements efforts center on interviews. Yes, interviews are very helpful for gathering requirements, but there are so many more effective techniques to elicit requirements from stakeholders than just interviews alone (see "Techniques to Elicit Requirements").

Techniques to Elicit Requirements

Interviews. Usually one-on-one or in small groups (three to four people) where key stakeholders are invited to attend a requirements elicitation meeting and asked questions to uncover needs, wants, and desires. The key questions should be determined prior to meeting with them based on the type of requirements you are looking for. There are a lot of ways to ask questions; generally, you want to ask open-ended questions to get them started and then follow up with probing questions to uncover requirements.

Facilitated discussions. Larger groups (five to ten people) come together with a facilitator trained on how to elicit requirements. The goal is to gather requirements faster than if you spent time interviewing each stakeholder independently.

Surveys or questionnaires. Used best when you want to gather requirements from a lot of stakeholders who are geographically dispersed. It is best if an expert in surveys or questionnaires develops the questions so you get the information needed.

Prototyping. Build an initial version of the solution; demonstrate it in the spirit of learning more about what features the users like and don't like. Use the feedback to make changes to the prototype, and show it to the users again. This repetitive process allows real-time feedback and a better understanding of the requirements.

Use cases. Think of use cases as stories that are easier for users to describe and understand. These are best used for the functionality (what the system must do) rather than how it will behave. Remember, use cases are *not* requirements; however, they are very helpful to get to detailed functional requirements.

Change requests. Ask to see the log showing what customers and end users are asking to be fixed on current solutions. System enhancements and bug reports are all good sources for possible requirements.

Observation. Watching users perform their jobs can be very helpful in understanding the current process from start to finish. Asking questions to better understand where they get frustrated or wish the solution did something different are excellent areas to find requirements.

Brainstorming session. Use a brainstorming session to help discover requirements for solutions that have not been developed before. Invite domain and subject matter experts into a session and let them brainstorm what they think the solution might look like. Allow enough time for the "good stuff" to come out, then ask them to prioritize those ideas they think are the best to go forward with. Use the highest-priority ideas as a basis for the initial set of the requirements.

Another very effective technique for gathering project requirements is the *voice of the customer* (VOC) technique. This term describes the process of capturing critical details regarding the desires, needs, and requirements of a given prospect, customer, or target group.[6]

There is no one homogenous VOC. Customer voices are diverse and reflect a variety of different needs. There are multiple customer voices within a single organization. For example, a product being considered for purchase within an organization may have the voice of the procurement department, the voice of the product end users, and the voice of the support and maintenance of the product. All of these diverse voices must be listened to, considered, and balanced to form the appropriate requirements for a truly successful solution.

VOC should be captured from both external and internal sources. External customers are those that purchase the output of a design and development effort such as new products and services. Internal customers are normally the receivers of newly developed systems, services, and other capabilities designed and developed by project teams for individuals within the same organization or company. Generally, internal and external customers exhibit the same traits. It therefore becomes imperative that the members of the team treat their search for identifying, understanding, and characterizing project requirements the same for both audiences.

Things to Avoid When Uncovering the Voice of the Customer

Engineering knows best. Don't let the technical implementation drive the requirements. What an engineer may find interesting may not be what the customers and end users find interesting.

Sales can be a proxy for the customers. What the sales team hears from customers can help guide the requirements elicitation process and discussion. Spend time validating what the sales force is hearing against what is known about the broader market and competition.

CEO said so. It can be very difficult to discount a pet feature, especially if it comes from the person at the top. Key executives may try to influence the requirements based on their own opinions. Gather facts to support why their requirements are good ones or not.

Who Are the Stakeholders We Need to Talk To?

Taking time to brainstorm a list of stakeholders the team wants to engage with is one of the most important aspects of planning the elicitation effort. A simple table with a column for the stakeholder's name, their role, possible techniques used during the elicitation effort, and what will be the output or result is effective for planning your requirements elicitation stakeholders (see Table 4.1).

Identify the Schedule and Resources

It is important to estimate the schedule and resources needed for the elicitation effort. Just like any other project task, requirements elicitation requires resources, time, and budget to complete. Include a measure of uncertainty for each key stakeholder.

Document the Assumptions and Risks

Take time to document all the assumptions made when determining the plan to gather requirements. Since assumptions by nature tend to be "big deal" items that many times are outside the influence or control of the project team, they can easily be converted into a risk mitigation table. Additionally, they should have a well-thought-out mitigation and management strategy. Focus on risks such as compressed schedule, resource constraints, and funding needed to travel to conduct focus groups or other

Table 4.1: Example List of Stakeholders

Stakeholder's Name	Current Role	Technique to Elicit Requirements	Desired Outcome
Jim Johnson	Business Analyst	Group Interview	Feedback on key features
Suzy Smith	Lead Engineer	Prototype	Validate usage is correct
Pat Pink	Marketing Lead	Observation	Gain "insider" viewpoint

costs associated with prototypes or participant observation (see "Tips for Mitigating Requirements Gathering Risks").

Tips for Mitigating Requirements Gathering Risks

Brainstorm as many assumptions as you can think of (e.g., will all the key stakeholders be available and willing to discuss the project?).

Share the list with key stakeholders to uncover new assumptions and validate any hidden costs associated with the brainstormed list.

When you believe you have a solid list, convert each assumption into a risk and document it as such in the elicitation plan.

Focus first on the highest risks associated with schedule, resources, and quality of the elicitation process.

Identify the possible impacts should the risk become a reality.

Document the likelihood (high, medium, low) of the risk happening.

If the risk becomes a reality, what is the magnitude of the impact to the project (high, medium, low).

What is the mitigation and management strategy should you need to employ it?

Using the Elicitation Plan

Taking time early in the project to fully understand and document an elicitation plan will save precious time. Investing time to discuss the process and strategy to elicit requirements, separate from the project plan, will give a next layer of detail to the overall time needed for the project. If you want to save time, this is one of the best ways to ensure that the time allocated for gathering requirements is used effectively. Often, teams spend too much time talking to the wrong people, gathering duplicate requirements from stakeholders from the same organization or group, or realizing late in the project they must allocate more time to talk to "new" stakeholders they forgot about because they began without a plan.

It is never too early to begin developing an elicitation plan. As soon as a project is assigned to a project manager, he or she should begin the process. Start by pulling two or three key contributors into a 90-minute to two-hour requirements elicitation kickoff meeting to discuss each of the sections of the elicitation plan. Write down what you hear. You will use this time to ensure that you and the team really understand what it will take to gather the project requirements.

Don't be constrained by the content areas shown in Figure 4.1 and described earlier. Customize your elicitation plan to meet your specific needs. Other information to consider in your plan might include:

- *Market and business context.* A high-level description of the market and how it interacts with the business goals for the project.

- *System domain*. Document the environment and conditions the solution will operate within. For example, if you wish to build a mobile communications device, you must understand the environments and conditions in which it will be used. To build a driverless car, you must understand the car industry, the manufacturing process, and how cars are used.
- *Outputs*. State the expected deliverables of the effort. For example, draft use cases, completed customer surveys, interview notes, and so forth.
- *Open issues*. Mark sections in the elicitation plan where the information is unclear or unknown. Or better, create a table at the end of the elicitation plan titled "Open Issues." Document all items under open issues so they are in one place that will make them easy to find, track, and resolve.

Once the sections are documented, have the project sponsor review the elicitation plan to approve the scope and validate any assumptions made by the team. Better to get the assumptions out on the table early so they can be addressed and the full scope of what it will take to successfully gather the requirements known and agreed to.

Benefits

Spending adequate time to get a complete understanding of what it is going to take to gather requirements for a project can be of significant value.

Following are four benefits gained by taking the time to develop an elicitation plan, even if it is only a few pages and at a high level:

- *Guarantee you will spend time with the "right" stakeholders*. Without a well-thought-out plan, you may begin gathering requirements, only to find out later that you have missed important stakeholders or gathered information from stakeholders who were not able to provide the right feedback.
- *Use the right technique to gather requirements*. Spending time to identify the correct elicitation technique (interview, focus group, prototype, etc.) will help ensure that you deliver the best possible set of requirements. Many requirements teams start with interviews with the existing users.[7] However, many unique solutions do not have current users, so there is a visualization leap required. When users are asked to help design a solution they have never seen or used, an interview may prove to be an ineffective way to gather a list of requirements. A prototype would produce better insight, but time to create, test, and then deliver a prototype requires planning to do it right. Without a solid elicitation plan, what was thought to take a few days could turn into several weeks.
- *Assign the right resources, armed with the right amount of time to gather requirements*. Often, gathering requirements is seen as a "second job" that has to be completed in conjunction with other responsibilities. Taking time to create a plan helps to ensure that the right people will be involved in gathering requirements, that their time will be used wisely, and that adequate up-front requirements gathering will be performed.

- *Gain agreement from top managers on the requirements elicitation effort (time, resources, and cost).* By documenting what resources and time it is going to take to gather requirements, senior leaders can guide the project manager on what compromises and decisions will need to be made. For example, if an elicitation plan shows five focus groups are planned at a cost of $50k, senior leaders can approve this cost *before* the focus groups are formed and scheduled, rather than be surprised at the end.

An elicitation plan can be as simple as one page. Or it can be quite lengthy (eight to ten pages) depending on how comprehensive and complex the project is. The main thing to remember is that it is a specific plan to gather requirements, *not* an overall project plan. Use this plan to augment the overall project plan with more details on the resources, time, and effort needed to effectively and efficiently gather project requirements.

REQUIREMENTS SPECIFICATION

Now that you have an understanding of the effort to gather requirements, it is time to talk to stakeholders. Before conducting the first interviews, it is good practice to develop a requirements specification and a consistent syntax to document the information from the stakeholders. A requirements specification is the process you will use to collect and document the project requirements. One of the primary purposes for developing a requirements specification is to achieve team consensus on how project requirements will be documented before making the more time-consuming effort of performing the work.

There are, of course, many requirements specification techniques to use. One of the best techniques was developed by Tom Gilb, who developed a simple, yet powerful end-to-end process for writing high-quality requirements. He calls this technique *Planguage*.[8]

The name Planguage is a combination of the words *planning* and *language*. Planguage is an informal, but structured, keyword-driven planning language. It aids in communicating complex ideas in terms any stakeholder can understand.

Planguage provides a standard format and vocabulary for each requirement. This helps reduce ambiguity, increase readability, and promote requirements reuse.

The keyword-driven syntax is a very effective way to document both functional (what the solution must do) and nonfunctional (how the solution will behave) requirements.

Specifying a Requirement Using Planguage

When specifying (writing) a requirement, it is much easier if you use a framework. As we have been discussing, Planguage is a keyword-driven syntax that provides an easy and effective way to ensure that you gather all the needed information from the stakeholder and can validate with other stakeholders a requirement has been written concisely, correctly, and completely.

Structuring Functional Requirements

The best way to use Planguage is to start with functional requirements. Table 4.2 lists a few basic Planguage keywords and definitions for writing functional requirements. Choose the Planguage keywords that will be most beneficial to your project.

Structuring Nonfunctional or Quality Requirements

The more challenging requirements to write are the nonfunctional or quality requirements. These requirements ensure that a solution behaves as expected. A few examples of nonfunctional requirements are level of reliability of the solution, scalability of the solution as it evolves and grows, and performance (how fast the system must be).

Table 4.2: Planguage Keywords for Functional Requirements

Basic Functional Keywords

Keyword	Definition
Tag	Unique, persistent identifier for traceability purposes
Gist	One-line description of the requirement or area being addressed
Source	Who provided the requirement
Functional requirement	The text detailing the requirement (The system shall ...)
Rationale	The reasoning that justifies the requirement, quantified if possible
Priority	Statement of priority and claim on resources
Author	The person who wrote the requirement

Additional Functional Keywords

Keyword	Definition
Stakeholders	List of parties materially affected by the requirement
Revision	A version number for the requirement (each requirement can have a revision history, not just an entire specification)
Date	The date of the most recent revision
Assumptions	All assumptions or assertions that could cause problems if untrue now or later
Risks	Anything that could cause malfunction, delay, or other negative impacts on the expected results
Subject matter expert	The person who is most knowledgeable (considered the expert) about the requirement
Dependencies	Anything that this requirement is dependent on (can't be implemented without)
Notes	Key information to assist in design, development, or delivery of the requirement
Defined	The definition of a term (better to use a glossary)

Table 4.3: Nonfunctional Planguage Keywords

Keyword	Definition
Ambition	A description of the goal of the nonfunctional requirement
Scale	The unit of measure used to quantify the nonfunctional requirement
Meter	The process or device used to establish the scale
Minimum	The minimum level required to avoid failure
Target	The level at which good success can be claimed
Outstanding	Stretch goal if everything goes as planned
Past	Previous results for comparison

When we state a nonfunctional requirement (or any qualitative statement), it is vital that we specify a few unique items not found in a functional requirement (see Table 4.3).

Don't forget, these Planguage keywords are used only when specifying *nonfunctional requirements*. Why? Functional requirements specify what a solution must do and are measured in "yes/no" terms. Most people capture these pretty well. Nonfunctional requirements are everything else associated with being measured on a scale other than as a simple "yes/no." The following section demonstrates how to use the Planguage technique to create high-quality functional and nonfunctional project requirements.

Using the Requirements Specificiation

The information captured in the requirements specification enables many things. Primarily, it helps with understanding requirements prioritization in a meaningful way. Rather than just calling something a high-priority requirement, the requirements specification also describes the impact of *not* implementing a requirement, as well as the rationale behind it. This enables sound decision making and prioritization throughout the project.

Functional requirements are usually easy to convert into a Planguage template, especially if the requirement is well known and has been successfully implemented before. The real value comes when the requirement is new, and asking a few more questions and documenting the answers significantly reduces the time to write a good requirement. Following is an example in order to illustrate how easy and effective it is to use this technique. While in a focus group discussion, an important stakeholder shared one of the key requirements as follows: *After every enrollment is completed, a confirmation has to be sent to the user.*

Is this a good requirement? As written, unfortunately, no, because critical information is missing. We don't know enough to be able to build and test this as stated. We would need to understand a lot more, such as:

- What does the system need to do? (Functional requirement)
- When does the system need to do this? (Functional requirement)
- Is there a time when the system *won't* do this? (Functional requirement)
- What is meant by completed? (Functional requirement)

- Why is this important for the system to do? (Rationale)
- How important is this in relation to other requirements? (Priority)
- If this is a high priority, what will happen if we don't implement? (Claim on resources)
- Who needs to know if this requirement is dropped? (Stakeholders)

Spending time to gain an understanding of the true requirement will help ensure that a complete requirement has been documented. Table 4.4 is an example of a complete functional requirement using Planguage.

Let's look closely at the functional requirement "confirmation" written previously using Planguage. Are there any obvious nonfunctional requirements associated with the functionality of the solution? To uncover a nonfunctional requirement, ask:

- Is there a degree in which the requirement can be measured? (Binary—yes or no; if yes, it is a nonfunctional requirement)
- How do you expect the system to behave? (Nonfunctional requirement)
- What is the goal? (Ambition)
- What will need to be measured? (Scale)
- What process or device can we use to measure? (Meter)
- What is the minimum level we need to establish? (Minimum)
- What is success? (Target)
- If everything goes as planned, what is a good stretch goal? (Outstanding)

Take the nonfunctional requirement information gathered and enter it into a table, database, or a spreadsheet (see Table 4.5).

A lot of information is now available in a very small space. Where most requirements specifications go wrong is in the description of how the nonfunctional requirements will be measured. Take time to talk to experts who do the testing and

Table 4.4: Example Functional Requirement Using Planguage

Keyword	Definition
Tag	Confirmation
Gist	Confirmation notification is key to enrollment process
Source	Chris Smith, during focus group discussions on July 2, 2015
Functional Requirement	After the user fills in all fields on the enrollment page and selects the enroll activation tab, the system shall send a confirmation to the user unless the health care provider is missing
Rationale	Task automation decreases error rate, reduces effort per enrollment, meets corporate business rule to confirm enrollment successfully
Priority	High; if not implemented, will cause business process reengineering and reduce program return on investment by $40m per year
Stakeholders	Intake analyst, marketing, health care provider
Author	Pat Jones
Date	July 5, 2015

Table 4.5: Example Nonfunctional Requirement Using Planguage

Keyword	Definition
Tag	Performance
Ambition	Ensure that enrollment into health care system is quick and efficient so users are not frustrated
Source	Chris Smith, during focus group discussions on July 2, 2015
Scale	Average time required for a novice to complete enrollment using only the online help system for assistance
Minimum	No more than 7 minutes
Target	No more than 9 minutes
Outstanding	No more than 5 minutes
Past	Recent site statistics show 11 minutes
Defined	Novice: A person with less than 6 months' experience with web applications and no prior exposure to our web site
Stakeholders	Intake analyst, marketing, health care provider
Author	Pat Jones
Date	July 5, 2015

validation of the solution to better understand how the requirement will be tested and validated. Many times, those who do the testing and validation know the most obvious, natural measurement because they are experienced in measuring quality and performance levels.

A project team should create its own set of keywords to meet the needs of the project. Some of the most creative keywords can be found in a concept glossary on Tom Gilb's web site.[9]

Simple Dos and Don'ts When Using Planguage

Do:

- Only use Planguage keywords that add value.
- Keep the Planguage list of keywords handy to help guide conversations with stakeholders.
- Spend time to understand how the nonfunctional requirements will be validated to ensure that minimum, target, and outstanding measurements are captured and discussed early in the requirements specification phase.
- Talk to a variety of stakeholders—especially the domain experts, subject matter experts, and testing team members—to understand how requirements are best validated.

(*continued*)

- Start with the scale first—figure out what you want to measure, and then associate the meter used to measure the scale.
- Use known, accepted scales of measure whenever possible.
- Begin with the standard Planguage keywords and create new keywords as needed

Don't:

- Guess; talk to testing and validation experts for their experience with similar measurements.
- Forget the meter must be employed before completion of the deliverable.
- Be fearful of Planguage; try it out on a small set of critical requirements that are new (never been done before) or are key to the success of the project.

Benefits

Most requirements specification tools focus on functional requirements and by nature mix in a variety of design and implementation solutions within the functionality of a solution. The keyword-driven syntax used by the Planguage technique is key to writing better requirements by pulling out a more detailed specification for both functional and nonfunctional requirements. This enables the project manager to provide improved prioritization of requirements, reduce design constraints, improve quality, and more effectively manage risk.

This in turn provides significant benefit to the consumers of the requirements specified such as a software developer who now knows what to design and develop, a tester who now knows what tests to run, and stakeholders who now know what they will be receiving.

THE PRODUCT REQUIREMENTS DOCUMENT

As requirements are written, it is very helpful to start to organize them into a document by logical groupings. Requirements documents can go by many names depending on your needs, such as a business requirements document, functional specification, system specification, or just the requirements document. For our purposes, we will use the term product requirements document (PRD). The PRD is used to document all requirements necessary to fully describe the features, functions, and capabilities required in the deliverables of a project. Even though it has *product* contained within its title, the PRD is used to define the requirements for any solution to be developed (see Figure 4.2).

The PRD is normally created in response to a marketing requirements document and should generally define the problems a solution is intended to solve. The PRD should *not* describe the solutions to the problems.[10] Solution development comes after the requirements are defined and documented.

PRODUCT REQUIREMENTS DOCUMENT

Project Name: Highlands **Rev#:** 1.0 **Date:** 11 Sept 2016

MARKET REQUIREMENTS

Market Drivers: Time to Market, Technical Leadership
Key Customers: Big Data LLC
Competition: ACME

PERSONAS AND USAGES

JOE:	**PAT:**	**HALEY:**
Long-time user	Geek	Power User
Daily use	Loves Technology	Customization
Traditionalist	Mobile Functions	24x7
Subject Expert	Extended Use	Millennial

REQUIREMENTS

- After the user fills in all fields on the enrollment page and selects the enroll activation button, the system shall send a confirmation to the user unless the healthcare provider information is missing
- System shall support 150 users simultaneously
- User shall be able to get to any screen within three (3) "clicks"
- Novice user shall be able to complete the enrollment form in 30 minutes maximum
- ....

RISK

Risk	Magnitude of Risk	Likelihood	Mitigation Plan
Technology not available for launch	High	High	Monitor monthly with technologist

Figure 4.2: Example Product Requirements Document

There are many PRD templates available. Feel free to tailor your PRD to suit the needs of your project. Rather than duplicating information that is already captured and still valid, reference it by the name of the document or by the tag.

Developing a Product Requirements Document

The purpose of the PRD is to clearly describe a solution's purpose, its features and functionality, how it is intended to be used, and what risks may be encountered.[11] The following guidelines are valuable when developing a PRD.

1. *Understand the product's reason for being.* In order to do this, you must spend time to understand the market requirements. This includes the product's objectives, your customers, and your competition. Every good project starts with an unfilled need that sparks the desire to develop a solution that addresses that need. It is critical to the success of the project that the project manager establishes a clear, concise value proposition for the creation of the product.

Few project managers stop long enough to create an "elevator pitch," which is three or four sentences (spoken in 10 to 15 seconds) that adequately describe the point of your product, why it is important, and how it would benefit the company.

2. *Classify the types of users and customers.* Once you have a clear understanding of the problem or unfulfilled need you want to solve, the next step is to gain an in-depth understanding of the target users and customers. These are many times called *personas:* fictional users who are realistic archetypes. They serve as a representation of the actual users and their expected experiences when using the product.[12]
3. *Define usages that will help users accomplish their goals.* Defining key usages per persona is important when discussing the user's main goals or objectives when using the product. This is the center of the product specifications process, and this is where creativity and innovation comes in. Again, avoid jumping to solutions, but rather focus on the user's problems and needs. Note that we have talked about goals and tasks but not features. Features will be linked to required tasks that map to specific user and customer's goals and objectives. If a feature cannot map to a goal, we have to ask if it is necessary.
4. *Use a format all stakeholders can understand.* Most requirements are written in a standard word processing application. The media and format are not as important as the content and the ability to find what the team needs and can be updated throughout the life of the project. Clearly articulate each requirement in a simple syntax, such as Planguage, which was discussed in the previous section.
5. *Brainstorm assumptions and risks to ensure success.* Once you think you understand their problem you are trying to solve, the users of the system, the usages, and the requirements, it is time to brainstorm all the assumptions made regarding the product planning, development, and launch. Spend time to question each assumption and comprehend the risks associated with the assumptions. What really matters is that the right product gets delivered, so don't let a wrong assumption derail your success.

   The key to developing a useful PRD is to be very clear what success looks like, and provide guidance to the project team so it is easy for them to make trade-off decisions when needed.

Identify the PRD Content

The next step is to determine what content you need in the PRD to ensure success. The PRD must be tailored to meet the needs of the project team. Following is a set of recommended content items to use as a reference for creating a customized PRD for your project.

1. *Project identifiers and purpose.* Project identifiers include the project name, the project manager, project sponsor, version number of the document, and the date the version was released.

The purpose statement describes why the document was created and what its intended use involves. An example purpose statement is:

> The primary purpose of this PRD is for the requirements team to ensure all aspects of the project requirements are easy to find in one place and provide the project team with the information necessary to understand and design the solution. It also provides the baseline for requirements change management.

2. *Market overview.* It is recommended that a brief description of the market be included early in the PRD. If a marketing requirements document has been generated, the market overview can usually be pulled directly from the document. The market overview provides a short description of the market, whether the market is new or an existing market, how the market is segmented, a description of the competing solutions, and a summary of the key market risks.
3. *Personas and usages.* This section of the document focuses on the customers, users, and how the product will likely be used. Customers are described through the use of *customer profiles.* Customer profiles identify the types of customers who are expected to purchase the product, any purchase decision information about the customers, and an overview of the customer's business model if it is a business-to-business-type purchase.

   Personas, also known as user profiles, describe the different types of users who will use the product. Users are many times a different set of people than those who will purchase the product. For example, parents (adults) normally purchase children's toys where the children are the users.

   Use cases describe how the product will be used. The best use cases describe how the user will interact with the various features of the product.
4. *Key features.* Provide a list of the key features that will satisfy the market objectives and meet the customer and user needs. Steer clear of too much detail when describing the features as the detail will be contained in the functional and nonfunctional requirements. The intent of this section is only to give the reader an overview of the key features being considered for the product.
5. *Functional requirements.* Organize the functional requirements in a way that supports their development. This usually means grouping them by functional area using subsections as required to present a logical breakdown of the functionality.

   The functional requirements should be entered in a table or a spreadsheet using keyword-driven syntax (Planguage) to ensure that clear, concise, and testable requirements are documented. Refer to Table 4.4 for an example of how to write a functional requirement using Planguage.
6. *Nonfunctional requirements.* Nonfunctional requirements should be organized by topics or requirement types. Example nonfunctional requirement types include performance, reliability, usability, security, maintainability, compatibility, interoperability, and customization requirements.

   The nonfunctional requirements should be entered in a table or a spreadsheet using keyword-driven syntax (Planguage) to ensure that clear, concise, and testable requirements are documented. Refer to Table 4.5 for an example of how to write a nonfunctional requirement using Planguage.

7. *Documentation*. Often overlooked are the documentation requirements associated with a solution and project. It is not uncommon for project teams to complete the design, development, and test of a solution and then come to realize that they had forgotten to create the documentation required by the customer and end users of the solution. Usually, this situation is caused by a missed documentation requirement.

   Examples of documentation that may be required include the following:

   - User documentation
   - Online help scripts
   - Labels and packaging
   - Technical documentation
   - Marketing collateral and sales tools

8. *Internationalization and location*. It is common for products and other solutions to be deployed on a worldwide scale, which many times require a number of customizations to meet market, customer, and user needs. All requirements that are specific to international use should be documented in the PRD. These may include requirements for unique user interfaces, customer or user documentation, special export requirements, and what languages must be supported.

9. *Legal requirements*. Although not commonly included in a PRD, legal requirements are included in this suggested list of content items to encourage project managers to at least investigate whether there are any special legal requirements that need to be documented and acted upon by the downstream recipients of the PRD.

   Legal requirements may include such things as trademarks and copyrights (do any components need to be registered or approved?), licenses that may be required, requirements for certificate of origin, and export and import requirements if the solution is going to be sold or used internationally.

Document the Requirements

Creating a useful PRD is much more difficult than just identifying the contents of the document. The hard work is defining and writing the requirements in terms that your project team can comprehend and take action upon. Before sitting down to write, start by taking the time to describe or draw your ideas out with a pencil and paper. This will help formulate your ideas, especially after talking to your stakeholders to understand their needs and desires (see "Golden Rules for Developing a Useful PRD").

Next, review the PRD template and eliminate sections that don't apply to your project. Begin filling in the template with the information that comes easy to you. Get past writer's block by ignoring the quality of your writing. Just write. You can fix the grammar, punctuation, and spelling later. It is very easy to unintentionally dive into documenting the actual solution rather than staying at the level of "what" versus "how."

Golden Rules for Developing a Useful PRD

1. Before writing a PRD, make sure you understand the user needs and problems you are trying to solve.
2. Use a PRD template to guide you through the process and prevent you from missing key elements.
3. Include the market requirements in the PRD so you stay focused and avoid scope creep.
4. "A picture is worth a thousand words"—complex ideas can be best conveyed with a picture, which makes it possible to absorb large amounts of data quickly.
5. Use Planguage to document the functional and nonfunctional requirements.
6. Focus on "what" the product must do, versus "how" to solve the problems.

Using the PRD

The best PRD is one that provides enough information so the downstream consumers (designers, engineers, validation team) can do their jobs, but not so much information that the important details get lost in the noise of long, rambling sentences and paragraphs.

Keep the PRD as simple and uncluttered as possible. Only have in the document what the design, developers, and testers really need. Customize the content items to meet the needs of your project to ensure that it will actually be read and used.

The key to ensuring usage of the PRD is for the author to ensure it is easy for the recipients to find what they need and understand what they need to do. Go talk to your team, ask them what they want, and give it to them. If your PRD isn't getting across the information they need, the most important thing is to be flexible and modify it to meet the needs of your audience.

Tips for PRD Use

Make sure the PRD is up to date, accessible, and easy to search so the consumers of the PRD can find what they need.

What is the right level of detail? It depends. Ask your audience what level of detail they need, and then provide it to them.

Spend time to educate your audience on Planguage so they understand the keywords and the information provided.

Whenever possible, include information on the market requirements so those using the PRD understand the market and strategy.

Think of the PRD as a compass for the team to ensure they are going in the right direction and not going down a dead-end road. It will be the document that ties all the downstream activities such as design, development, and testing efforts together. If no one uses the PRD, the team is setting off on a long, uphill hike without a map and no idea of which direction they are going.

Benefits

Will anyone really read the PRD? You must hope so! The PRD will be a lifesaver when the size and complexity of a project is fairly large. If the complexity is low, the PRD may only be two to four pages. If the complexity is high, the documentation may be several hundred pages (including Planguage keyword tables).

Documenting the product requirements is very helpful in getting the team focused. It ensures that everyone has a clear understanding of the product being developed and where they fit into the process. It is very common for a requirements team to become distracted by an exciting feature they like (not necessarily a feature a customer or user asked for), rather than staying focused on the problem they are trying to solve.

One of the biggest benefits of a PRD is that it gets the entire project team involved. Avoid at all costs writing a PRD alone without any input from stakeholders. Include key partners such as designers, developers, and testers to write and review as you go. This will spark new questions and discussions about risk and assumptions, as well as encourage the team to talk about the right level of detail and inspire new thoughts and new ideas, which will only make the product better.

The bottom line is that a PRD helps reduce ambiguity and provides the team with the details to actually build and test a product.

REQUIREMENTS AMBIGUITY CHECKLIST

After several requirements have been documented, and *before* development of the requirements begins, set up a series of requirements review sessions for key stakeholders to check for clarity, completeness, and quality.

The goal of a requirements review is to ensure that the team has developed the requirements to their highest level of quality and gain agreement among the stakeholders that the requirements are correct, complete, and unambiguous.

To help this process go faster, use a requirements ambiguity checklist to expose the most common mistakes requirements authors make.[13] Once the stakeholders agree and sign off on the requirements, development or design can begin. See Table 4.6 for an example checklist.

Review Each Requirement for a Single Interpretation

Take time to meet with the requirements team to confirm that each requirement is clear, concise, and coherent for the intended audience, and that each requirement possesses a single interpretation. There are three areas the team should discuss to ensure that there is only one interpretation of the requirement:

Table 4.6: Example Requirements Ambiguity Checklist

Status	Checklist Items
☑	Requirements are clear to the intended audience, with only one interpretation.
☑	Terms are defined in a glossary and used consistently.
☑	Poor grammar that causes unintended meaning (or no meaning at all) has been removed.
☑	All "dangling else" statements have been removed.
☑	Terms relying on personal opinion, such as *fast, easy, state-of-the-art, obvious, simple, relatively, as appropriate,* and *excellent,* have been removed.
☑	Words without precise meaning, such as *soon, virtually, more, most, some, similar, eventually,* and *recent,* have been removed
☑	Unbounded lists, such as *TBD, etc., at least, including, such as,* and *at a minimum,* have been eliminated.
☑	Optionality words, such as *should, could, possibly, if possible,* or *may,* have been eliminated.

1. *Completeness.* Are there any placeholders ("TBD," question marks, use of etc., or notes) to add information that has not been addressed? Are there any unbounded lists? Look for statements such as *at least, including, such as,* or *at a minimum,* to name a few.
2. *Optionality.* Are the correct imperatives used? Remove any use of the words *should, could, would, can, may, might, possibly,* or *if possible.* Preface functional requirements with the imperative *shall* (or *will*), and nonfunctional requirements with *must.*
3. *Vagueness.* Are there any words without a precise meaning, such as *friendly, similar, some, few, eventually, recent, more,* or *most*? Time references, such as *before, after, current,* or *simultaneous,* can be tricky. Personal opinion, such as *excellent, fast, easy, state-of-the-art, hardly ever, by and large, more or less, obvious,* and *simple,* are ambiguous. Double check to make sure there aren't any double-negative statements and that proper grammar, punctuation, and spelling is used.

Ensure that All Terms Are Defined and Used Consistently

The requirements team should have all acronyms and specific terms defined in a glossary and available to refer to when there is a question on how a specific term is being used.

Tips to Ensure that Terms Are Defined

Take time to define terms that are not universally known by the target audience.

Highlight all instances of defined terms with some visual trait (e.g., color, font, capitalization, or boldface) so they stand out from the other text.

Use a glossary to define words used in the document to safeguard consistent usage and understanding.

Review Each Requirement for Weak Words

Weak words are notorious for being vague or ambiguous. Replace any term that is qualitative rather than quantitative, as it is most likely weak. There are too many weak words to list them all. Richard Bender has a comprehensive list of words in his whitepaper titled "The Ambiguity Review Process" that helps point out potential ambiguities when writing a PRD.[14] He describes a trap he calls the "dangling else," where words such as *must be, will be, is one of, should be, could be,* and *can be* are ambigous. Also review the words listed above under completeness, optionality, and vagueness. Replace or remove them to improve the quality of your requirements.

Include Alternative Methods to Describe Requirements

If possible, include a different way to describe a requirement, especially if a diagram, algorithm, table, use case, user stories, or a link to a prototype would increase understanding. To reduce ambiguity and increase comprehension, show process and data flows with a diagram or algorithm. Any complicated logical flows, dependencies, and time indicators are good candidates.

When there are many categories or attributes within a requirement, use a table. State tables are very effective when several conditions or complex logical structures govern the responses of the system under specific definitions.

Use cases describe the user experience of the end solution, which is an excellent way to document system usage and provide context to the requirements. User stories are typically very short—one to two sentences to enable quick updates to the project based on user specifications as they are known.

Using the Requirements Ambiguity Checklist

As stated previously, the requirements ambiguity checklist is best used as a guide when requirements authors, contributors, and stakeholders are reviewing requirements to ensure that all statements are unambiguous to the intended audience. The burden of clarity is on the author, not the reader.

The ambiguity review process is best used as a two-step process:[15]

Step 1: *The initial ambiguity review.* This initial review is conducted by someone who knows little to nothing about the domain. The intent of this review is to ensure that the logic and structure are clear and understandable by a layperson. Since they are not a domain expert, they can't read into the requirement facts that aren't explicitly written. Identify three to five key requirements you want reviewed and encourage them to identify generic ambiguities such as unclear references.

Step 2: After the issues raised in the initial ambiguity review are resolved, the next step is to have a domain expert use the ambiguity checklist to review for content. Focus on ensuring that the requirements are complete and correct; terms are defined and used consistently; there are no "weak words"; and that diagrams, tables, use cases, or prototypes are present to improve clarity.

If the requirements are in a word processing application, then the ambiguities are documented in either a copy of the requirements or in a separate document.

Depending on the accessibility of the requirements and logistics of making the requirements available to the reviewers and if the comments made by one reviewer can be seen by the others, keep the timing of providing feedback relatively short. For instance, if you hold a review session on Tuesday, ask for feedback to be sent back by Thursday. That way, you can follow up on Friday to ensure that all review comments are collected and can be addressed in a timely manner.

It is common on the first review to find about 10 to 15 ambiguities per page of detailed requirements. The cost associated with fixing the error after release of a solution can be over four to five times as much as one found during design, and some numbers show up to 100 times more expensive to correct than an error found during the maintenance phase. This is best illustrated in a chart. See Figure 4.3 to better see the cost comparison.[16]

After a series of reviews with both domain experts and those who are not domain experts, ambiguities per page drop by a factor of 10x. We find that once a requirements author realizes that the reviewers are looking for dangling else terms such as *must be* and *will be* and vague terms such as *friendly, similar,* and *more,* the next requirement they write will be void of the same error. The author gets better at writing, and the reviewers will find substantially fewer errors.

Tips for Using the Ambiguity Checklist

Use the checklist to help remind you of weak words to avoid.

Do a search to remove any instances of "dangling else" before stakeholders provide feedback.

Remember, fixing requirements defects during the planning phase costs substantially less than waiting until your customer finds the error in the released solution.

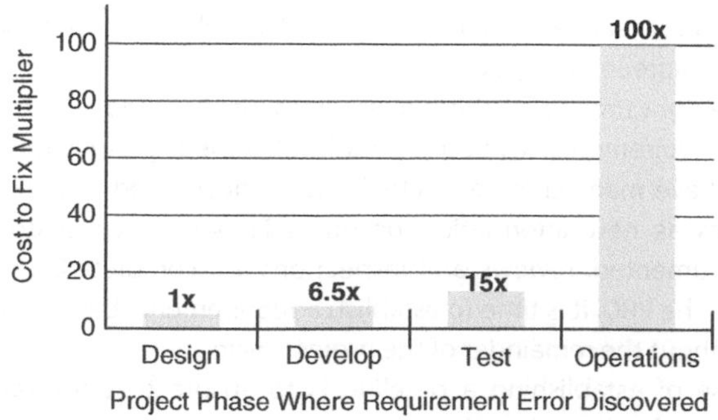

Figure 4.3: Relative Costs to Fix Requirements Errors

Benefits

The requirements ambiguity checklist will improve the readability and provide clarity to individual requirements as well as the entire requirements document. Use it while writing requirements and every time stakeholders, especially domain experts, conduct reviews to provide a framework to help ensure the team is spending time to discuss the right things and fix common problems that occur in writing requirements.

The key benefits are:

- Domain experts provide feedback and insight early in the documentation of the requirements, which results in higher-quality requirements.
- Errors are fixed earlier so the cost is less.
- Breaking the requirements into smaller pieces provides focus and timely feedback.
- One clear, concise, and measurable set of requirements reduces rework and unnecessary features creeping into the product.

REQUIREMENTS BASELINE

After the ambiguity reviews are complete and all requirements have been formally approved (usually via an inspection process), the stage is set for moving into the requirements management process.

It is estimated that few initial requirements documents capture more than half of the eventual requirements. So change will happen! Unmanaged and poorly managed changes to requirements lead to mistakes, unnecessary features, and expensive rework. Managing changes is a very important part of a successful project. Equally so, is the establishment of a requirements baseline. A requirements baseline is a snapshot in time that represents the agreed-upon, reviewed, and approved set of requirements for a project.[17]

Once approved by all the stakeholders, the baseline becomes the foundation of the requirements management process. Any changes needed after the baseline is set are subject to stringent change control. This is because the baseline is a finite agreement between the project team and their sponsor and stakeholders. Any changes required will be subject to agreement by both sides.

When is the right time to establish a requirements baseline and to start managing the changes to requirements? While in the midst of gathering and writing requirements, it is normal to have many changes. Actually, it is encouraged to make modifications to requirements as new information becomes known. Once customer needs are adequately documented, reviews and inspections are completed, and requirements are approved in the PRD, it is time to establish a requirements baseline and control the changes throughout the remainder of the project cycle.

The purpose of establishing a baseline is to ensure that the requirements are clear, common, and coherent to the downstream consumers of a requirements document. Spending adequate time with internal business partners, customers, and other stakeholders to determine if the requirements are complete, necessary, correct, unambiguous, consistent, modifiable, traceable, and testable will pay larger dividends as changes to the requirements begin to occur.

Establishing a Requirements Baseline

You may have heard the phrase "freeze the requirements." This really means it is time to "establish a requirements baseline."[18] The Institute of Electrical and Electronics Engineers (IEEE) Standard Glossary of Software Engineering Terminology defines a baseline as:

> A specification or product that has been formally reviewed and agreed on, that thereafter serves as the basis for further development, and that can be changed only through formal change control procedures.

As we have been discussing, the requirements are the foundation to a project and the lack of a requirements baseline can give free rein to constant change in requirements and therefore project scope. A baseline is about spending time to come to a common agreement between stakeholders. It essentially involves setting the right expectations, including responsibilities, risks, assumptions, and approaches. Once an agreement is reached, the requirements document is under change control to manage the baseline going forward.

Typically, a requirements change will affect the scope of a project (commonly called scope creep). Scope creep is generally not a good thing. It can drive costs, resources, and schedules beyond their thresholds, and when features are implemented without complete understanding of how they impact other aspects of the project, it can cause confusion throughout the project.

Once the requirements have become stable, it is time to establish the baseline and implement a formal requirements change control process (see Chapter 8). Controlling the project requirements baseline is essential to project success.

Establish a Requirements Change Board

Karl Wiegers defines the requirements change control process as "all the activities that maintain the integrity and accuracy of the requirements as the project progresses."[19] In order for the requirements to be managed effectively, a change control board (CCB) is formed. This team is usually a small cross-functional team that can represent viewpoints of all project stakeholders and is given the authority to manage the new requirements baseline using a well-defined, well-documented (and communicated) change management process.

The CCB owns the implementation of a formal change management process that begins as soon as the baseline is established and continues throughout the project life cycle. Normally, the CCB is a small team who meets regularly to ensure standard methods and procedures are used to efficiently handle any changes to the requirements. This is best described as a formal, repeatable process to reduce the risk of introducing change into the project scope without assessing and controlling the requested change. A simple but robust change management process is presented in Chapter 8.

Will there be changes on every project? The answer is yes. Do we want a change after the baseline has been established to be as painless as possible? The answer is, again, yes. Early in the definition of the project we want changes to be easy to implement because the elicitation efforts will continue to uncover new requirements. However, once we have an agreed-upon scope, an approved budget, and a schedule established, new requirements and modifications to requirements can be very disruptive and cause the project to derail. Changes to requirements will happen; you just have to be ready for them when they do.

Benefits

Setting a requirements baseline and establishing a change control board to manage requirements changes is a good practice. The benefits of establishing a baseline and then managing changes to that baseline are many; following are a few of the crucial benefits:

- Establishing a requirements baseline helps to ensure all requests for change are evaluated and managed by the right stakeholders at the right time.
- Changes are planned, managed, and communicated so that resources, budget, and schedule can be aligned to support the change.
- An effective CCB will assess the overall impact and risk associated with the change and have the authority to approve, defer, or deny based on criteria set for the project.
- An updated and maintained requirements baseline ensures that the project team is designing, developing, and testing the approved set of requirements, which will reduce rework, errors, and unnecessary features.

Final Thoughts

Requirements are a primary factor in project success or failure. We know and understand that one size does not fit all, so feel free to tailor the five requirements tools outlined in this chapter to meet the needs of your organization and project.

Spend time at the very onset of a project to outline an effective elicitation plan so you can ensure that you have a clear understanding of the voice of the customer by gathering requirements from many stakeholders—and a comprehensive plan to do so.

Every business situation is different; some are more complex than others, and your stakeholders will all have different perspectives. Use Planguage to help simplify by breaking down each requirement into keywords that any stakeholder, from the technology geek to a finance wizard, can understand.

Organize your requirements into a document such as the product requirements document so the downstream consumers of the document can easily find what they need and understand what they need to do. Pick and choose what your project needs—at a minimum the functional and nonfunctional requirements.

Use an ambiguity checklist when stakeholders review requirements to ensure that they are replacing weak requirements with clear, concise, and measurable ones. The common mistakes can be avoided by providing guidance to your reviewers.

Establish a requirements baseline after the reviews are done so the team members know what requirements have been approved. When changes occur, use the baseline as the basis for requirements change management.

References

1. Project Management Institute. *Pulse of the Profession: Requirements Management—A Core Competency for Project and Program Success.* The Pulse of the Profession, In-Depth Report (Newtown Square, PA: Project Management Institute, 2014).
2. Ibid.
3. Rierson, Leanna. *Developing Safety-Critical Software: A Practical Guide for Aviation Software and DO-178C Compliance* (Boca Raton, FL: CRC Press, 2013): 99–100.
4. Wiegers, Karl, and Joy Beatty. *Software Requirements*, 3rd ed. (Developer Best Practices) (Redmond, WA: Microsoft Press, 2013): 4.
5. Hass, Kathleen, and Rosemary Hossenlopp. *Unearthing Business Requirements: Elicitation Tools and Techniques* (Vienna, VA: Management Concepts, 2008): 76–77.
6. Roman, Ernan. *Voice-of-the-Customer Marketing: A Revolutionary Five-Step Process to Create Customers Who Care, Spend and Stay* (New York, NY: McGraw-Hill, 2011): 3.
7. Robertson, Suzanne, and James Robertson. *Mastering the Requirements Process: Getting Requirements Right*, 3rd ed. (New York, NY: Pearson Education, 2013): 102–108.
8. Gilb, Tom. *Competitive Engineering: A Handbook for Systems Engineering, Requirements Engineering, and Software Engineering Using Planguage.* (Burlington, MA: Elsevier Butterworth-Heinemann, 2005).
9. Web site: www.gilb.com. Under Resources for Requirements, see concept glossary. Accessed October 30, 2014.
10. Wikipedia: http://en.wikipedia.org/wiki/Product_requirements_document. Accessed February 2015.
11. Cagan, Martin. *How to Write a Good PRD* (white paper) (Silicon Valley, CA: Silicon Valley Product Group, 2005). http://www.svpg.com/assets/Files/goodprd.pdf
12. Wiegers and Beatty, 2013.
13. Ibid.
14. Bender, Richard. *The Ambiguity Review Process* (white paper). http://www.benderrbt.com/Ambiguityprocess.pdf
15. Robertson and Robertson, 2013, 388.
16. Soni, Mukesh. "Six Sigma. Defect Prevention: Reducing Costs and Enhancing Quality." http://www.isixsigma.com/industries/software-it/defect-prevention-reducing-costs-and-enhancing-quality/. Accessed December 18, 2014
17. Wiegers and Beatty, 2013.
18. Ibid.
19. Ibid.

5

SCOPE PLANNING

Project planning begins with identifying what work needs to be completed for a project to be successfully implemented. Regardless of the project management methodology used, it will require you to first and foremost define the scope of the project. Thus, it will require the project manager to engage in scope planning activities.

Scope planning is a collaborative process between the project manager, the project sponsor, and other key stakeholders who can help shape the work that must be completed in order to achieve the business goals driving the need for a project.

The purpose of scope planning is to ensure that all the required work and only the required work is clearly identified, that the deliverables and outcomes are documented, and that the boundary conditions are adequately defined to complete the project successfully.

Scope planning involves identifying your goals, objectives, tasks, resources, budget, and timeline. This chapter includes a number of high-impact tools that can be added to a project manager's PM Toolbox to assist with the process of scope planning.

Used in conjunction with tools for schedule, budget, resource, and risk planning, project scope planning tools eventually lead to project plan development. Later, during project implementation, the scope baseline will be a vital foundation for disciplined scope control and change management, creating a barrier to unnecessary scope creep.

THE PROJECT SWOT ANALYSIS

The project SWOT (strengths, weaknesses, opportunities, threats) analysis is a technique used by project managers to develop project execution positioning. By positioning a project to take advantage of its particular strengths and opportunities while minimizing weaknesses and threats, the project SWOT analysis helps identify a sound strategy for project execution. The analysis is performed to gain an understanding of and act upon the project's capabilities and environment.[1]

Project capabilities—expressed as the project's internal strengths and weaknesses—tell us what our project can and cannot do well. At the same time, our assessment of the project environment indicates what opportunities and threats are presented by the environment within which the project operates. Information about the environment, combined with the knowledge of the project's capabilities, enables project teams to identify factors needed to meet the business and project requirements (see Figure 5.1).

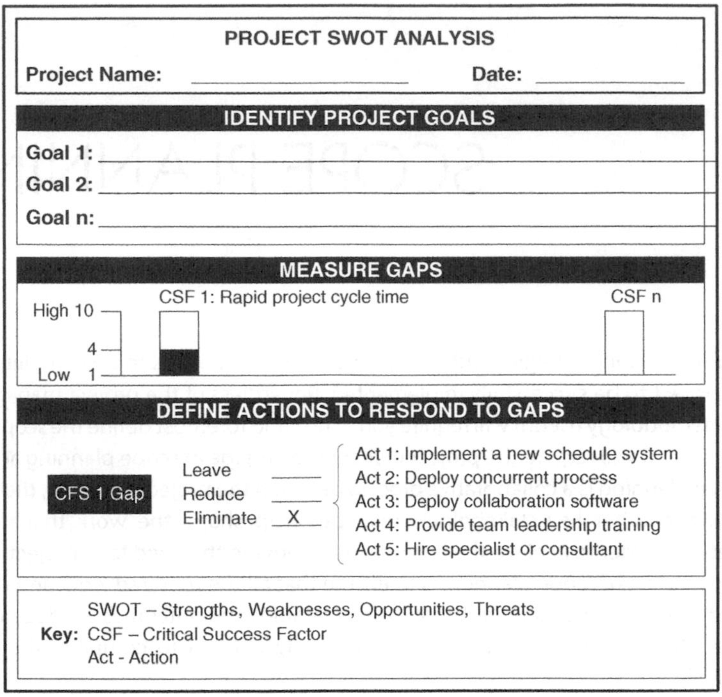

PROJECT SWOT ANALYSIS
Project Name:
Date:
IDENTIFY PROJECT GOALS
Goal 1:
Goal 2:
Goal n:
MEASURE GAPS
CSF 1: Rapid project cycle time
CSF n
High 10
4
Low 1
DEFINE ACTIONS TO RESPOND TO GAPS
CFS 1 Gap
Leave
Reduce
Eliminate X
Act 1: Implement a new schedule system
Act 2: Deploy concurrent process
Act 3: Deploy collaboration software
Act 4: Provide team leadership training
Act 5: Hire specialist or consultant
SWOT – Strengths, Weaknesses, Opportunities, Threats
Key: CSF – Critical Success Factor
Act - Action

Figure 5.1: Project SWOT Analysis Template

Measurement of where the project stands regarding these factors provides clues about execution gaps, prompting the team to consider strategies and actions to address the gaps. Awareness of the gaps and a clearly defined response allow the team to formulate a realistic project scope and related strategies for attaining the project goals. In short, the project SWOT analysis is an effective tool to include in a project manager's PM Toolbox because it identifies areas of strengths to leverage and execution gaps to resolve in developing an effective project plan.

Performing a Project SWOT Analysis

Much of the information needed to perform a SWOT analysis at the project level can be collected through tools described previously. For a good start in performing the project SWOT analysis, three information inputs are of vital importance:

1. The project charter with its supporting detail.
2. The project business case.
3. The project requirements.

While the project charter provides knowledge about the fundamental boundaries and constraints of the project, the supporting detail (strategic and tactical plans, project selection criteria, and project mission) helps determine the context in which the boundaries were drawn. The project business case describes the business environment in which the project will exist, thus providing vital information about capabilities needed internally and external to an organization. Exactly why the project requirements are so relevant for the analysis will become clear in the first step in performing a project SWOT analysis.

Identify Project Goals

Projects are implemented to create value for a firm. Consequently, the project goals define what constitutes value when the project is completed. When it comes to the project SWOT analysis, focus only on the critical goals, those that can make or break the project. For example, if a firm is heavily involved in a time-to-market race with their competition, they may have a requirement that a project shave off 30 percent of what is considered a typical delivery time for a project of that type. This is a significant challenge for the implementing company and its project team, who may have little experience in fast-track projects. Because the company's management may view this project as an opportunity to enter into a new market, the project needs to be successful. But what does it take to be successful? The answer is in identification of the project success factors.

Select the Project Success Factors

Fundamentally, project success factors (commonly called critical success factors) are areas in which a company must do well in order to be successful.[2] So what are these areas? The areas come from two major domains. The first domain is the business success factors that are identified in the project business case. The second domain is the goals of the project, which support the business success factors but are execution focused.

The project success factors are those that are necessary to meet the goals of the project. Begin, then, by asking what you need to do well within the project in order to meet or exceed the project goals. In the example stated earlier, rapid project cycle time is identified as a primary project success factor that supports the business goal of early entry into a new market in order to capture market share. That is a very complex project success factor, requiring the synchronization of several components, including concurrent engineering, the use of collaboration software, cross-functional teams with interpersonal skills, and integrated scheduling. Of course, more may be necessary for rapid project cycle time, but these four components are good examples.

Concurrent engineering is about overlapping project activities to speed up the project pace.[3] Its crux is reciprocal dependencies between the activities that exchange incomplete information, making it more difficult but faster to work.[4] In the product development context, that exchange is significantly more efficient if performed in a collaborative manner supported by collaboration software, which commands significant resources and skills. Rapid product development also demands flexible cross-functional teams, fully equipped with soft, interpersonal skills to handle conflicts and negotiations germane to fast-track projects.[5] On top of all of this, scheduling this project calls for the ability to juggle multiple critical paths, perhaps including 30 to 40 percent of all activities, with a constant need to use schedule crashing or fast-tracking techniques.

When you are finished analyzing the project capabilities domain, look again at the project requirements, particularly the requirements focused on customer needs and desires. In which areas in the project environment must you do well to meet or exceed the requirements? You may need to include a first-tier vendor involved in the project execution, as this has been shown to be a proven technique to speed up projects.[6] Certainly, possibilities of external critical success factors (CSFs) abound, but the checklist shown in the example titled "Screening Your Environment for Possible Critical Success Factors and Gaps" offers a few ideas of where to look.

Screening Your Environment for Possible Critical Success Factors and Gaps

Here is a short, general checklist of areas where you can find possible CSFs and related strategic gaps:

- Stockholders
- Customers
- Governments
- Competitors
- The general public
- Creditors
- Suppliers/Vendors
- Unions
- Local communities

Again, possibilities are limitless, but note that these environment-related CSFs are more challenging to deal with because, unlike internal project capabilities, they are external to the project and less controllable. Brainstorm with the team to identify ten or so for each domain, internal and external. Rank them and focus on the critical few. Keep in mind that identifying CSFs without understanding their measurement criteria, dynamics, and interactions is fruitless.

Measure the Gaps

After you have selected project execution CSFs, the next step is measuring the gaps. A gap is the difference between the ideal and actual level of a CSF. If the CSF in our example from Figure 5.1—rapid project cycle time—were ideal, it would draw a perfect scoring on all of its four components, including concurrent engineering, collaboration software, interpersonal skills, and integrated scheduling. That would mean that the entire CSF and its four components are at a level ideal for meeting the project goals. The actual level, however, is where you believe you currently stand with regard to that CSF and its components.

A measurement scale is necessary to identify the magnitude of the gaps. Depending on the degree of desired rigor and the time available for the project SWOT analysis, you can choose from among several alternatives. For example, smaller and simpler projects may do just fine by having a straightforward scale of small, medium, and large gaps.

Other organizations categorize gaps by urgency, perhaps using colors—for instance, green for "no gap," yellow for "caution," and red for "danger." This allows them to send an immediate visual message that describes the gap.

In contrast, a perceptual scale spanning from 10 (the ideal level in example SWOT Analysis) to 1 (the widest gap) more precisely measures a gap. How do you identify a gap with this scale? One way is to assign a narrative description of the state of a CSF's component to each level in the scale (1 through 10). Then, after a team discussion, each team member assesses the actual level of a component. Further, the actual level of each component can be averaged for the whole team. For example, the actual level of rapid project cycle time for Figure 5.1 would be 4. Comparing the actual level with the ideal one will yield the gap—in our case, a large gap of six units. Whatever gap measuring method you choose, keep in mind that measuring a gap is a subjective judgment, not an exact science.

Having a large gap on internal capabilities is an apparent weakness, while a small gap is a strength. Similarly, where we find a small gap in assessing the project environment, we can view it as an opportunity. Along the same line, a sizable external gap poses a threat. Expressing gaps as potential threats can help you relate the problem to the project sponsor, top managers, and other stakeholders who may be instrumental in getting the resources you need to address the gaps.

Decide How to Respond to the Gaps

Identifying a gap brings you to another decision: what to do about it. Generally, you have three options: leave it as is, reduce the gap, or eliminate the gap.[7]

As Figure 5.2 indicates, when your project has few or no gaps in internal capabilities and project environment (upper left-hand corner), your best option may be to leave things as is.

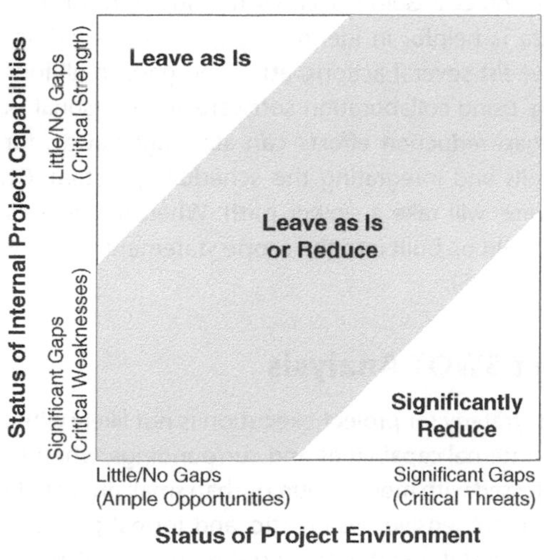

Figure 5.2: Project SWOT Analysis and Strategies to Act on Gaps

Why would anyone select this option? A lack of time or an insignificant impact of eliminating gaps may be reasons for opting to leave small gaps alone. However, those who decide to eliminate small gaps may find the motivation for doing so in the ease of closing them or in their drive for perfection. With larger gaps in internal capabilities or project environment (the shaded area in Figure 5.2), the case may be more complex. First, acting on gaps usually requires resources, typically controlled by management. If your knowledge of the style of management and resource availability tells you that your action may not be fruitful, and therefore not worth the effort, leaving the gap as is may be understandable. However, it is often worth the case to management, especially if you use a language that management reveres: impact to business results.

Take, for example, our case of the six-unit gap in CSF 1 in Figure 5.1. Not acting on this gap may prohibit the company's management goal of entering a new market of time-to-market focused competitors, resulting in loss of expected sales and profits. Given the strategic and profit significance of our case, management sees that our gap deserves an action of eliminating the gap. However, if the project was resource or funding constrained, we might choose the *leave as is* option.

Receiving an infusion of resources, though, is not an easy scenario in today's organization, where various projects often brutally compete for their share of the resource supply. Rather, it may take preparing well, presenting the case for the scarce resources, and fighting for them with a passion based on facts. Once the resources have been obtained, developing a plan to reduce the gap is important.

Figure 5.2 also shows that in a situation burdened with large gaps in internal and external capabilities (lower right-hand corner), perhaps the most viable option is to act to significantly reduce gaps; eliminating them may be too challenging.

Define Specific Actions to Respond to Gaps

Reducing or eliminating project execution gaps is a good decision, requiring specific actions to deploy resources. Going back to the structure of a CSF, its components, and the actual state is helpful in identifying actions needed. In our project SWOT analysis example, we list several actions, attacking gaps in all four components: concurrent engineering, using collaboration software, interpersonal skills, and integrated scheduling. Some gap reduction efforts can start right away, for example, training on interpersonal skills and integrating the scheduling system. Others, like installing collaboration software, will take a slower path. When the actions are identified and understood, they should be built into the scope statement and, possibly, into the work breakdown structure (WBS).

Using a Project SWOT Analysis

Identifying a sound strategy for project execution is not likely without a critical evaluation of the project's internal capabilities and surroundings, whether the project is large or small. Typically blessed with more resources than small projects, large projects should strive for a more comprehensive, systematic, and formal project SWOT analysis, preceding a detailed scope statement. Applying the project SWOT analysis in an informal manner is a common approach for smaller projects. Faced with little time or resources to do a detailed analysis, project managers for small projects should make the analysis part of their mental process, constantly questioning their project capabilities and surroundings. They should not be concerned that the analysis must be written; it does not have to be (see the example titled "Ten Minutes Can Do It").

Ten Minutes Can Do It

Jeffrey D'Esposito, project manager for Procter & Gamble, had this to say about using the project SWOT analysis:

> When I first ran into the project SWOT analysis tool, I was very pleasantly surprised and, frankly, very proud of myself. For a long time I was doing a SWOT analysis on my projects without knowing there was actually a formal tool. As a manager of multiple small projects, I never had time to do a formal, written analysis. Rather, I was doing it informally, verbally in ten or so minutes with my team members. We simply called it risk assessment. But it worked, and it worked very well.

Jeffrey's story is not a solitary case. Many project managers do the same thing. Know your gaps before you venture into the project. Make them visible. Ask your manager to help reduce or close the gaps. And if it only takes ten minutes, you can find the time in your busy schedule.

Benefits

The most successful ventures are built on the ability to utilize one's strengths, moderate weaknesses, seize the opportunities, and neutralize threats.[8] Pressured for time on many concurrent fronts, project managers all too often do not build their projects on this premise. Rather, they dive into detailed project planning without taking stock of their project's strengths, weaknesses, opportunities, threats, and related gaps. This is where the project SWOT analysis comes in; it takes stock and offers a clear picture of project gaps. The value of the analysis, therefore, is in enabling the project to do the following:

- Position itself in the best possible way to take advantage of its particular strengths and opportunities within an organization, while minimizing the weaknesses and threats.
- Uncover strengths that have not yet been fully leveraged and identify weaknesses that can be corrected.
- Bring to the attention of the senior management team significant gaps that can jeopardize the project and get their support to close the gaps and reduce the risk of failure.
- Continually focus actions on business success factors and project requirements to ensure alignment of project capabilities.

The proactive nature involved in the use of the project SWOT analysis helps build the mentality that it is never too early to bring a defensive shield up. Facing project execution gaps early helps to define alternative project scenarios and prepare for major project danger zones.

THE SCOPE STATEMENT

The scope statement is a written narrative of the goals, work, and outcomes of a project. It defines the project and becomes the basis for making decisions and decision trade-offs during project planning and execution. A good scope statement is necessary to guide a project to successful completion.

Scope statements take many forms, depending on the project type as well as the nature of an organization. They should be viewed as a dynamic tool that contains the best information available during early project planning, and as the project progresses, it should be modified to reflect significant changes incurred.

The scope statement is effective in establishing the project baseline and boundary conditions, which cannot be compromised without consent of the approving managers. It also means that within the project boundaries, the project manager and his or her team are empowered to operate and make appropriate decisions.

Developing a Scope Statement

The fundamental premise in developing a scope statement is that the statement must be as change resistant as possible (see the example titled "Innovative Ways to Develop a Change-Resistant Scope Statement"). Note that this premise does not have to do with successful control of changes through change management systems such as a change control plan and scope control. Rather, the premise here is very different and rooted in a set of principles that help minimize the impact of changes from the environment surrounding the project.

Innovative Ways to Develop a Change-Resistant Scope Statement

The application of the following principles in scoping projects can help define projects with increased resistance to change later in the implementation:

Principle 1: Reduce project complexity. Adding work elements to a project means that you create more interactions between the elements. Each new element adds complexity by increasing the number of interactions, so when a change hits, more elements need to be changed or redone. In contrast, scoping projects with fewer work elements reduces the number of interactions and increases the project's resistance to change.[9]

Principle 2: Design robust project outcomes. Some project outcomes are designed to perform within a narrow range of conditions. Others, however, are designed for a wider range of conditions and are said to have a robust design. When a change in the range of conditions occurs, a project scoped with a wider range of project outcomes becomes more resistant to change. In contrast, even a slight change in the conditions can cause ripple-effect changes in projects with narrowly designed outcomes.

Principle 3: Freeze the scope early. When you freeze the project scope early, you establish an early baseline, which helps to manage change more effectively. Late changes typically impact a large portion of the project, impeding the progress and imposing delays and cost overruns. Therefore, an early freeze may force earlier decision making and enable a faster completion of the project. This circular reaction, early freeze–faster completion–fewer changes, offers you advantages in defining a change-resistant project scope.

The quality of a scope statement hinges in many ways on the quality of the input information. Specifically, the following inputs carry great weight in developing a scope statement that has value:

- Business success factors
- Project requirements
- Project charter
- Project SWOT analysis

Truly, a project's reason for existing is in helping an organization meet its business goals. Recognizing this, we reviewed earlier in the book several tools for capturing the business goals and project requirements with the purpose of internalizing the goals and requirements to incorporate them in the project. If you prepared any of the tools, the time is now to put them to use in developing the scope statement, such as the example shown in Table 5.1.

Identify Business Purpose

Long gone are times when projects were viewed only as technical ventures whose major purpose was to provide a certain capability. Today, in addition to producing a capability, a project must support the delivery of business goals: desired profits, market share, competency building, customer satisfaction, productivity gains, and so forth.[10] Therefore, we begin the development of the scope statement with the project's rationale—what is its business purpose? What business goals will the project accomplish? What business plans will it support? For a traditional project manager, this thinking of projects in business terms is not easy; rather, it is both challenging and demanding. But so are today's projects. The demand is for them to be faster, cheaper, and better than those of the previous year. This is so much the case that some experts believe the project business purpose should be called the "project passion statement" and answer the question: What unique and distinct value will your project create for the customer and the company's business?

Table 5.1: A Simple Scope Statement

Scope Statement

Project: Jumpin Jive

Revision #: 1
Date: 17 December 2017

Business Purpose:

Improve our customer service satisfaction rating by 5% through the development and deployment of a new customer relationship management software solution

Project Goals:

- Completion Date: July 22, 2019
- Estimated Budget: $166,000
- First-Year Customer Adoption: 1,200 users
- Minimum Satisfaction Score: 87%
- Maximum System Downtime: 4 hours per month

Project Work Statement:

Work to be completed on this project includes an analysis of the customer relationship workflow, design the software solution, develop a prototype solution, perform customer usage tests, develop a final solution, conduct quality and acceptance tests, and release the software to the production platform.

Primary Deliverables:

- Workflow analysis diagram
- Workflow gap report
- Prototype SW design files
- Customer usage test plans
- Customer usage test report
- Production SW design files
- Quality and acceptance test plan
- Quality and acceptance test report
- Release plan
- Production SW build

Key Milestones:

- Workflow analysis complete: 9/30/18
- Prototype SW complete: 12/15/18
- Usage tests complete: 1/31/19
- Production SW design complete: 6/15/19
- Quality and acceptance tests complete: 7/10/19
- Production SW available: 7/15/19
- SW solution visible to market: 7/22/19

Constraints:

Our key developers will not be available in June because of travel to our European division.

Project Work Exclusions:

This project excludes work associated with customer- or customer segment–specific customizations, field testing, next revision needs analysis, and user training.

Define Project Goals

As compelling and inspirational as the business purpose may be, it is still no more than a broad direction, lacking details about specific targets for the project. Those targets are defined by project goals for time, cost, quality, and business success. By specifying when you will finish the project—for example, by July 22, 2019—you set your time or schedule goal. To attain the schedule, you must determine what cost budget you need—for instance, "the budget for this fab upgrade is $400m." Projects in industries not using cost budgets may indicate the desired number of resource hours, such as 1,200 resource hours. Unlike the schedule and cost/resources goals, expressing quality goals in a specific and measurable fashion may be a challenge. Since quality is about meeting or exceeding customer requirements, usually expressed by certain standards, a sound strategy is to define the project quality goal by referring to a certain standard agreed upon with the customer. For example, "This project manual will be in line with the Project Management Body of Knowledge (PMBOK Guide)."

Describe the Project Work Statement

What work will you do in this project to deliver the project outcome and support the business goals? Can you express that very succinctly in a sentence or two? For example, the work statement for the erection of an optics plant may read: "Design an optics plant, procure it, erect it, and commission it." Or, for a software project, "Define workflow, configure software, develop training plan, develop prototype, train personnel, and release software."

The idea here, again, is to identify *major* elements of project work, which will be described in greater detail in the supporting specifications, project WBS, and other documentation. Also, as you read the example statement, you might have noticed the way the work statements are structured: the verb (define) followed by the noun (workflow), the verb (configure) followed by the noun (software). This method of writing has a very specific purpose of reducing the activity orientation of the work statement and highlighting its goal orientation. Of course, you may choose to write work statements differently.

Identify Deliverables

Performing the work described in the work statement must result in major deliverables or outcomes. Take the preceding example of the work statement for the software project. Its *major* deliverables are workflow gaps report, software design files, test plans, test reports, and production software. A closer look at this set of deliverables reveals several guidelines for identifying the deliverables. First, there is almost one-for-one correspondence between elements of work statement and deliverables. "Customer relationship workflow" (element of the statement) produces a "workflow gap report" (deliverable). Also, major elements of work lead to major deliverables. Focus your identification of deliverables in the scope statement on deliverables that you can make Level 1 or 2 elements in your project WBS. It is in the WBS that you will identify *minor* deliverables—or subdeliverables—(Levels 2, 3, and so on), not here. Another guideline evident from the example set is that your deliverables may include interim deliverables—for example, products of early project stages (workflow report)—as well as end deliverables (software release). Finally, deliverables may be of the product (e.g., machine, facility, report, study) and service (e.g., training). Identifying deliverables should be followed by an additional step: Define each one in terms of how much, how complete, and in what condition they will be delivered. This is typically done in the supporting specifications and other documentation. Should a WBS dictionary be available, it can take on this role.

Select Key Milestones

Milestones are major events and points in time, indicating the progress in implementing your work statement and producing deliverables. As a crucial part of the scope statement, you need to identify major milestones in your project. Consider again the example of the preceding software project. The project team identified these key milestones:

- Workflow analysis complete: 9/30/18
- Prototype SW complete: 12/15/18
- Usage tests complete: 1/31/19
- Production SW design complete: 6/15/19
- Quality and acceptance tests complete: 7/10/19
- Production SW available: 7/15/19
- SW solution visible to market: 7/22/19

Note the correlation between deliverables and milestones. This is a personal choice designed to provide full consistency from work statement to deliverables to milestones. This correlation becomes even more critical if payments from a customer are tied to deliverables and milestone events. Although there are many other ways of identifying major milestones, they share some common ground. They focus on several major milestones defined and understood by project stakeholders, with a clear date and consistent with the list of deliverables.

Define Major Constraints

All projects face constraints that may change the way project work is defined, deliverables produced, and milestones met. These constraints may be physical, technical, resource, or any other limitation. Think of climbing Mount Everest as a project. The physical constraint here is the climate, which limits the time the project can be executed to certain months. In another example, management ordered a project manager to deliver a software package per a fast-track schedule. Because of the lack of resources for fast software quality testing, the project milestones had to be changed and pushed out by several months. These examples provide a crisp message: Identify major constraints and build your scope and project plan around them, or face the derailment of your project during execution.

To manage constraints means to clearly identify them, size them up, and formulate the scope statement around them. As the project unfolds, you will need to revisit them and verify whether they still exist. Considering that they are a foundational element of the scope, you need to rescope the project if the constraints change.

Determine Exclusions

Habits die hard, as the following anecdote vividly points out. In the early 2000s, a contractor involved in the transfer of some technology to Africa scoped its projects to include computer centers with office furniture. It took the owners several years to realize that they would be better off buying the furniture locally than having the contractor import it from Europe. Consequently, the owners asked the contractor to leave the furniture out of the scope. The contractor's office was properly notified but to no avail; the furniture arrived at the offices of the African owners. Why? The furniture used to be a habitual part of the scope, and the exclusion notification was neither strong nor visible enough.

To avoid a situation like this, you need to specifically identify project work elements that will be explicitly excluded from the project scope. State specifically what is not part of the scope. In the previous example scope statement, the project specifically excluded work associated with customer-requested customizations, field testing, next revision needs analysis, and user training.

Include Supporting Detail

To be clearly directive and crisp, the scope statement should be succinct, possibly written in the imperative mood (see the example titled "Simple Dos and Don'ts in Preparing the Scope Statement"). Technical and other details usually have no place in the statement; rather, they should be attached in the supporting documentation.

Simple Dos and Don'ts in Preparing the Scope Statement

Do:

- If the statement is longer than a page, split it into two tiers: a one-page summary (as in example scope statement) and the supporting detail.
- Avoid repeating in one tier what is mentioned in the other tier.
- Use active rather than passive language, and spell out acronyms.
- Include functional groups providing resources in writing the statement.
- Have the statement approved by managers.

Don't:

- Write a scope statement without having a system to structure it.
- Write it as a purely technical statement of the project.
- Use ambiguous language (e.g., "nearly").
- Mix major and minor goals, deliverables, and milestones.
- Proceed without having an independent party review the scope statement.
- Forget to include constraints and exclusions.

Evaluate and Fine-Tune the Scope Statement

There are at least two levels of evaluation that deserve your attention. First, check the statement for completeness, comparing it against the information inputs and the information requirements that we have discussed here. Did you include all primary project requirements? Did you define the project goals? Did you identify major constraints? Has any specifically excluded element of scope been identified?

Second, you should assess the quality of certain information, for example, schedule and cost aspects of the project goals. The scope statement is used again for detailed schedule and cost planning. However, the detailed numbers are integrated and may be different from your schedule and cost numbers in the scope statement. What should you do? You can replace the numbers in the scope statement with those integrated numbers from detailed planning, or you can reduce the scope and cut the integrated numbers to comply with the original numbers from the scope statement. Whatever you choose, it is obvious that the scope statement is only the first step in this iterative cycle of project planning, and for that reason, you have to fine-tune it as you go through the planning cycle.

Using a Scope Statement

Not only is the scope statement a must-have tool to include in any PM Toolbox, but it is one of the most indispensable tools available. Every industry and every project family can use it. Except for highly repetitive projects that can utilize the same informal scope statement, every project regardless of the size and complexity would greatly benefit from having a formal written scope statement. History has seen successful projects without formal scope statements, but researchers have found that having a well-defined idea of what you want to do in a project (the scope) is a critical factor in project success.[11] Therefore, if you strive for the heights of success, make sure you define the project scope and control its implementation properly.

Variations

The scope statement that we have described is designed to be a cross-industry tool to serve as many project audiences as possible. Still, almost any industry or project family may find it useful to have a variation of the tool. Take, for example, product developers. Having a product statement included in the scope statement (after the project goals section) is a regular practice. Such a statement typically spells out the following:[12]

- Specification of the target market: exactly who the intended users are.
- Description of the product concept and the benefits to be delivered.
- Delineation of the positioning strategy.
- A list of the product's features, attributes, requirements, and specifications (prioritized: "must have" versus "would like to have").

Others, such as those involved in government contract projects or contractor/subcontractor arrangements prefer the use of a statement of work (SOW). A project scope statement and SOW serve related and often overlapping functions. Both set expectations and parameters for the project. The SOW, however, provides a detailed breakdown of the project and deals with the nuts-and-bolts details of how the project team will accomplish the project goals.

There are three basic types of SOWs:[13]

1. *Design/Detail SOW*. This SOW type tells a supplier or vendor *how* to do the work. The statement of work defines buyer or client requirements that control

the processes of the supplier—such as measurement, tolerances, and quality requirements. In this type of SOW, the buyer or client bears the risk of performance.

2. *Level of effort or time and materials SOW*. This SOW type can be used for nearly any type of service. The real deliverable under this type of SOW is an hour of work and the material required to perform the work.
3. *Performance-based SOW*. This SOW defines work agreements based on what has to be accomplished. Desired outcomes from the work are described in clear, specific, and objective terms that can be reasonably measured.

A statement of work typically includes the following information about a project:

- *Project purpose*: Describes why the organization is investing in the project.
- *Objectives*: Describes what the project aims to achieve once it is complete.
- *Location of work*: Describes where the work will be performed.
- *Period of performance*: Specifies the allowable time for projects. This can be represented in start and finish dates, number of hours that can be billed, or anything else that describes scheduling constraints.
- *Deliverables*: Describes the tangible outcomes from the project.
- *Deliverables schedule*: Describes when each of the deliverables is due.
- *Value of work performed*: Describes the estimated total cost of the project.
- *Specifications and standards*: Describes any specifications or industry standards that need to be adhered to in fulfilling the work obligation.
- *Method of measurement of acceptance*: Describes how the buyer or receiver of the project outcomes will determine if the project outcomes are acceptable.

Whether a project manager chooses to use a scope statement, an SOW, or some other variation, every project will benefit from the use of one of these tools. They provide the foundation for project planning as well as the overall vision of what the project end state will look like.

Benefits

Project managers value the scope statement for what it is—a first-step tool for planning for a project that directs all subsequent tools in the planning and control effort. It captures the fundamentals of a project, bonds them, and displays them for everyone to see. Setting the scope baseline helps project teams stay focused, allowing them to be guided by the project vision it creates. Once this baseline is available, the project team should consider establishing a change control plan (see "Change Control Plan Helps Set the Direction for Controlling Scope").

The scope statement is also comprehensive, simple in format, and easily adaptable. It includes a look at all major project dimensions, providing the project manager with a comprehensive vision of what encompasses a project, but also simple enough that it enables an easy grasp of versatile dimensions of project assignment. With little effort it can be adapted to specific industry or company needs, taking out or adding new elements as needed.

Change Control Plan Helps Set the Direction for Controlling Scope

Working on a significantly large project will almost inevitably call for some sort of change control plan, typically part of the project plan. Although smaller projects usually cannot afford such a level of documentation, they still need to set a clear direction for controlling changes. Therefore, both smaller projects informally and larger projects formally need to address the following issues:

- *Who has the change approval authority?* If the authority is to rest with the change review board, then chairman and board members need to be appointed for the project. Clearly define their responsibilities. In some cases, especially in smaller projects, project managers may have full authority to make changes and will not use a board.
- *How is the scope of the change authority defined?* For example, a board may be authorized to deal with major changes substantially affecting the scope. For the sake of responsiveness, the board chairman may have the approval authority, with other members reviewing (not approving!) their areas of expertise. Changes with small or no impact on scope may be in the purview of the project manager.
- *What is the change request procedure?* The change control plan needs to describe the change request procedure and any forms or documentation to be used for submittal to the change authority.
- *How do we ensure the implementation of the approved changes?* For example, nominating an administrator for this role is a possible solution.
- *When in the project life cycle do we begin using and stop using the change request procedure?* The change control plan should address this issue.

THE WORK BREAKDOWN STRUCTURE

A project work breakdown structure is an outcome-oriented grouping of project elements that organizes and defines the total scope of the project—work not in the WBS is outside the scope of the project.[14] When presented in a graphical format, it becomes obvious why the WBS is often described as a project tree diagram, hierarchically displaying project outcomes, which are further broken down into more detailed tasks (see Figure 5.3). This same project tree analogy helps to visualize project outcomes at one level as parents to those from the next-lower level, who then become parents to the next-lower level outcomes, and so on.

The project WBS provides the project manager the means to divide a project into manageable increments, helping to ensure the completeness of all work that is required for successful completion of a project. Projects are planned, organized, and controlled around the lowest level of the WBS or work packages that are assigned to a project team member for responsibility for completion.[15]

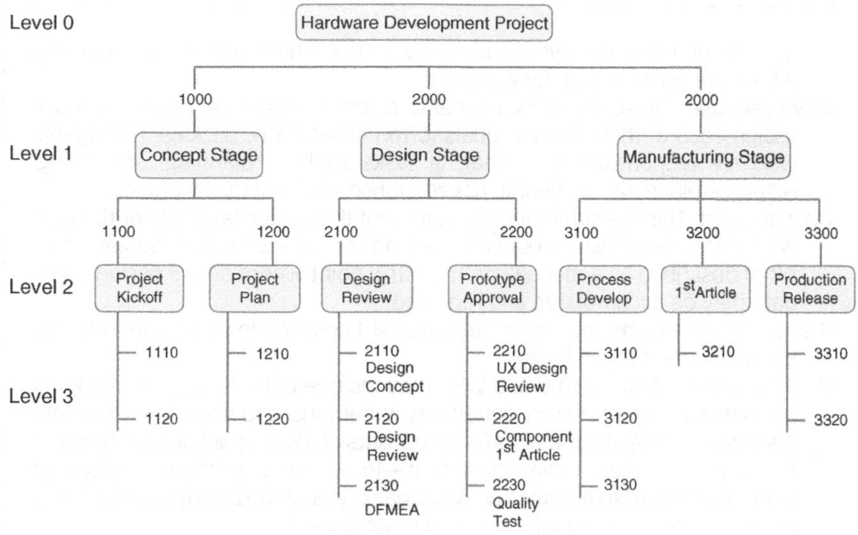

Figure 5.3: An Example Hardware Development Project WBS

Don't confuse the project WBS with an array of intimidating acronyms such as CWBS, BOM, or OBS. These tools are as logical and conceptually simple as the WBS, but they have different purposes. For example, CWBS (contractual WBS), which is less detailed than a WBS, is used to define the level of reporting that the project contractor will provide to the owner in larger, contractual projects. Widely applied in manufacturing industries, BOM (bill of materials) is a hierarchical representation of the physical assemblies, subassemblies, components, parts, and so forth that are necessary to produce a product. Finally, an OBS (organizational breakdown structure) indicates which organizational units are responsible for which work elements from a WBS (for basic terminology, see "WBS Language").

WBS Language

Work elements. Any project outcome in the WBS is called a work element, consisting of an item of hardware, software, service, or data. While some elements are the direct outcome of work, others are the aggregation of several logically grouped deliverables.

WBS level. This is the hierarchical location of a work element in the WBS. Work elements at the same stage of structuring are on the same level. There is no universal system for numbering levels. We number the overall project level as 0 and subsequent levels as Level 1, Level 2, and so forth. Using

(continued)

level numbers enables you to uniquely code each work element, providing a basis for cost control, for example.

Work package. These are work elements at the lowest level of the WBS. We assign each of them to individuals (often called work package managers), who are responsible for managing tasks such as planning, scheduling, resource planning, budgeting, risk response, and quality assurance.

Cost account. This is a summary work element that is one level higher than the work package. A cost account includes one or several work packages and is often described as a management control point where actual performance data may be accumulated and reported.

Branch. All work elements underneath a Level 1 deliverable constitute a branch. Branches may vary in length.

WBS dictionary. At a minimum, a WBS may include brief descriptions of work packages, along with entry conditions (what inputs to a work package are necessary) and exit conditions (what outputs of a work package are required to call it complete). Adding more to it—for example, schedule dates, cost budgets, staff assignments for work packages, and descriptions of other work elements—may make sense in large projects.

Constructing a Project WBS: A Top-Down Approach

There are two basic ways of developing a project WBS: top-down and bottom-up. In this section we detail the top-down approach, which is a very convenient approach for project managers and teams with experience in project work, have knowledge of the project deliverables, and are comfortable with a systems approach to organizing the work to be performed on a project.

The development of a WBS is likely to be an easier and more meaningful exercise if you are equipped with information about the following:

- Project scope
- Project workflow
- Project requirements
- Project situation

The project scope statement provides an understanding of what the project will produce. You first need to know "what you will produce" (scope) before you decide "how you will produce it" and depict it in the WBS. Note the practice of some experienced project teams who prefer to develop the scope statement and WBS in parallel rather than sequentially or actually include the WBS as part of the scope statement.

In constructing a WBS, the knowledge of the project workflow is crucial. For instance, to develop a meaningful WBS for a software development project, you need to understand the process of software development. Knowledge of the process will indicate which activities are necessary to produce the required project deliverables.

Finally, specifics of your project situation are known to influence the anatomy of a WBS. Take, for example, the case of an enterprise software development company where each department had its own WBS when working on a project until it was acquired.

The new owner immediately ordered all projects to develop an integrated WBS for each project, mandating that a large project's WBS include a project management branch.

Select WBS Type

After acquiring all necessary information about WBS shaping factors, you have more choices to make before being able to construct a WBS. Which method will you use to structure your WBS? Consider the three major ones: WBS by project life cycle, WBS by system, and WBS by geographic area (see Table 5.2).

The underlying principle of the WBS by project life cycle method is self-explanatory: You break the overall project into phases of its life cycle on Level 1 of the WBS. This principle of conveniently following the natural sequence of work over time is widely popular in some industries. A good example is a software development project consisting of phases such as requirements definition, high-level design, low-level design, coding, testing, and release.

In contrast, users of a WBS structure by geographic area segment the project work by major geographic sites or regions such as in building construction. It is not unusual that literal geographic regions—for example, northwest site, southwest site, southeast site, and northeast site—are used as the element for Level 1 of the WBS.

You have probably noticed that our discussion about the three structuring methods was limited to Level 1 of the WBS. What about Level 2 and other lower levels? Each of the three can continue with the underlying principle of further dividing the work into the lower levels. For instance, WBS by geographic area can have work elements on Levels 2 and 3 that could be composed of work elements that will be performed in a specific geography identified at Level 1. Many practitioners, however, find hybrids more practical, combining two or three methods in the same WBS. For example, they may have a WBS by project life cycle on Level 1, systems on Level 3, and a geographic breakout of work at a lower level.

Which method of WBS structuring is right for you? Before answering this question, you should know that structuring a WBS is not an act of science. It is rather an act heavily influenced by a company's culture, shaped by top managers with the purpose of determining "how we get things done around here." If there is no previous history of using a WBS in your company, you should probably follow your industry norm. This does not rule out the use of a different type of WBS structuring, but the bar of resistance to a new WBS structure will be lower if you go with the standard approach. Should you decide to go against the grain, be prepared to engage in some organization transformation discussions with others in the organization.[16]

Table 5.2: Methods for Structuring the WBS

	Method of WBS Structuring		
WBS Level	**Life Cycle**	**System**	**Geographic Area**
0	Project	Project	Project
1	Phase	System Component	Area or region

Establish the WBS Level of Detail

How many levels should a project manager use to structure a WBS (see "Too Many Levels Can Create Turmoil")? The answer to this question will determine the total number of work packages. Considering that the number of work packages influences the necessary time and cost you need to manage a project, you need this number to align with what your available time and budget can tolerate.

Too Many Levels Can Create Turmoil

"How many WBS levels do we want our projects to have?" This was a question that designers of the PM process in a power distribution organization asked themselves. To prepare a good answer, they first looked at the projects they manage: 10 to 15 projects per year, ranging from \$100k to \$5m, mostly involving the design and construction of electrical substations. After some benchmarking, the designers made their decision: Each project shall have a five-level WBS.

Soon after the deployment of the PM process began, a silent rebellion in smaller projects began, followed by outright refusal of their project managers to use the process. The justification was simple. The amount of work on scheduling, budgeting, and control of 250 work packages in a five-level WBS drove smaller projects to a halt. As a result, project managers went to the old, ad-hoc way of management. The moral of the story: Match the size and structure of WBS with the size and structure of the project. The smaller projects likely could have been adequately organized and managed with three WBS levels.

As explained earlier, the work package is the central point for managing to the WBS, as illustrated in Figure 5.4. Simply said, work packages are discrete tasks, or combination of tasks, that have definable end results—outcomes—that assigned organizational units "own" and need to create. When using them for the integration of project planning and control in a very detailed WBS, for each work package, you will assign responsibility, develop the schedule, estimate the resource and cost, develop plans for risk response and other planning functions, and measure performance. Obviously, as the number of work packages grows, so does the necessary time (and cost) to perform project planning and execution. Reaching a point when there may be too many work packages renders their management impractical and cost prohibitive. Closely related to the number of work packages is their average size. Clearly, work packages need to be small enough that they are manageable.

In summary, establishing the level of detail of the WBS includes determining the number of levels in the WBS, the number of work packages, and the average size of the work package that are compatible with your tolerance level and the industry practice. Table 5.3 shows WBS data from some actual projects.

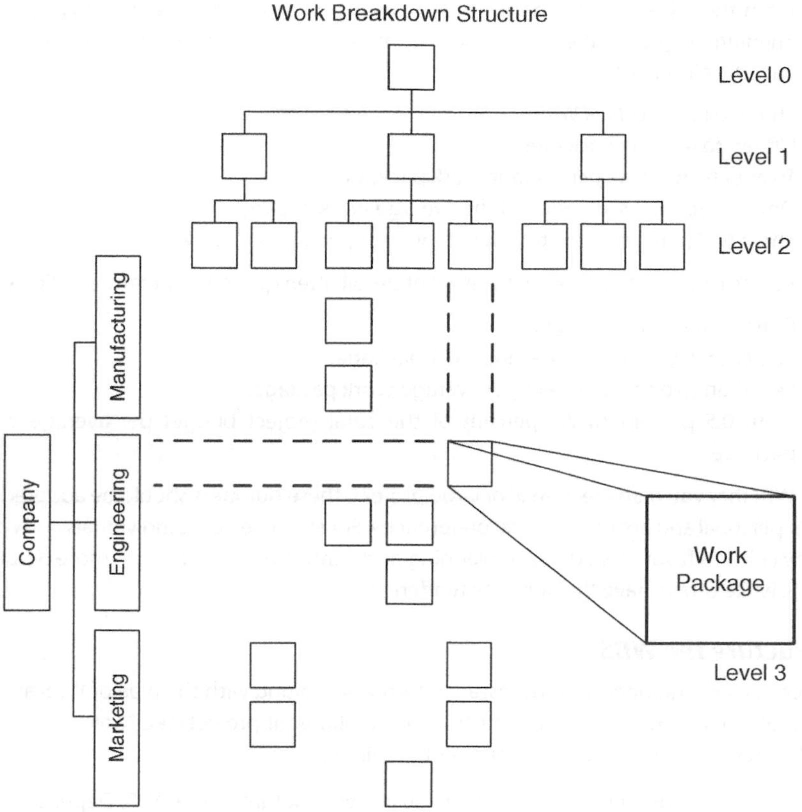

Figure 5.4: Work Package, a Critical Link in Managing the WBS

Table 5.3: Examples of the Level of Detail of the WBS

(1) Project	(2) Project Duration (days)	(3) Project Budget (person-hours)	(4) # of Levels in WBS	(5) # of Work Packages	(6) Mean # of Hours/ Work Package (3) / (5)	(7) Mean # of Days/ Work Package* (2) / (5)	(8) Mean % of Budget/ Work Package [(6) / (3)] × 100
IT infrastructure	90	500	3	15	33	6	6.6
Selecting a billing platform	180	1200	4	36	33	5	2.7
S/W development	270	1200	3	25	48	11	4
Hardware development	365	500	4	29	17	13	3.4

*Assuming all work packages are sequential without any overlapping.

From the table, you can derive the following guidelines for the majority of small and medium projects in the field of information technology, software development, and product development:

- Three to four levels of WBS.
- Fifteen to 40 work packages.
- Twenty to 50 hours per average work package.
- One to two weeks' duration of the average work package.
- Three to 7 percent of the total hours budget per average work package.

For larger projects, however, the level of detail often quoted in literature indicates[17]:

- Five and more levels of WBS.
- Eighty to 200 hours per average work package.
- Less than two to four weeks per average work package.
- From 0.5 percent to 2.5 percent of the total project budget per average work package.

Whether you manage a small or large project, these numbers should be adapted for your personal and possibly cultural preferences. For example, some individuals and company cultures favor more detailed planning and control, and accordingly, more detailed WBS, while others have the opposite tendency.[18]

Structure the WBS

Once you are equipped with necessary information, along with the type of WBS and its level of detail, you are ready to construct a WBS for your project (see "Golden Rules of WBS Structuring"). The steps are outlined as follows:

Step 1: Start by identifying the major construct of structure of the WBS. Depending on the type of WBS you selected, these may be phases, systems, geographic areas, or a combination. One useful approach here is called the *scope connection*. In particular, when developing the scope statement, you identify major deliverables, which could be borrowed to serve as the major elements of the WBS. This helps integrate the scope statement with the WBS, linking business and project goals—via major deliverables—with lower-level deliverables and work packages.

Step 2: Divide the major WBS elements into smaller, more manageable outcomes, level by level, until a point is reached where the outcomes are tangible, verifiable, and defined to the level of detail that enables them to be used for the integration of project planning and control activities.

Step 3: How will you represent your WBS? In the case of smaller projects, drawing a WBS as a tree diagram is a very visual and preferential way (see the upper part of Figure 5.4). As your number of WBS levels grows, so does the complexity, making it more difficult to stick with the tree format.

An alternative may be in the TOC (table of contents) format. For example, Figure 5.5 shows the TOC format for the WBS from Figure 5.4. To some extent this WBS construction process looks like a random process. Bringing more order to it is possible by following several guidelines (a shortened version appears in the "Golden Rules" example).[19]

Hardware Development Project

1000	**Concept Stage**		
	1100	Project Kick-off	
		1110	Materials Preparation
		1120	Kick-off Meeting
	1200	Project Plan	
		1210	Integrated Plan Preparation
		1220	Project Plan Approval
2000	**Design Stage**		
	2100	Design Review	
		2110	Design Concept
		2120	Design Review Meeting
		2130	DFMEA
	2200	Prototype Approval	
		2210	UX Design Review
		2220	Component 1st Article
		2230	Quality Test

Figure 5.5: Table of Contents WBS Structure

Step 4: Make sure the WBS is outcome oriented. Since the WBS is about outcomes, activities and tasks should be relegated to the lowest levels of the WBS.

Step 5: Be sure that the WBS includes all project work. What is left out will not be resourced and scheduled, which is a risky proposition.

Step 6: Make each work element relatively independent of others on the same level.

Step 7: Keep breaking the work down into work elements until you reach a level at which there is a method in your organization capable of producing the element. Stop there. It is an acceptable practice to have WBS branches of unequal length.

Step 8: Produce a WBS that integrates work elements or separate levels to the point that their aggregate is an equivalent of the project completion.

Golden Rules of WBS Structuring

- Focus on documenting the project outcomes.
- Show all project work.
- Make objectives relatively independent.
- Use symmetrical branches when justified.
- Build the WBS as an integrated effort.

Evaluate the WBS Structure

Since the WBS development lacks the rigor and discipline of the scientific approach, there is no single correct WBS structure. Rather, there can be many different but equally good WBS structures. To make sure that your WBS is sufficient, you should evaluate it against the preceding guidelines. If there is a need for some revisions, making them will ensure that you have created a WBS capable of acting as a framework for integration of project planning and control.

Increase Productivity through WBS Templates

Having each project team develop a WBS from scratch can create several problems. First, WBS development consumes resources. Also, when each project uses a different WBS type, the benefit of comparing various projects and drawing synergies is lost. These problems have been successfully cured by the use of WBS templates.

Specifically, this means adopting templates for certain project families. Highway construction projects are an example of a project family. Other families may include software development projects, manufacturing process projects, and hardware development projects. Generally, a family of projects is a group of projects that share identical or sufficiently similar project assignments. When the templates are developed and adopted, the development of a WBS for a new project is reduced to adapting the template. This saves time, produces a quality WBS, and enables interproject comparability. In a nutshell, templates increase productivity.

Constructing a Project WBS: A Bottom-up Approach

Brainstorming all project work that needs to be performed and then organizing it into a WBS hierarchy is at the heart of the bottom-up approach. This approach, essentially an application of the affinity diagramming method, is beneficial to those without much project experience or those who do not prefer to take a systems approach. Projects developing or deploying novel technologies, typically fraught with high uncertainty and lacking precedents, can benefit from this approach as well, even if the project team is experienced. Despite its brainstorming nature, the bottom-up approach may be preceded by collecting necessary information for the WBS development, selecting the type of WBS, and establishing the level of its detail—in other words, some of the steps taken in the top-down approach. Other steps in the bottom-up approach follow.

Generate a Detailed List of Outcomes

This step includes having each project team member brainstorm what is going to be created and delivered in the project. Each outcome may be recorded on a sticky note and posted in a visible place, with the number of generated outcomes ranging from 40 to 60—a good number for a small or medium-size project. Higher numbers may be required for large projects. In the process, adhering to the brainstorming principle of not critiquing the ideas is crucial.

Sort Deliverables in Related Groupings

The result of this step is the grouping of related outcomes. The aim may be to create groupings of five or so outcomes. Screen them carefully for relationships and group them into new groupings, striving to have three or four levels of groupings for small and medium projects, possibly more for larger projects.

Create Duplicates and Consolidate Outcomes

Project team members may have different and conflicting ideas about grouping outcomes. If that occurs, create duplicates of the outcomes and post them in the groupings suggested by the team members. Develop the discussion to understand the rationale for the conflicting groupings, and try to reach a consensus. If this is not possible, use a preferred type of voting to decide on the final grouping. Also, consolidate similar outcomes and eliminate redundancies. This should give you the preliminary WBS hierarchy.

Create Names for the Groupings

The hierarchy needs names for groupings/outcomes on different levels. As much as possible, the idea here is one of garnering consensus among team members. Spending enough time on naming the groupings/outcomes is useful for a good understanding of what is going to be delivered on the project, as well as for creating buy-in.

Evaluate WBS Structure

Similar to the top-down approach, bottom-up WBS development lacks the rigor and discipline of the scientific approach, leaving room for mistakes. Therefore, this is the time to evaluate the WBS developed against the guidelines for structuring a WBS. Again, useful revisions and corrections are welcome, aiming at the improvement of the WBS to the level it needs to perform as a framework for integration of project planning and control.

The bottom-up approach is a good method for novice project managers and for use on unfamiliar projects. For full evaluation of its potential, we should recognize that it provides an easy start, encourages strong involvement, and downplays vocabulary issues. Its ease of use may give it an edge over the top-down approach, which requires more time to start and a shared vocabulary, and also limits team participation.

Using the WBS

The WBS was initially used to bring order to the integration of management work faced with large and complex projects in the government domain. Logically, then, most of the "science" of the WBS was conceived in governmental agencies and is very well reflected and covered in the works of prominent PM books.[20] What is not well covered is how to adapt the science for what is the dominant stream of projects in today's business world—dynamic small and medium projects. For these, the WBS is a must-have tool. Whether you are in software or hardware development, marketing or accounting, manufacturing, infrastructure technology, or construction, practically any area and industry, small and medium projects benefit from a WBS to structure the project work. It is possible to run a successful project without using a WBS, but the point is that probability of success is higher when a good WBS is used, as opposed to having an inappropriate WBS or not having one at all (see "Tips for WBS Use").

Tips for WBS Use

- Adopt a template WBS for each project family.
- When developing the template, begin with few levels. Add more if project team members ask for it.
- Build "blank" work elements in the template to be used by unusual projects.
- Develop a WBS in each project, small or large, starting from a template WBS.
- Allow smaller projects to use a fewer number of levels in the template.

For each work element, assign an individual responsible for producing the outcomes listed under that work element. If, for example, you identify 20 elements in the WBS, each work element would have an "owner" responsible for the element: its schedule, cost, risk response, performance measurement, and project control. A convenient tool for assigning responsibilities in this manner is the responsibility matrix (Chapter 3). Listed on its vertical axis are work elements, while across the top are persons involved in the project. In the cell at the intersection of a work element row and a person column, different responsibilities are assigned for the accomplishment of the work element.

The second planning action made possible by the WBS is scheduling the project work. Here, the development of a project schedule begins at the work package level. In particular, once the schedule for each work package is developed, the schedule for a work element on the next level is no more than a summarization of schedules of its work packages with resource constraints resolved. Similarly, the schedule for any higher-level work element is the summary of schedules of its constituent work elements.

In a way similar to schedule integration, the WBS provides a formal structure for resource estimating. Again, resources are estimated for work packages, and their aggregation up the WBS leads to total resource requirements. When resources for a work element are spread over the element's schedule, you will obtain a time-phased resource plan, a great baseline against which to compare the actual performance and strategize corrective actions, if a variance occurs. The reason we emphasize resources is that the majority of small and medium projects prefer the use of resource-based estimates to the cost-based estimates. Should you prefer the latter, multiply resources by their hourly labor rate to obtain cost estimates.

Planning for other management functions, such as risk, quality, and change management, should also be performed around the skeleton provided by the WBS. Take, for example, risk response. The basic place for a risk plan including risk identification, analysis, and response is the work package. Summing up risk plans for work packages that belong to the next-higher-level work element will yield a risk profile for the element. Continuing with this summation up the WBS hierarchy will produce a risk profile for the total project.

Benefits

There are two reasons why the value of the WBS cannot be overstated: It helps organize the project work, and it creates a framework from which management of a project can be fully integrated. In particular, the WBS enables a project team to organize the work into small outcomes that are manageable, making the assignment of responsibility for each of them easy. Because the outcomes are relatively independent, their interfacing with and dependence on other outcomes is reduced to a minimum. Still, they are integratable as you move up the WBS so the team can get a view of the total solution. This extraordinary capability of the WBS to help organize the project work is an enabler of what is an even higher value of the well-constructed WBS; it serves as the framework for integration of project planning and control functions. Some call it the single most important element in project management.[21]

At the heart of the significance of the WBS is its ability to serve as a project planning and control framework, enabling the achievement of the following fundamental project management actions:

- Assign the responsibility for the project work.
- Schedule the project work.
- Estimate the cost or resources to complete the project work.
- Develop the response to risks associated with the project work, and perform other planning functions such as quality planning.
- Manage changes to project scope.
- Control the project work effort to accomplish project goals.

Finally, the WBS provides a strong visual impact. As a practicing project manager recently commented, "the WBS brings order to disorder in a visual way." At a glance, the WBS turns the dlsorder of the verbal and obscure scope statements into the order of a clearly structured hierarchy of work.

THE PRODUCT BREAKDOWN STRUCTURE

For those project managers involved in product development, planning activities involve the combination of product-based planning and project-based planning.[22] This type of planning is based on the premise that the product solution is developed first, and *then* the project activities, tasks, deliverables, and resources required to create and deliver the product are planned.

The product breakdown structure (PBS) is a critical tool for product-based planning and should be an essential part of any product development project manager's PM Toolbox. Essentially, the PBS disaggregates a final product into its constituent components.

Like the project WBS, the PBS is a visual aid that represents the relationship between a product and its components. However, the project WBS and the PBS are used for different purposes. The main difference between the two is that the PBS focuses on the product, whereas the WBS focuses on the work required to create the product.

Constructing a Product Breakdown Structure

When creating a PBS, it is common to utilize the knowledge of the team that will be designing and developing the product. This usually involves a group of cross-functional product specialists. A structured brainstorming session, complete with a whiteboard, sticky notes, pens, and brainpower, is all you need to construct a PBS.

Start with the End in Mind

It is helpful to begin the PBS construction process by first diagramming the product that will eventually be created and developed. We recommend doing this by diagramming the *whole solution* that fulfills the customer's expectations.[23] For example, if we purchase a laptop computer, we wouldn't consider it acceptable if we received a box of circuit boards, a second box that contained the enclosure, another box containing the peripheral devices such as memory and network adapters, and finally an envelope containing a CD with the computer software applications and operating system. Rather, unless we're a computer hobbyist or a system integrator, we *expect* to receive an integrated laptop that we can unpack, plug in, and begin using immediately. The point we're making is that the whole solution may include not only the core product, but also a number of enabling elements of the product that are needed in order to fulfill the customer's expectation.

It is helpful to think of the whole solution consisting of two parts: (1) the core components of a product and (2) the enabling components of a product. The core components are the tangible elements that when integrated, constitute the physical product developed. In systems language, these are the subsystems of the integrated product. The enabling components are the additional elements needed to ensure the product meets customers' expectations. Both the core and enabling components of a product need to be included in the planning of a project.

Let's look at an example to illustrate. Imagine that you are in charge of leading the project commissioned to create the next-generation smartphone for a phone manufacturer. Your whole product solution diagram may look something like the one illustrated in Figure 5.6.

The product solution begins with the core components that consist of the physical elements that make up the phone such as the digital circuitry, the embedded software, the radio device, and the enclosure packaging (keep in mind this is a simplistic view for discussion sake).

The product solution also includes other important elements needed such as a software application development platform, interface to the wireless communication infrastructure, manufacturing of the product, quality assurance, and customer support for the users of the product. These are the enabling components of the product that are needed to ensure complete customer and user satisfaction when the phone is delivered.

By diagramming the product and its components you are beginning to structure your project. In effect, you are creating the project architecture, where the term *architecture* refers to the conceptual structure and organization of a system.[24] The process of creating a PBS now has an excellent starting point.

Choose the PBS Structure

Once you understand the core components of the product you intend to develop, you are ready to construct your PBS. The first step in this process is to select the type of structure you wish to use.

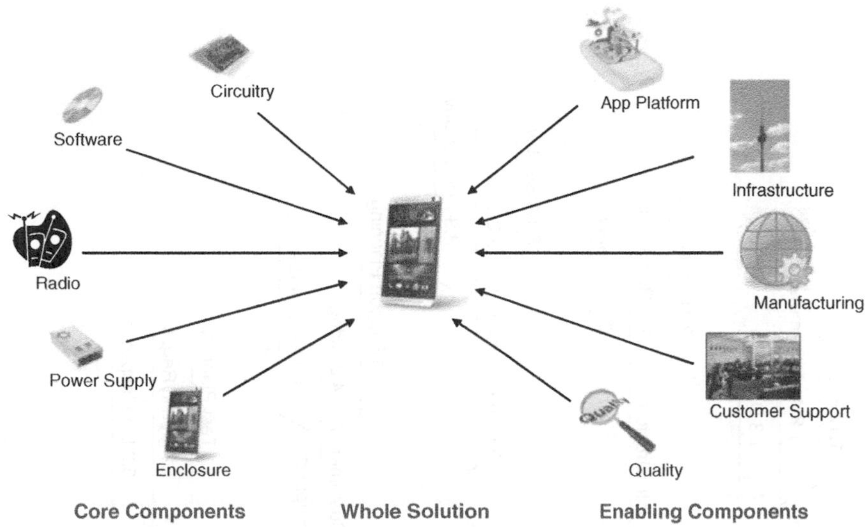

Figure 5.6: Example Whole Product Solution Diagram

Like the project WBS, multiple structure types are available for the PBS: the hierarchical tree design, the table of contents design, and the mind-map design to name three. The hierarchical tree and table of contents designs are identical to those described for the project WBS. If you create a diagram of the whole product solution as recommended in the previous section, the mind-map design is a good complementary PBS design to choose. Figure 5.7 illustrates an example of the mind map PBS structure for the cell phone product discussed previously.

Decompose the Product

With the PBS structure decided upon, the next step is to begin decomposing the product into its components and subcomponents. At the top level (or in the middle of the mind-map structure) is the integrated product that will emerge from the product development process. In the example used earlier, the cell phone is the top-level product in the PBS.

Next, decompose the top level, or integrated product, into its primary components. As illustrated in Figure 5.6, you may have a combination of core components and enabling components if you take a whole solution approach. You will want to include both component types in your PBS. Continue decomposing each component into its subcomponents until a logical point of decomposition is reached (Figure 5.7). This is normally the level where the subcomponents can be described as a set of project deliverables.

Validate the PBS

Working your way up from the lowest level of the PBS, validate that the subcomponents at each level will integrate to create the subcomponent or component at the

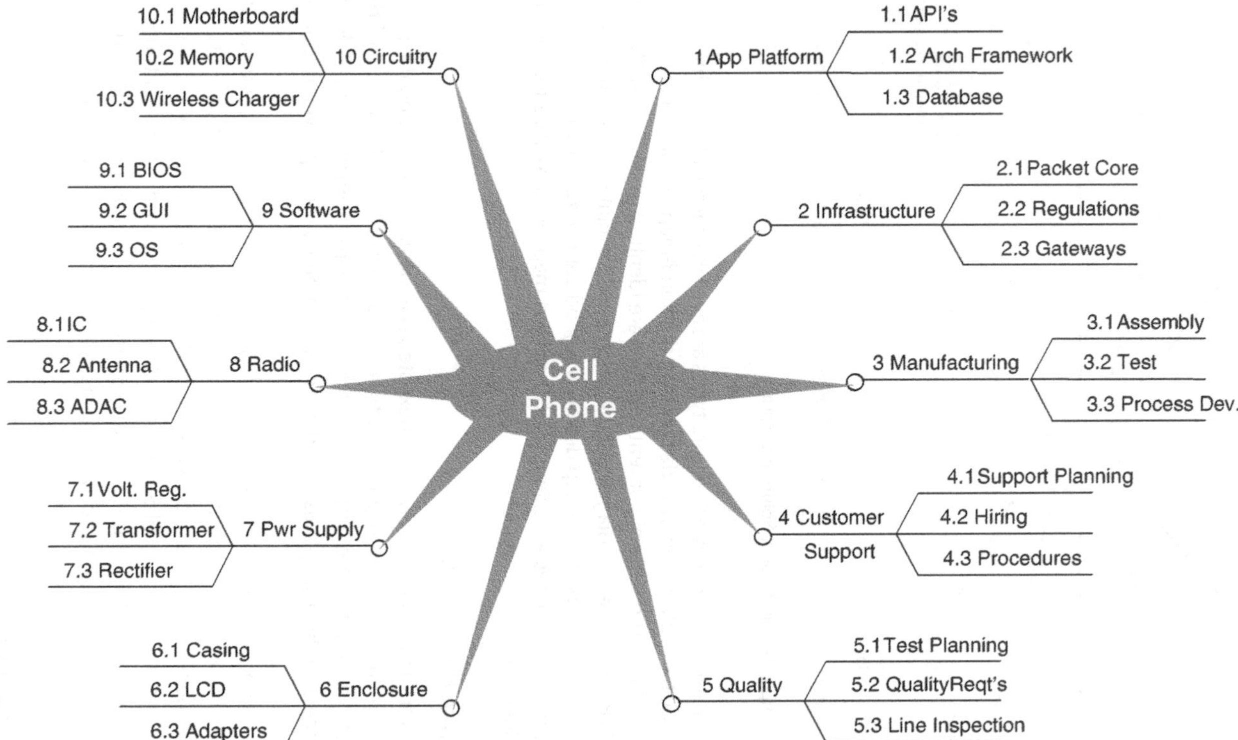

Figure 5.7: Mind-Map PBS Design Example

next-highest level. Continue this process until the top-level product is validated. If gaps are discovered during the validation process, go back to the decomposition step to correct for the gap.

Using the PBS

As stated previously, the PBS is primarily used for product development projects where product planning and project planning are often occurring in a concurrent manner. The PBS is used to bring these two planning efforts together by providing a visual representation of the product and its components, as well as the relationship between the components, in order to facilitate the project planning process.

The PBS is used to describe *what* the project work effort is meant to create. This should be established before full-scale project planning begins, where project planning focuses on *how* the product will be developed. Obviously, *what* should come before *how*.

We found it interesting to witness a recent debate on a professional networking site concerning the value of using a WBS versus a PBS.[25] We had no idea this was the basis of such a hotly contested philosophical war between factions. Not only did we find the debate interesting, we found it a bit disturbing. As is usually the case in such mental exercises, the key point of interest was lost. The discussion should not have been about the virtues of using one tool versus the other, but rather about how the two tools can be effectively used in tandem.

Since planning on a product development project includes both product-based planning and project-based planning, once a PBS is created that decomposes a product into its components to the level of deliverables, the components can then become the elements contained within the WBS. Decomposition of the work necessary to create each product component can now take place using the WBS. Used in this manner, the PBS and WBS become a set of integrated tools, which demonstrates both the decomposition of the product as well as the decomposition of the work necessary to create the product.

Benefits

Since the planning process on a product development project involves concurrent planning of product and project, the opportunity is great for these two efforts to become disconnected and even in conflict with one another. This of course can be a disastrous situation for the project manager as well as the project sponsor. If used, the PBS is an effective tool that can help to keep this from occurring. It focuses the planning effort on both what will be developed as well as how it will be developed. Especially if used in conjunction with the project WBS.

The PBS is a very visual tool that helps the project manager visualize the various components of the product and project. This visual representation aids in establishing structure and order to very complex projects. By engaging the sense of sight in the project structuring process, we are gaining another tool to establish order out of complexity.

An intangible, but important benefit gained from the use of the PBS is an increase in team cohesion. How is this? The PBS is normally created early in the project planning process, a time when a project team is beginning to form. Since development of the PBS is a collaborative effort between the cross-functional product specialists, much is learned about the various specialties, and a realization that the product development project will only be successful through a collaborative effort normally sets in. Nothing drives team cohesion faster than having to collectively solve problems and create a solution—in this case, the PBS.

References

1. Stroh, P. J. *Business Strategy: Plan, Execute, Win!* (Hoboken, NJ: John Wiley & Sons, 2014).
2. Alleman, G. B. Performance-Based Project Management: Increasing the Probability of Project Success. New York, NY: AMACOM, 2014).
3. Handifield, R. B. "Effects of Concurrent Engineering on Make-to-Order Products." *IEEE Transactions on Engineering Management* 41 (4): 384–393, 2004.
4. Reinertsen, D. G. *The Principles of Product Development Flow: Second Generation Lean Product Development* (Redondo Beach, CA: Celeritas, 2009).
5. McDounough, E. F. I., and G. Barczak. "Speeding Up New Product Development: The Effects of Leadership Style and Source of Technology." *Journal of Product Innovation Management* 8 (3): 203–211, 1991.
6. Handfield, R. B., Gary L. Ragatz, Kenneth J. Petersen, and Robert M. Monczka. 2009. "Involving Suppliers in New Product Development." *California Management Review* 42 (1): 59–82.
7. Cleland, D. I. *Project Management: Strategic Design and Implementation* (New York, NY: McGraw-Hill, 2006).
8. Thompson, A. T., and A. J. Strickland. *Crafting and Implementing Strategy*, 19th ed. (New York, NY: McGraw-Hill, 2013).
9. Reinertsen, 2009.
10. Berkun, Scott. *Making Things Happen: Mastering Project Management* (Sebastopol, CA: O'Reilly Media, 2008).
11. Kahn, K. B. *The PDMA Handbook of New Product Development*, 3rd ed. (Hoboken, NJ: John Wiley & Sons, 2012).
12. Cooper, R. G. *Winning at New Products: Creating Value Through Innovation*, 4th ed. (New York, NY: Basic Books, 2011).
13. Web site: www.purchasing-procurement center.com. Accessed October 15, 2014.
14. Project Management Institute. *A Guide to Program Management Body of Knowledge*, 5th ed. (Newtown Square, PA: Project Management Institute, 2014).
15. Ibid.
16. Hammer, M., and J. Champy. *Reengineering the Corporation* (New York, NY: Harper Business, 2006).

17. Kerzner, H. R. *Project Management: A Systems Approach to Planning, Scheduling, and Controlling* (Hoboken: NJ: John Wiley & Sons, 2013).
18. Schneider, A. "Project Management in International Teams: Instruments for Improving Cooperation." *International Journal of Project Management* 13 (4): 247–251, 2005.
19. Department of Defense. *MIL HDBK-881A; Department of Defense-Work Breakdown Structure* (Washington, DC: Department of Defense, 2005).
20. Department of Energy. *DOE G 120.1-5, Performance Measurement Systems Guidelines* (Washington, DC: Department of Energy, 1996).
21. Kerzner, 2013.
22. TSO. *Managing Successful Projects with PRINCE2* (London, England: TSO, 2012).
23. Martinelli, Russ, James Waddell, and Tim Rahschulte. *Program Management for Improved Business Results*, 2nd ed. (Hoboken, NJ: John Wiley & Sons, 2014).
24. Ibid.
25. Website: https://www.linkedin.com/groups/What-is-Work-Breakdown-Structure-35313.S.5879948537357152257?trk=groups_search_item_list-0-b-ttl&goback=%2Egmr_35313%2Egna_35313. Viewed on October 8, 2014

6

SCHEDULE DEVELOPMENT

Project scheduling involves the planning of timelines for completing the work identified and establishing dates during which project resources will be needed to perform the work.

The project schedule is the cornerstone of project work and, as such, serves as the working tool for project planning, execution, monitoring, and control. By developing a project schedule, a project manager is *planning* the time element of the project. By authorizing work according to the tasks contained within the schedule, the project manager triggers the *execution* of the project. By comparing the actual execution dates of the tasks with the scheduled dates, a project manager *monitors* the project and when actual performance deviates from the scheduled dates to an extent that corrective action must take place, a project manager is using the schedule to exercise *control* measures.

The process of schedule development involves the integration of multiple aspects concerning a project, including the estimated duration of tasks, the constraints imposed by the availability of resources and budget, and expected due dates. The goal of schedule development is to be able to answer the following questions:

1. If performed to plan, when will the project be completed?
2. At what dates should each task begin and end?
3. Which tasks are more critical to ensure timely completion of the project?
4. Which tasks can be delayed, if necessary without delaying the project completion date?
5. When are project resources needed and when will they be released?
6. Which resources are most constrained according to their task load?

The project schedule can be presented in various ways, as demonstrated in this chapter. The schedule type is often driven by the preferences and needs of the various project players. A functional manager, for example, may be interested in a schedule type that show the resource allocation requirements; the senior sponsor may be interested in a schedule that shows only major project events and milestones; and a project manager may need a detailed schedule showing each work breakdown structure (WBS) element.

Since project managers have a set of stakeholders with varying scheduling needs, they need to have various schedule types in their PM Toolbox. We will start by describing the most widely used project schedule type—the Gantt chart.

THE GANTT CHART

Using bars to represent project activities, the Gantt chart is a scheduling tool that shows when the project and each activity within the project start and end against a horizontal time scale. Even though the Gantt was developed around 1917 and is the oldest formal scheduling tool, it is still the most widely used tool.

Having a Gantt chart helps ensure that project participants have the necessary time allocated on their calendars and are available to perform their activities. Figure 6.1 illustrates a simple example of the Gantt chart.

Developing a Gantt Chart

Developing a Gantt chart takes several steps. Though the first step—determine level of detail and identify activities—is normally part of scope planning, we include it here in order to provide an integrated procedure.

The quality of the Gantt chart is firmly rooted in quality inputs about the following:

- Project scope
- Project WBS
- Responsibilities
- Available resources
- Schedule management system

Information about the project scope and the breakdown of project outcomes and activities from the WBS helps schedulers analyze and understand the project

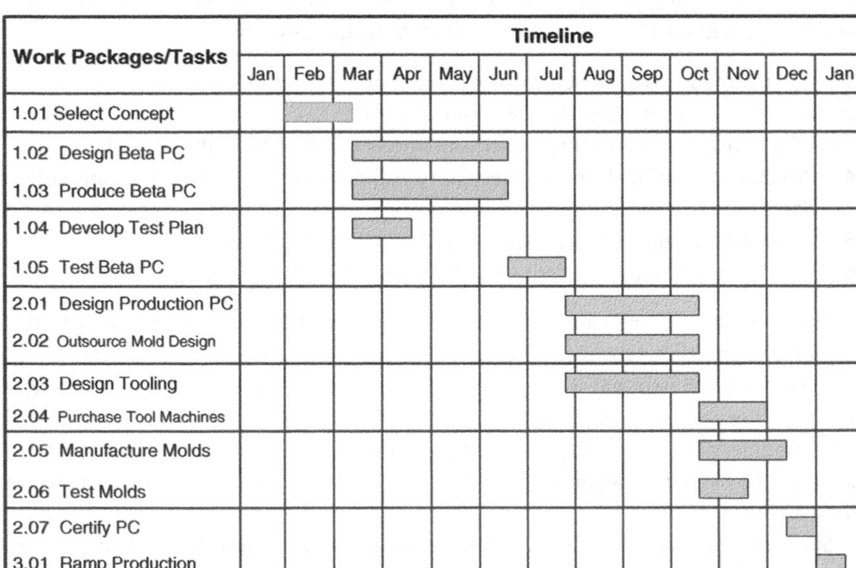

Figure 6.1: An Example Gantt Chart

activities that are being scheduled. Naturally, those responsible for the execution of the activities should be in the best position to schedule them because they are the best source of knowledge about the activities. Part of their knowledge relates to priorities, activity sequencing, and duration estimating. Another part of their knowledge relates to what resources are available and in which time periods. Still another part of their knowledge is the schedule management system, a method leading companies deploy to ensure their schedules are systematically developed and used (see "Schedule Management System").

Determine Level of Detail and Identify Activities

How many line items should there be in the Gantt chart? Twenty-five? Fifty? Seventy-five? The answer will be determined by the level of detail used in the WBS and the number of activities required to create each project outcome. Consider, for example, a practice used by an information technology (IT) organization in a financial services company. For a certain type of project profile, the team's decision is that a typical Gantt chart will have around 25 activities, with no activity longer than three weeks and shorter than one week. This provides guidance for the next step and ensures that the chart is sized appropriately, neither too cumbersome and time consuming nor too high level and lacking necessary information to manage. Specifically, the amount of detail should be enough for the intended user to monitor progress and for coordination.

Brainstorming and breaking the project down into constituent outcomes and activities that need to be performed in order to complete the project is next. You can enlist the help of the WBS (Chapter 5) for this purpose, identifying activities necessary to achieve work packages in the WBS. At this time, you don't need to worry about how big the activities are; rather, the emphasis is on making sure that all necessary activities are identified. Once that point is reached, refer to the level of detail established in the previous step. If it is determined that there are too few activities, breaking down some of them to hit the desired level of detail is a natural choice. Or, if there are too many activities, it is helpful to combine activities to arrive at the desired level.

Schedule Management System

Scheduling a project is rarely a matter of sitting down and developing a single schedule before the project starts. The process of developing and using schedules needs to be planned and organized.[1] To ensure such an approach, some companies tend to resort to a schedule management system that helps to determine the following:

What schedules are needed? Depending on the size of the projects, you can use a hierarchy of schedules for larger projects (see the Hierarchical Schedule section later in this chapter) or a single schedule for small projects, as well as preliminary and final schedules.

(*continued*)

How will the schedules be used? Some schedules, such as a summary schedule, can be used for management oversight, while a detailed schedule can be used for organizing and controlling all work.

How much detail is needed? Limiting the number of activities in a schedule, such as up to ten in a milestone schedule, is an attempt to preempt unnecessary detail and eliminate wasted time.

What tools will be appropriate? Any of the tools in this chapter are appropriate, if used for their designed purpose.

When will the schedules be prepared? Before the project starts. For projects with a high level of uncertainty, a rolling-wave concept may be applied: Prepare a front-end schedule for the first 60 days of a project, then develop a more detailed schedule as the project unfolds.

How will the schedules be monitored and updated? Frequency and schedule control tools need to be defined in tune with needs of a company.

Scheduling can be overwhelming unless approached in a systematic way. Developing a schedule that is more detailed than necessary is as useless as the schedule that the management team cannot follow or understand. Inappropriate scheduling emphasis like this may easily turn people against the use of formal scheduling techniques. It is the job of the schedule management system to prevent this.

Sequence Activities

Sequencing activities involves arranging them in a logical order of execution. This requires a good knowledge of the work flow and priorities of the project and ensures that we first perform activities that need to produce outputs necessary for work on the subsequent activities. Illogical sequence of activities is destined to cause rework and additional cycle time.

Estimate Activity Durations

Resources, human and material, drive the process of estimating activity durations. Begin the process by asking, "What resources do I need to successfully complete this activity?" The answer should provide the names of resources and work time for each one of them to complete the activity: for example, 100 hours of work from a software programmer. Next, knowing the availability of a resource and using the company's work calendar (e.g., no work on Sundays), convert the work time into calendar time. For instance, because she is involved in multiple projects, the programmer's 100-work time hours will have to be spread over 12 weeks to get her job done. This should be repeated for each activity.

Draft and Refine the Gantt Chart

Drawing a Gantt chart requires a sheet or form with a horizontal time scale and the list of activities across the vertical axis (Figure 6.1). Adding up calendar times for project activities gives a rough idea of the total time the time scale needs to show. This is a good approach for projects in which all activities are sequential. Should there be overlapping activities, you can reduce the duration of the time scale accordingly.

Next is listing all activities on the vertical axis per determined sequence, followed by these steps:

Step 1: Draw a bar representing each activity, with its length proportional to its duration on the time scale.

Step 2: For multiple activities that form a phase of work, add a summary bar, called a *hammock activity* or simply a *hammock*, just above the first activity. A hammock begins when the first of the activities begins and ends when the last of the activities ends. A reasonable measure is to have a hammock activity for every four to ten detailed activities that relate to each other. Because management needs a big-picture view of the project, the hammocks' summary level of detail makes them very convenient for that purpose.

Step 3: Look again at the whole chart. Are all necessary activities there? Logically sequenced? With appropriate timescale and reasonable durations? Make any changes necessary to finish refining the chart and get ready to use it.

Using the Gantt Chart

The Gantt chart is an effective tool for smaller and simpler projects, where there is less need to show dependencies between activities, since they are well known to all involved.[2] As the project size and complexity increase, the Gantt chart becomes less applicable. Simply, the chart gradually loses the ability to handle the increasing number of activities, data, and dependencies between activities involved. In large and cross-functional projects, using the Gantt chart as the primary scheduling tool is impractical and ineffective (see "Tips for Gantt Charts").

In contrast, in large and complex projects, the complementary use of the Gantt chart and critical path schedule may be a very smart strategy. The latter effectively copes with number of activities, data, and dependencies between activities. It, however, does a poor job of showing in a simple and visual manner what activities to work on, in what order, and for how long. That is where the Gantt chart may add value. Extracting from the sizable critical path schedule activities due in the next week or two and then presenting them in the Gantt chart format provides clear and practical short-term schedule focus. The project manager is still responsible for coordinating the interfaces between the owners of the project activities and outcomes.

Depending on one's knowledge and experience, a 20-activity Gantt chart can be developed in anywhere from 10 to 40 minutes. Some experienced project managers use the rule "activity per minute," meaning it should take one minute for each activity in the Gantt chart. Note that the more people involved in the Gantt chart development, the more time may be needed.

Benefits

Having a Gantt chart helps ensure that everyone understands the timeline for project activities. Then, project participants will have the necessary time allocated on their calendars and be available to perform their activities.

The Gantt chart brings value to the project manager through its visual and simple design that has lasted for decades. It creates a pictorial model of the project that makes it both an excellent management and communication tool. Its simple design enables a user with little or no instruction the ability to create and read a project schedule.

Finally, the Gantt chart is a very useful tool for resource planning and allocation. By defining the number of resources needed for each project activity, and then adding them up for each time period, the total resource requirements over a defined period of time for each activity as well as for the project as a whole can be established.

Tips for Gantt Charts

- Rely on the Gantt chart as long as it has fewer than 100 activities.
- Use a single Gantt chart as a primary scheduling tool in small, simple projects.
- Don't use a single Gantt chart as a primary scheduling tool in large, complex, cross-functional projects. Rather, utilize multiple, integrated Gantt charts.
- Team-developed Gantt charts lead to higher quality, better buy-in, and stronger commitment of team members.

THE MILESTONE CHART

This scheduling tool shows milestones against a time scale in order to signify key project events and to draw stakeholder attention to them (Figure 6.2). A *milestone* is defined as a point in time or event whose importance lies in it being the climax point for many converging activities. For instance, "Requirements Document Complete" is a distinctive milestone for software development projects, and "Market Requirements Document Complete" is a characteristic milestone for product development projects. While these milestones relate to the completion of key deliverables, other types may include the start and finish of major project phases, major reviews, events external to the project (e.g., trade show date), and so forth.

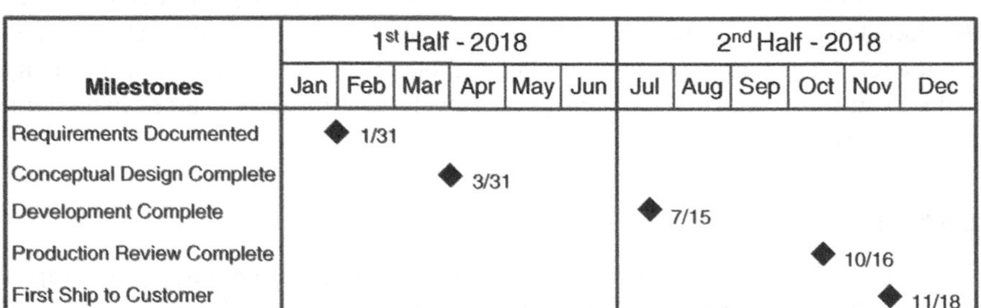

Figure 6.2: Example Milestone Chart

Developing a Milestone Chart

Traditionally, the milestone chart has been used to focus senior managers on highly important project events, whether the project is large or small. It can also help strengthen the emphasis on goal orientation while reducing focus on activity orientation. Its value can be additionally improved if the chart is customized to specific project needs.

A relatively simple procedure for developing a milestone chart includes several steps that build heavily on a schedule with activity dependencies that is constructed in a separate procedure. We nevertheless include it here in order to present an integrated picture of milestone chart development.

A milestone chart is as good as its inputs. Having a solid definition of the project scope provides those who schedule a good understanding of the milestones being scheduled. The quality of the chart is certainly expected to be higher when the owners of milestones are responsible for their scheduling (see "Who Is Involved in Scheduling?") and follow guidelines established in the schedule management system. If scheduling of milestones is based on a previously developed detailed schedule, the quality of the chart is bound to further improve.

Who Is Involved in Scheduling?

The involvement of project participants in developing schedules hinges to a great extent on the organizational strategies for project management. In the matrix environment, for example, many players are involved—team members, project managers, functional managers, the project office, and executives.

Team members typically own work packages and tasks, reporting their completion and estimating how much time is necessary to complete each unfinished work package or task. While they have to know some scheduling terms, such as start date, finish date, data (reporting) date, resource availability, and so on, there is no need for extensive knowledge of scheduling theory.

As providers of resources to a project, functional managers care about the accuracy of the estimates and availability of resources when projects need them.[3] Like team members, their knowledge of scheduling theory is basic.

Project managers are the ultimate users and owners of the project schedule. They facilitate schedule development and monitor the data furnished by team members for completeness and feasibility. They then use the schedule to monitor and control progress to plan, working with the functional managers to make schedule modifications when needed. Project managers need a decent amount of knowledge about scheduling theory.

The project office (or the scheduling group) should have scheduling experts who are capable of designing and maintaining a project scheduling system that all other players utilize. Also, their knowledge in running scheduling software and checking time, cost, and resource estimates in order to support the system and individual projects is essential.

(continued)

> The job of executives in project scheduling is not about scheduling theory, tools, or software. Rather, their focus is on asking questions, reading reports, directing project-related personnel, and providing overall support. Like a well-conducted orchestra of master musicians, these players need to synchronize their actions to produce a meaningful scheduling concert.

Select Milestone Chart Type

You may choose to develop a high-level milestone chart with only a few high-profile milestones intended to inform managers or other stakeholders. Another option may be a milestone chart designed to help manage the work necessary to accomplish a project deliverable or complete a project phase. Which one is better? It depends on the situation. As an example, for a certain type of project, a company can use high-level milestone charts with five standard milestones identified as decision points between project phases. At these milestones, top management reviews the project status and decides to either continue or terminate the project. In addition, the company may also use a more detailed chart with 10 to 15 milestones, which the project team uses to review major project deliverables. Selecting the type of chart based on its purpose and use is an important step in building the milestone chart.

Identify Key Milestones

In this step, we let the type of the chart chosen guide the selection of the desired milestones. Consider all types of milestones—key deliverables, start and finish of the project and its major phases, major reviews, important events external to the project, and so forth. Which ones are key to monitoring the project progress? If a company uses standard milestones, the answer is simple: those same standard milestones. If this is not the case, focus on the major synchronization points and key decision points in the project.

Sequence the Milestones

Sequencing milestones is about studying dependencies between activities and comprehending how their outputs will converge to a culmination or synchronization point—the selected milestone. Marking their position in the detailed schedule will provide their sequence, indicating which activities have to be started or completed to proclaim that a milestone is accomplished.

Draft and Refine the Milestone Chart

Once milestones are marked on the detailed schedule with activity dependencies, the milestone chart is essentially drafted. This is the time to ask the following questions. Are all necessary milestones there? Are they logically sequenced? In appropriate positions on the schedule? It is important to ensure that a sufficient number of milestones have been chosen so as not to have a prolonged period between milestones. It is easy to select all milestones at the project beginning and end, where many activities start and complete. However, that would leave the middle void of milestones, reducing the ability to control project progress. As these questions get answered, the information will be created to make any changes and distill the chart.

With milestones marked on the schedule, two points should be discernible. The first is the laborious work that must be done. Second, the selected milestones that signify keys to the project progress must stand out so as to avoid getting lost in the details of the laborious work. In short, not only should the trees (details) be evident, but the forest (milestones) as well.

Finalize the Chart

Quite understandably, managers don't necessarily like to bother with detailed schedules. Rather, they will ask for a chart showing only milestones so they can glance and discern key data on project progress. You should use information from the detailed schedule with milestones—time scale, milestone names, and time positions of milestones—in order to prepare the milestone chart. To do so, list milestones on the vertical axis of a sheet, draw the time scale across the horizontal axis, select symbols for milestones (e.g., diamond), and place the symbols across the time scale (see "Tips for Milestone Charts").

Tips for Milestone Charts

- Provide adequate time spacing between milestones.
- Use the chart for both key events and detailed milestones.
- Use the charts in both large and small projects to demonstrate actual progress against plan.
- Use the chart in conjunction with another schedule that show activity dependencies.
- Team-developed milestone charts lead to higher-quality schedules, broader buy-in, and stronger commitment of team members.

Using the Milestone Chart

Traditionally, the milestone chart has been used to focus management on highly important events whether projects are large or small.[4] As a result, when used for this purpose, the chart typically shows a few key milestones. In other words, use the chart to provide key data pertaining to planned and actual project progress. When the WBS is used in the project, these highly important events and key data are usually related to Level 1 in the WBS.

Recent developments have seen the increased use of multiple milestone charts at one time, where each chart corresponds to a certain level of WBS. Consequently, a chart on Level 4 of a five-level WBS, for example, may easily have a couple hundred milestones, each one tied to a work package. The chart is then used in conjunction with a detailed network diagram chart so that the effect of dependencies on milestones is discernible. This practice is well accepted in technology organizations that compete on fast project cycle times (see the "Hierarchical Schedule" section later in this chapter).

Benefits

A chart with a few key milestones—related to Level 1 of a WBS, for example—is in a solid position to capture and enjoy management attention and time with high-profile project events (see "The Lack of Milestones May Kill Thousands").[5] Not so with charts including many milestones linked to work packages. The gain from such charts is the ability to increase the emphasis on goal orientation ("milestone accomplished" or "milestone not accomplished"), while reducing focus on activity orientation ("I am working on it").

The visual nature of the milestone chart also provides value to both the project manager and his or her stakeholders. It creates a pictorial model of the project ideal for effective communication between project team and stakeholders as well as within the project team. In this manner, the milestone chart serves as an effective project tracking tool as well as an effective planning tool.

The Lack of Milestones May Kill Thousands

Napoleon, a master of warfare, never made more than the sketchiest plans even for his great projects.[6] Those sketchiest plans probably resembled milestones to signify key events in a campaign. When he invaded Egypt, he did so making his expectations clear to each of his generals. In contrast, when Napoleon invaded Russia with his 400,000+ troops in June 1812, he chose not to communicate the plans. Despite his generals' desire to have the plans communicated because of the harsh Russian winter, Napoleon never shared them. In December of 1812, Napoleon's defeated army left Russian soil, this time numbering only 20,000+ troops. The lack of a clear milestone schedule might have contributed to the finale of Napoleon's Russian campaign—from the zenith of glory to ashes.

THE CRITICAL PATH METHOD DIAGRAM

The critical path method (CPM) tool is a network diagram technique for analyzing, planning, and scheduling projects. It provides a means of representing project activities as nodes (see Figure 6.3) or arrows, determining which of them are "critical" in their impact on project completion time, and scheduling them in order to meet a target date at a minimum cost.[7]

Originally developed for use in large, complex, and cross-functional projects, the CPM tool is now employed in smaller projects as well. Having a CPM schedule helps the project manager to see the total completion time, understand the sequencing of activities, ensure that resources are available when needed, monitor those that are critical, and measure progress.

Activity	Description	Immediate Predecessor	Duration (days)
a	Project Kick- off		0
b	Get materials for a	a	10
c	Get materials for b	a	20
d	Manufacture a	b,c	30
e	Manufacture b	b,c	20
f	Customize a	e	40
g	Assemble a & b	d,f	20
h	Release	g	0

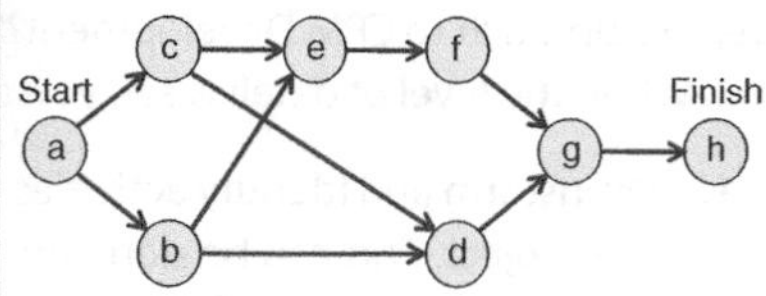

Figure 6.3: Example Critical Path Method Diagram

Constructing a CPM Diagram

Constructing a CPM diagram is an exercise in patience and discipline that involves proceeding through several major steps. In it, as with all schedule development tools, a crucial step is to determine the level of detail and to identify activities. Although this step usually belongs to the scope planning process, we include it here because it helps explain the tool development in an integrated way.

The process of building a CPM schedule is destined to produce a better product If quality information about scope, responsibilities, resources, and the overall schedule management system is available.

Project scope provides schedulers with the knowledge of the project activities that need to be scheduled. Clear definition of responsibilities—who does what in the project—points to who has the best information about the individual activities and should therefore schedule them. To develop realistic schedules, these "owners" of the activities also need to know which resources are available and when. Finally, the schedule management system will direct schedulers in developing and using the CPM.

Determine the Level of Detail and Identify Activities

How can large or small individual activities influence the number of activities in the CPM? A rule at one company may help clarify this point. Large-fab construction projects run at around 2,000 activities, lasting from two to four weeks. This helps everyone realize what level of detail is acceptable and what is unacceptable. The goal for scheduling is to account for the complexity and size of the project in a way that gives the team enough information—not too little and not too much—to direct the daily work, identify interfaces between workgroups, and monitor progress at an effective level (see "Why a Team Approach to CPM Development?").

When the level of detail is set, you are ready to perform the following activities:

1. Brainstorm and identify activities that are necessary to complete in order to finish the project. This can be done by means of the WBS, perhaps the most systematic and integrated way of activity identification (see the Work Breakdown Structure section in Chapter 5).
2. Refocus the attention on the established level of detail. If the number of activities is lower than the intended number, continue breaking down larger activities. If the number of activities is over the target number, combine related activities to reach the desired detail.

Why a Team Approach to CPM Development?

Using a project team to build a CPM diagram is perhaps the most effective way of doing it. Here is why:

- Team members are usually the best source of knowledge about their piece of the schedule.
- Each team member can see where and why he or she is critical to the success of the project.
- The team can find creative ways to best sequence and shorten the duration of activities and the total project.
- As a unit, the team can focus its energy and mind on mission-critical activities.
- Involvement of team members enhances commitment and a sense of ownership of the project.

Sequence Activities

Sequencing is about identifying dependencies between activities by determining an activity's immediate prerequisite activities, called *predecessors*. A portion of the dependencies will be arranged in pure "technological order." These are termed *hard* or *logical* dependencies, meaning that the technology of work mandates such sequence. An example is that one must write the code before testing it; the other way around is not possible. Disregarding hard dependencies may lead to rework and project delay. But not all of the dependencies are hard; some of them are soft or preferential. They are not required by the work logic but set by choice, reflecting one's experience and preferences in scheduling. For example, we may decide to write a piece of software code, test it, write another piece, test it, and so forth. Dependencies may also be dictated by availability of key resources. If two activities require the same resources, one will have to follow the other. Once the dependencies are established, they can be recorded, as we have done in Figure 6.3.

Assign Resources and Estimate Activity Duration

The age-old rule of scheduling is that people and material resources get the work done. As a result, it is logical to estimate an activity's duration by identifying resources necessary to successfully complete it. Consider, for example, 100 hours of work from a business analyst. This is the work (effort) time, which in the case of mature work technologies is calculated by dividing the amount of work by the production norms.[8] With the information that the analyst splits her work time between this and three more projects, and knowing the company's work calendar (50 hours per week only; no work on Saturdays and Sundays), she may need eight weeks to get it done. This is the calendar time. Write the calendar time in the fourth column of the table in Figure 6.3.

Draft a CPM Diagram

Each activity is drawn on the network diagram as a circle or rectangle, with identifying symbols and duration within the circle or per convention chosen. This format is called AON (activity-on-node). Later in this chapter, we discuss another format of drawing network diagrams, AOA (activity-on-arrow). To pursue the AON format, indicate sequence dependencies by arrows connecting each circle (activity) with its immediate successors, with arrows pointing to the latter. For convenience, connect all circles without predecessors into a circle denoted "Start." Similarly, connect all circles without successors into a circle marked "Finish."

Identify the Critical Path

Normally, the diagram shows a number of different paths from Start to Finish, defined as sequences of dependent activities. To calculate the time to pass through a path, add up the times for all activities in the path. The critical path is the longest path (in time) from Start to Finish. It indicates the minimum time necessary to complete the entire project. Essentially, the critical path is the bottleneck route, and the highest priority set of activities to manage.

There is another way to calculate the critical path: using the forward/backward pass procedure.[9] While adding up activity times is simpler for smaller projects, it is too cumbersome and difficult for larger projects. Rather, the large projects use the pass procedure. Say, for example, you have the start date for a project. Then, for each activity there exists an earliest start time (ES). Assuming that the time to finish the activity is t, then its earliest finish time (EF) is ES + t. Figure 6.4 shows how to go through the forward pass to calculate ES and EF for each activity. The process, from left to right, is as follows:

- ES is the largest (or latest) EF of any immediate predecessors.
- EF is ES + time to complete the activity (t).

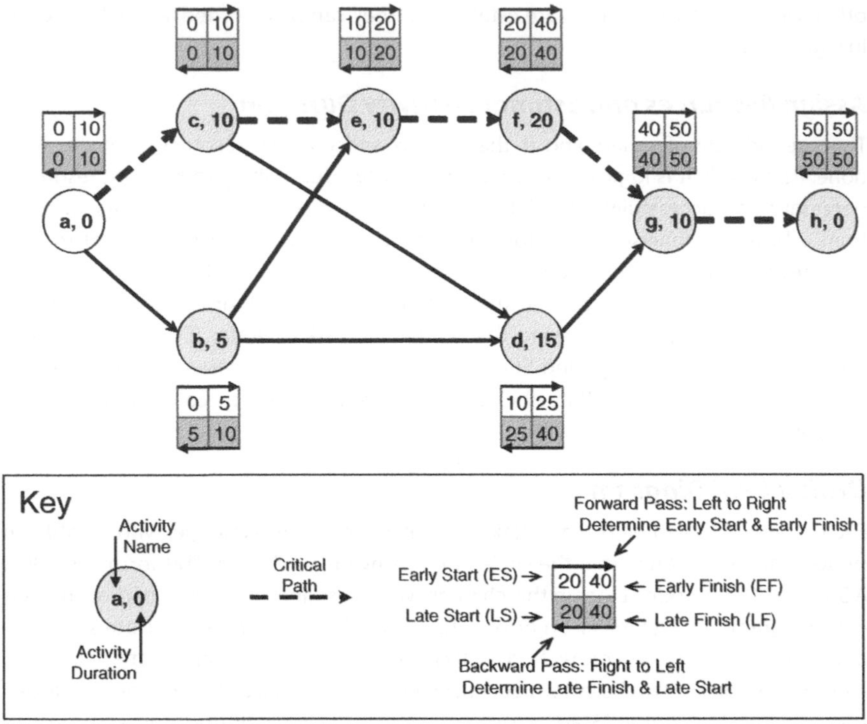

Figure 6.4: Forward and Backward Pass

Suppose now that you want to finish the project by the time that is equal to the EF for the project. If so, you can define the concept of late finish (LF), or the latest time that the project can be finished, without delaying the total project beyond EF. Thus, LF is equal to EF.

Similarly, you can define late start (LS) as LF − *t*, where *t* is the activity time. Building on these concepts, we can go through the backward pass, from right to left, to calculate for each activity (see Figure 6.4):

- LF is the smallest (or earliest) LS of any of immediate successors.
- LS is LF – Time to complete the activity (*t*).

Now that the forward/backward pass is finished, note that Figure 6.4 indicates that in some activities, early start is equal to late start, while in some it is not. The difference between an activity's early and late start (or between early and late finish) is called *total float*. Total float is the maximum amount of time you can delay an activity beyond its early start without delaying the project completion time. *Free float* is another kind of float equal to the amount of time you can delay an activity without delaying the early start of any activity immediately following it. While an activity with the positive total float may or may not have free float, the latter never exceeds the former. The formula to calculate free float is the difference between the activity's EF and the earliest of the ES of all of its immediate successors.

In our example in Figure 6.4, activities *b* and *d* have free float of 5 days and 15, respectively, while all other activities have zero free float. Activities on the critical path have zero total float and are called *critical activities*. They are shown in Figure 6.4, with thick arrows connecting critical activities on the only critical path. It is, however, legitimate to have multiple critical paths, a common situation in fast-tracking projects.

An activity with zero total float has a fixed scheduled start time, meaning that ES = LS. Consequently, to delay the start time is to delay the whole project, which is why such activities are called *critical*.

In contrast, activities with positive total float offer some flexibility. For example, we can relieve peak loads in a project by shifting activities on the peak days to their late starts. That won't impact project completion time. But this flexibility may vanish quickly. Consider a path with a very small total float, which we call *near-critical path*, the second-highest priority to pay attention to. If we let an activity on the near-critical path slip, its small total float may be gone and it becomes critical path. In case of the free float, we can delay the activity start by an amount equal to (or less than) the free float without affecting the start times or float of succeeding activities.

Review and Refine

Look closely at the drafted diagram and ask the following questions:

- Has any important activity been left out of the schedule?
- Is the activity sequencing logical?
- Are durations of activities reasonable?
- Is the project schedule time-constrained or resource-constrained? (See "Time- or Resource-Constrained? Or Both?")

Time- or Resource-Constrained? Or Both?

Although Intel is in a constant time-to-market race with its competition, project managers in one of its divisions understand the relationship between time and resources when developing their schedules. The problem is that no matter how fast they want to complete their projects, the availability of resources is limited. This helps classify schedules into two categories:[10]

1. Schedules under a time-constrained situation. The project must be finished by a certain time (called the drop-dead date in Intel language), with as few resources as possible. Here, it is time, not resources, that is critical. These are typically the highest priority projects.
2. Schedules under a resource-constrained situation. The project must be finished as fast as possible without exceeding a certain resource limit. Here, it is resources, not time, that is critical. Projects like these are usually of lower priority.

(*continued*)

Between these two extremes are resource-leveling projects with their medium priority. With these, once a schedule is developed, project tasks are shifted within their float allowances to provide smoother period-by-period resource utilization. As long as management clearly communicates in which category each project is, project managers face no problem—most commercially available software they use already have algorithms to develop schedules for any of the three situations. The trouble arrives when project managers are told their projects are time-constrained but with limited and insufficient resources. Faced with such systems constraints, they know they have to find resources on their own. So what do project managers do? They make do. They utilize overtime, work long hours, convince team members to do the same, and navigate their way through. Most of the time they succeed. After all, Intel's culture is all about performance.

This is the time to answer the questions and, if the answers require, to make necessary corrections.

If a company competes on time, a project manager must check if it is possible to reduce the project duration. The only avenue to do that is to find ways to shorten activities along the critical path. This is possible by fast tracking, or schedule crashing, or a combination of the two approaches.[11] Note that fast-tracking or crashing noncritical activities is irrelevant, because it does not reduce the duration of the critical path. Fast-tracking means changing the hard and soft dependencies—in other words, changing the logic of the diagram by obliterating previously established dependencies and creating new ones by attempting to overlap certain activities on the critical path. In the process, neither activity durations nor resource allocations will be changed.

"Crashing" means shortening the duration of activities along the critical path without changing dependencies. The way to do this is by assigning more people to the activities, working overtime, using different equipment, and so on. The crucial question is whether the gains from the reduction of project duration exceed the costs of acceleration. For the majority of time-to-market projects the answer to the question is yes. For more details, see the Schedule Crashing section in Chapter 12.

Using the CPM Diagram

The CPM tool was originally developed for large, complex, and cross-functional projects. This is still the primary use of CPM diagrams because it can easily deal with a large number of activities and their dependencies, directing our attention to the most critical activities. However, with the dissemination of CPM knowledge, it is not unusual to also see CPM used for smaller projects.

A fine application of CPM can be found in conjunction with the Gantt chart. In short, extracting from the sizable CPM diagram activities due in the next one or two weeks, presenting them in the Gantt chart format, provides clear and practical partial short-term outlook schedules.

Benefits

Having a CPM diagram helps the project manager to see the total completion time, understand the sequencing of activities, ensure resources when necessary, monitor those that are critical, and measure progress (or lack of it). This is easier to accomplish if certain rules are followed (see Tips for CPM Diagrams under "Three Don'ts for CPM Diagrams").

The CPM diagram also offers the project manager value through its graphical structure. The CPM diagram is easily explainable by means of the project network diagram that clearly charts the technological order of work. Data calculations are not difficult and can be handled readily and quickly by personal computers.

The greatest benefit to the project manager is arguably the focus on priority that the CPM diagram provides. It pinpoints attention on the small group of activities that are critical to successful project schedule completion. This focus greatly adds to higher accuracy and, later, precision of schedule control.

Three Don'ts for CPM Diagrams

1. Don't let the CPM chart control you. It is just a schedule and won't make a decision for you. You will.
2. Don't consider it gospel. If there is a better way to schedule, go for it!
3. Don't throw it aside when your project starts slipping. Review it, update, and improve, then use it again!

Tips for CPM Diagrams

- If you need to accelerate the schedule, do it by fast-tracking and crashing.
- Watch out! Accelerating the schedule may increase the number of critical activities. While in earlier times 10 percent of all activities were critical, in today's fast schedules, you often see 40 percent to 50 percent of activities being critical.
- Sprinkle major milestones over your CPM diagram. It helps you see the forest (milestones) and the trees (activities).
- Color-code activities performed by various resource providers in order to identify their interfaces and provide their coordination.

Variations

The CPM diagram discussed here is of the AON format. Other formats include CPM with AOA, Program Evaluation and Review Technique (PERT), and precedence diagrams. In AOA, activities are shown as arrows, and the arrows are connected by circles (or dots) that indicate sequence dependencies. In that way, all immediate predecessors of an activity lead to a circle at the beginning point of the activity arrow, while all immediate successors stem from the circle at the arrowhead. Thus, a circle becomes an event, where all activities leading to the circle are completed.

The CPM diagram is very similar to the PERT diagram, but while CPM's activity duration estimate is deterministic, PERT uses a weighted average to calculate expected time of activity duration as follows:[12]

$$TE = (a + 4m + b)/6$$

where

a = optimistic time estimate
b = pessimistic time estimate
m = most likely time estimate

The PERT diagram has been used primarily in research and development projects, while CPM, which was originally developed for construction projects, has spread across other industries.

The precedence diagram is an AON network that allows for leads and lags between two activities (see the Time-Scaled Diagram section in this chapter for more details on leads and lags). This makes it easier to portray rich and complex dependencies for real-world projects, giving the precedence diagram a wider application across industries and an edge over the CPM and the PERT. These two methods allow for leads and lags only by splitting activities into subactivities, leading eventually to a significant increase in the number of activities in the network, and making it more complex and difficult to manage.

THE TIME-SCALED ARROW DIAGRAM

The time-scaled arrow diagram (TAD) is used to analyze, plan, and schedule projects in order to meet a target date at a minimum cost. In the process, the TAD helps to determine which project activities are "critical" in their impact on project completion time to provide focus for the project team. As its nominal definition states, TAD is an activity on arrow tool, and is the only critical path method that is displayed against a time scale (Figure 6.5).

Like other schedule development tools, the TAD helps you identify the total completion time, understand the sequencing of activities, identify when resources are needed, monitor activities that are critical, and assist the project manager in measuring progress.

Developing a TAD

Building a TAD is an exercise that requires endurance and discipline, unfolding through several major steps. As with all schedule development tools, the first thing to do is to determine the level of detail required and identify the project activities. Typically, this is part of scope planning. We include it here in order to provide an integrated view of this tool's development.

The TAD's quality is heavily dependent on solid information about project scope, team member responsibilities, availability of resources, and the higher-level schedule management system.

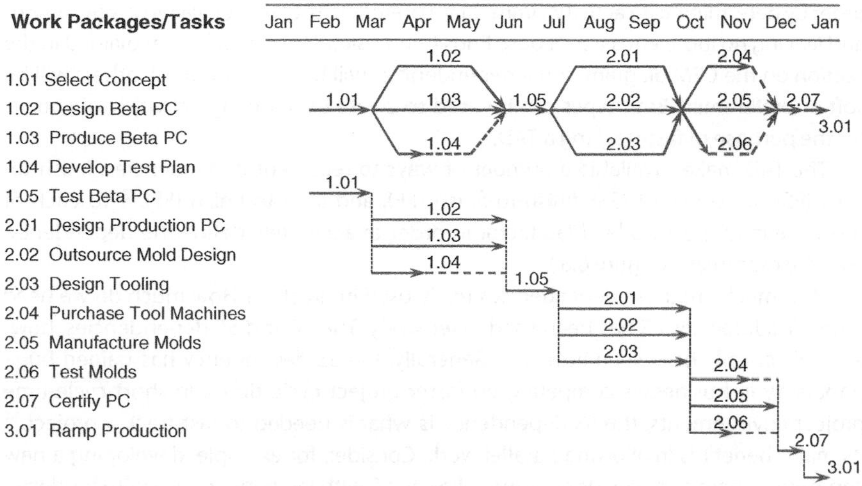

Figure 6.5: Example of a Time-Scaled Arrow Diagram

Clearly, you need to understand the project scope in order to schedule the project activities. The purpose of knowing who is responsible for certain activities is to indicate who will schedule and then manage the activities. For this to be possible, project schedulers need information on resource availability. Guidelines for how to develop and maintain a TAD will come from an organization's scheduling management system.

Determine the Level of Detail and Identify Activities

How large or small a project is influences the number of activities in a TAD. An example may shed more light on this issue. Large new product introduction projects typically include 300 to 500 activities, with durations between three and five weeks. The intention is to provide just enough information—neither more, nor less—than what one needs to direct and monitor project work of a certain size and complexity. Providing more or less information than necessary may either overload the team or deprive it of essential information, respectively.

Once the level of detail is chosen, identify activities that have to be performed to complete the project. As is the case with all types of scheduling tools, an excellent way to do this is to employ the WBS, perhaps the most systematic and integrated way of activity identification. Activities necessary for developing a TAD are those needed to produce work packages, the lowest-level elements of the WBS.

When the activities are identified, concentrate on the level of detail that was selected earlier. If the actual number of activities is smaller than the targeted number, resume breaking down larger activities. Should the actual number exceed the target number, some similar activities can be merged to reach the desired level of detail.

Sequence Activities

Sequencing involves determining the logical flow of project activities and establishing the dependencies between the activities. This means putting activities into a specific order by determining an activity's immediate prerequisite activities, called predecessors, and leaving no loose ends (see "Loose Ends May Mislead the Team"). As explained in the section on the CPM diagram, some dependencies will be hard or logical; others will be soft or preferential. Both types can be used to create overlapping activities, of course, for the purpose of fast-tracking a TAD.

The TAD makes available a number of ways to represent dependencies: finish-to-start (FS), start-to-start (SS), finish-to-finish (FF), and start-to-finish (SF).[13] To each of those we may specify a lead/lag factor in order to accurately define the dependency. (See the examples in Figure 6.6.)

How much are these dependencies really used in practice? How much do we need them? Traditionally, FS has been used extensively. The FF and SF dependencies, however, historically have seen little use. Generally, the SS dependency has gained huge popularity in businesses competing on faster project cycle times. In short-cycle-time project environments, the SS dependency is what is needed to fast-track a project.[14] Its major benefit is in allowing parallel work. Consider, for example, developing a new computer, where software development has an SS with lag dependency with hardware

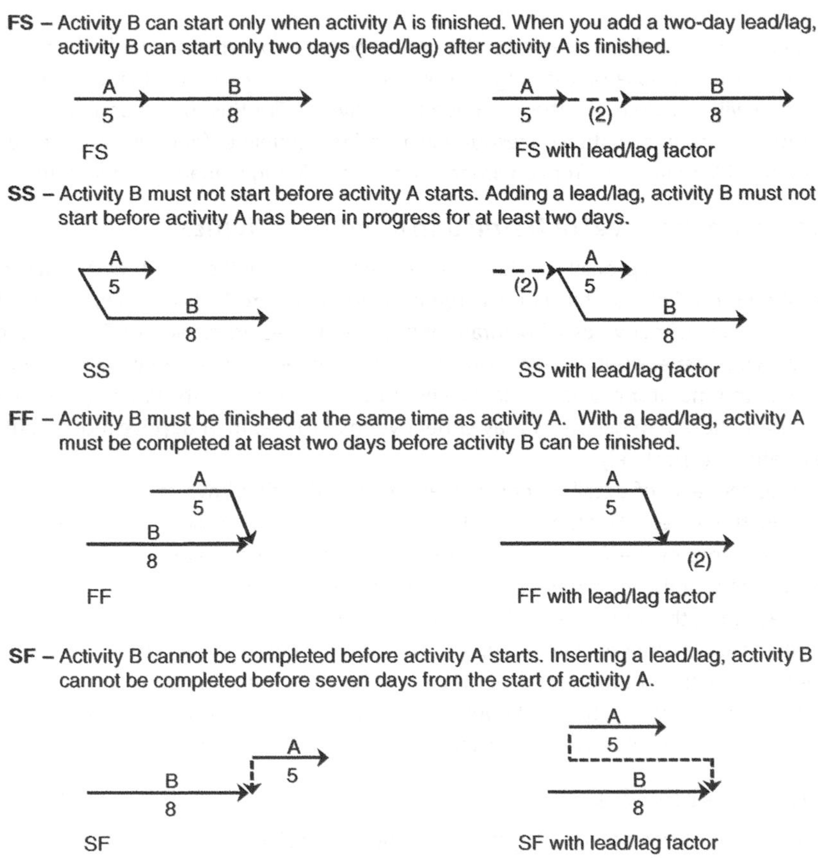

FS – Activity B can start only when activity A is finished. When you add a two-day lead/lag, activity B can start only two days (lead/lag) after activity A is finished.
A
5
B
8
FS
A
5
(2)
B
8
FS with lead/lag factor
SS – Activity B must not start before activity A starts. Adding a lead/lag, activity B must not start before activity A has been in progress for at least two days.
A
5
B
8
SS
(2)
A
5
B
8
SS with lead/lag factor
FF – Activity B must be finished at the same time as activity A. With a lead/lag, activity A must be completed at least two days before activity B can be finished.
A
5
B
8
FF
A
5
(2)
FF with lead/lag factor
SF – Activity B cannot be completed before activity A starts. Inserting a lead/lag, activity B cannot be completed before seven days from the start of activity A.
A
5
B
8
SF
A
5
B
8
SF with lead/lag factor

Figure 6.6: Types of Dependencies between Activities

development. To start its work, the software team needs at least a hardware design, and once they get it, they can carry on with their work in parallel with the hardware development.

Loose Ends May Mislead the Team

We frequently see TADs with loose ends, including arrow tails and arrowheads that are not connected to other activities. When we ask project managers why, we often get an answer like, "I only want to show the critical path and dependencies on it. Other paths and their dependencies are not important to me." This is a risky practice. To determine the critical path, the team has to evaluate all paths with properly connected activities. If there are loose ends, the team may not see the real critical path. Then the whole purpose of having a TAD—focus on really critical activities—is defeated.

Assign Resources and Estimate Activity Duration

The heart of schedule development is resource allocation and scheduling. Although it was touched on in the CPM section, we repeat it here. The first rule of resource allocation is to identify resources necessary to successfully complete activities. For example, you may need a cost estimator and precisely 80 hours of his work (effort) time. Assume now that the estimator is shared by this and two other projects. Look at the company's work calendar as well (50 hours per week maximum; no work on Saturdays and Sundays). All this information tells that the estimator will need to spread his 80 hours' worth of work over 10 calendar weeks (calendar time). Reiterate steps of identify resources, determine work time, and convert it to calendar time for all remaining activities. Estimating activity durations can be tricky, especially in multiproject environments (see "Switchover Time Adds to the Schedule Inaccuracy").

Switchover Time Adds to the Schedule Inaccuracy

Some 90 percent of projects are implemented in a multiproject management environment. This means that a practice of having project managers run multiple projects at a time, anywhere from two to ten, is widely accepted.[15] While such an approach provides outstanding benefits in terms of better management, it also generates a unique problem that calls for very meticulous scheduling of the projects. In particular, the problem here is the switchover time. When the project manager switches from one project to another, she needs switchover time to align her thinking and get into the new project, physically and mentally.[16]

(*continued*)

Since team members also operate as members of multiple project teams, they suffer from the same problem. As the projects grow in complexity, so does the switchover time.[17] Clearly, this time is a loss in a busy day of a project manager and team member. For example, some experts indicate that this loss may be up to 20 percent of the project manager's or team member's time, when involved in four projects at a time. The real problem then is that this switchover time loss typically is not taken into account in scheduling multiple projects. As a consequence, project schedules are notoriously optimistic and inaccurate.

There are at least two strategies to deal with the problem. One is to reduce the available monthly work hours of a multiproject person used for scheduling purposes by the corresponding switchover time loss. Another is to increase the estimate of a multiproject person's work hours for a specific project by the corresponding switchover time loss. These strategies are not attractive, but they are necessary for the realistic development of any project schedule or dependency diagram.

Draft the TAD

Draw activities as arrows, connecting them to one another—arrowhead to arrow tail to indicate the sequence of dependencies (see Figure 6.6). In that manner, all immediate predecessors of an activity lead to the beginning point of the arrow tail, while all immediate successors stem from the arrowhead. Thus, the beginning point of the arrow tail becomes an event, where all activities leading to the point are completed. Obviously, a TAD can be drawn in two different formats (see Figure 6.6 and the example that follows, "Cascade- versus Spine-Formatted TAD").

Cascade- versus Spine-Formatted TAD

Cascade

One zone, one activity. A zone is a horizontal swath or strip across a TAD printout. The cascade format allows only one activity per zone.

Why called cascade? A well-arranged succession of activities, one per zone, appears like a cascade of events.

Less complex. The cascade resembles a Gantt chart, a simple-looking tool, which creates a sense of lower complexity that is easier to apply.

Less practical. Because of one activity per zone, a larger TAD may require many sheets to print it and large wall space to post it.

Spine

One zone, multiple activities. The TAD printout allows multiple activities per zone.

Why called spine? Activities are symmetrically arranged around a central path, usually the critical path, resembling a spine of the network.

More complex. The appearance of the spine is much like any other network, something that looks scarily complex to some project managers.
More practical. Because of multiple activities per zone, you can print a larger TAD on a single sheet of paper and post on it on a small wall space.

Identify the Critical Path

Normally, a TAD shows a number of different paths—that is, sequences of dependent activities. The paths can be used in two ways to find the critical path. First, you can visually find a path composed of activities without float—no complex calculations necessary. Among all network users, this convenience is available only to those who practice TAD. Adding up the times for all activities in a path (as we did with CPM) will tell how long it is. As a reminder, the critical path is the longest path in a TAD, which indicates the minimum time necessary to complete the entire project. Second, you can find the critical path with the forward/backward pass procedure and calculate total and free float, as explained in the CPM section featured earlier in this chapter.

Using a TAD

Like any network diagram, the TAD's original targets were large, complex, and cross-functional projects. The TAD is well suited to such projects because of the ease with which it handles a large number of activities and their intricate dependencies, directing our attention to the most critical activities. While it is still applied for this purpose, a large number of project managers have used the TAD for medium-size and small projects (see "Tips for TAD"). In this case, a TAD is typically drawn in a cascade and often called a "Gantt chart with links." Perhaps more than anything else, this format facilitated the growing popularity of the TAD.

A sizeable TAD can be used along with a Gantt chart to provide focus on the day-to-day project work. In particular, we can take out from TAD those activities due in the next week or two, show them in the Gantt chart format, and have their "owners" use them as short-term outlook schedules. This provides a balance in focusing on both the big picture of the project with TAD and daily work details with Gantt charts.

Tips for TAD

- If you need to fast-track your schedule, use SS with or without lags. Be prepared to see and manage 40 to 50 percent critical activities in the schedule.
- Building on the similarity between a cascade-formatted TAD to a Gantt chart, spread the use of the TAD in all smaller projects. This will significantly enhance the quality of scheduling.
- Add major milestones to a TAD to help vital events serve as beacons in the sea of activities.
- Insist on the use of template TADs to boost quality and productivity of scheduling.

Benefits

The TAD offers a unique benefit not available to any other network diagram: the ability to read directly off the schedule's time scale when the project and each activity starts and ends, as well as the total float. Like other network diagrams, the TAD helps identify the total completion time, understand the sequencing of activities, ensure resources when necessary, monitor those that are critical, and measure progress (or lack of it).

The TAD's graphical appeal and intuitive logic provide additional utility to a project manager. The TAD's unambiguous sequence of work, supported by the timescale, is easier to clarify than any other network diagram. Data calculations are not difficult and can be handled readily and quickly by personal computers. Additionally, the TAD exhibits the dependencies between constituent activities of a project simply and directly. This helps fathom the order of activity execution.

Perhaps the highest value comes from the TAD's focus on priorities; TAD directs our mind on the vital few activities of critical importance to the project completion date. The outcome is higher accuracy and, later, precision of schedule control.

THE CRITICAL CHAIN SCHEDULE

Introduced in 1997, the critical chain schedule (CCS) is a relatively new tool to the world of project managers.[18] A CCS is a network diagram that strives for accomplishment of drastically faster and more reliable schedules (see Figure 6.7). It uses several unique

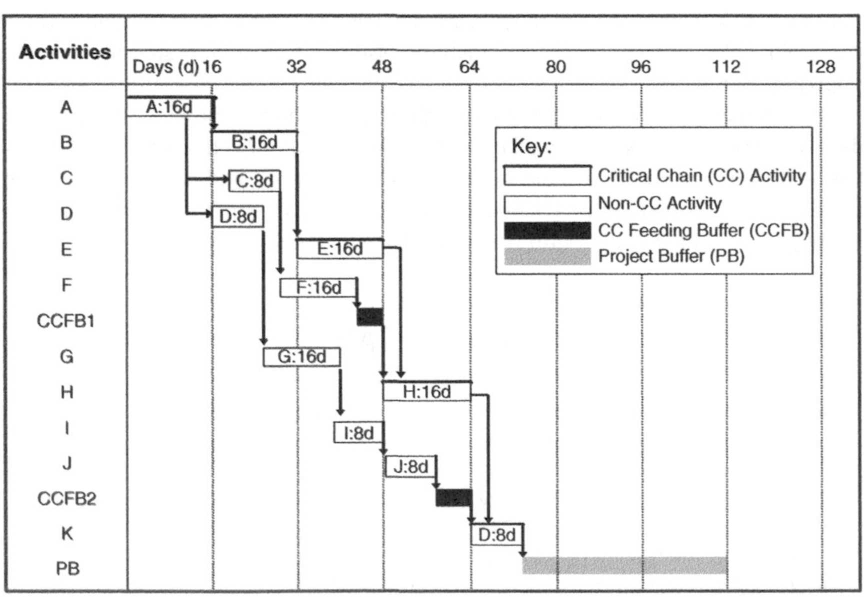

Figure 6.7: Example Critical Chain Schedule

approaches. First, the CCS focuses on the critical chain, the longest path of dependent events that prevents the project from completing in a shorter time. Unlike the critical path, the critical chain never changes. Second, its activity durations are estimates with 50 percent probability. For this reason, they are significantly shorter than those used in other scheduling tools, which are often with 95 percent probability. Third, in contrast to the critical path, the critical chain is defined by the resource dependencies. Fourth, buffers are built to protect the critical chain during the course of project implementation.

Developing a Critical Chain Schedule

Because of the inherent challenge of faster schedules that it seeks to make possible, a CCS's quality is even more dependent on the depth and degree of definition of inputs than other schedule development tools.

While the scope, responsibilities, and schedule management system will provide information about the what, and the who elements, as well as how to schedule project activities faster, the real emphasis is on the CCS's requirement for dedicated team resources, meaning that team members work full-time on one project only. Because of this, the logic goes that members of the dedicated project teams are more productive than members who are shared by multiple project teams. A reason for this is that the switching time cost created by one's work in multiple projects is eliminated as discussed in the previous section. Although this is generally true, there are some exceptions. A study found that when a team member who is focused on a single project is assigned a second one, productivity often increases a bit because the team member no longer has to wait for the activities of other members working on the initial project (see Figure 6.8). Rather, the team member can shift his or her attention to the second project.[19] When a third, fourth, and fifth project is added, however, the productivity plummets rapidly, and the team member becomes a bottleneck of all projects he or she is involved in. This is why the CCS approach insists on using dedicated teams.

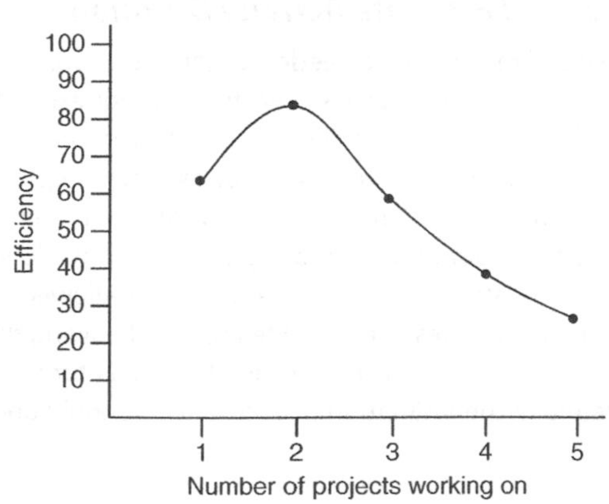

Figure 6.8: Productivity of Multiproject Team Members

Determine the Level of Detail and Identify Activities

The *number* of activities in a CCS is closely related with the *size* of activities. Therefore, choosing to have approximately 100 or 300 or 500 activities will help determine how large individual activities will be. To illustrate this point, consider one company's golden rule: Large projects, including between 5,000 and 10,000 person-hours, will have around 500 project activities that range in duration from two to four calendar weeks. Not only does this clearly tell everyone that neither 180-activity schedule nor 15-week activities are tolerated, but it also spells out the company's belief about the right level of detail. Given the complexity and size of the project, such level of detail provides sufficient information to manage the project without making it unnecessarily burdensome and time demanding.

Once the decision has been made about the level of detail, these actions should be taken:

- *Brainstorm and identify activities that are necessary to complete in order to finish the project.* As with other scheduling tools, resort to the WBS for the activity identification. In this process, disregard how large the activities are; rather, ensure that all necessary activities are on the list.
- *Go back to the chosen level of detail.* If the list of activities is below the intended number of activities, continue breaking down larger activities; if it is over the target number, combine similar activities to reach the desired goal.

Sequence Activities

Sequencing means arranging the activities in a logical workflow and identifying the dependencies between the various activities. Deep knowledge of workflow is a prerequisite here. The principle of sequencing is to know that a preceding activity produces outputs that become inputs to an activity that follows. If the obtained diagram fails to observe the principle, it is likely that we are missing the logic of the project work, resulting in rework and delays in project execution.

Assign Resources and Estimate Activity Durations

Since people and material resources are needed to complete project activities, they dictate activity durations. Therefore, a natural starting point for estimating the durations is: "What resources do I need to successfully complete this activity?" The answer should provide the names of resources and work time for each to complete the activity—for example, 100 hours of work from a programmer. The key point here is that CCS uses a unique technique of activity duration estimating that does not allow for contingencies (see "When Estimating Durations, No Contingency Safety Allowed"). Considering that the critical chain approach requires dedicated teams, and knowing the company's work calendar—5 days a week, 10 hours a day—those 100 hours turn into 10 workdays or 14 calendar days. Naturally, the estimation of each activity should undergo this process.

Identify the Critical Chain

The critical chain is the longest path in the network diagram, considering activity and resource dependencies. Stated a different way, it is the sequence of dependent events that keeps the project from completing in a shorter time.

Add the Resource Buffers

Critical chain schedules always consider the resource constraints and include the resource dependencies that define the overall longest path. Practically, this is handled by adding resource buffers to protect the critical chain from unavailability of resources. Resource buffers are added to the critical chain only, do not take any time in the critical chain, and are termed *resource flags*. For example, any time a new resource will be used in a critical chain activity, we will add a resource buffer. This signals to the project manager and resource provider when to make the resource available to work on a critical chain activity. Since timely resource availability is critical to the rapid execution that CCS advocates, some companies use incentives to reward behavior of early delivery of activity outputs and standby time of resources.[20]

Create a Project Buffer

Unlike other schedule development tools, the CCS uses a novel concept of the project buffer. Its purpose is to protect the project completion date by aggregating risk contingency time in the form of the project buffer at the end of the critical chain (for management of the buffer, see the Buffer Chart section in Chapter 12). There are several methods to determine the buffer duration. One of them is to divide the duration of the critical chain by two (called the "50 percent buffer sizing rule"). The buffer is used to absorb uncertainty or disruptions that may occur on the critical chain and has no work assigned to it (see Figure 6.7).

Create Feeding Buffers

Protecting the critical chain with the project buffer is not enough. There is a significant risk that activities that are not on the critical chain but feed into it may slip to the point of pushing out the critical chain. To protect the critical chain from the risk, we can aggregate contingency time at all points where noncritical activities feed into the critical chain (see Figure 6.7). These contingency times are termed *critical chain feeding buffers*. During the project implementation, these buffers will be used to absorb uncertainty or disruptions that may occur in noncritical chain activities. To determine these buffers, use one-half of the sum of the activity durations in the chain of activities preceding the buffer. No work is assigned to the buffers.

When Estimating Durations, No Contingency Safety Allowed

Most project managers tend to include contingency time into each activity estimate without specifying it. The reason is simple: add the safety time. The CCS strives to eliminate the safety. Here is what it means. Figure 6.9 shows a typical distribution of activity time performance. The solid line (the left ordinate) tells us the incremental probability of a given activity duration time on the *x*-axis. The dotted line indicates the cumulative probability (the right ordinate) that the activity will be finished in a time less than or equal to the activity duration time on the *x*-axis.[21]

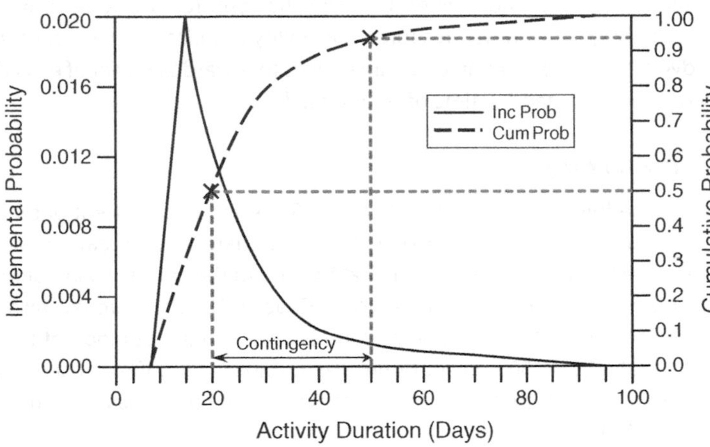

Figure 6.9: A Typical Distribution of Activity Time Performance

When project managers include contingency time within an activity, they really go for 95 percent probable estimate (cumulative probability). As the cumulative curve shows in the figure, it is duration equal to or less than 50 days. Without the contingency time, the duration is less or equal to 20 days. That is a 50 percent probable estimate. The difference between the 95 percent probable estimate and the 50 percent probable estimate is contingency time, 30 days in this example. To avoid excessive activity duration, and speed up the schedule, the CCS eliminates the contingency time, using only 50 percent probable estimates.

Using the Critical Chain Schedule

The most appropriate application of the CCS is for a dedicated project team seeking a significant reduction of the project cycle time in a company with an outstanding performance culture. The only job of this team is their project. Equipped with all necessary resources, the team operates in a company whose performance culture focuses on exceeding its customer expectations, creating maximum value for its shareholders, and providing strong growth opportunities to its employees (see "Tips for Critical Chain Scheduling").

Tips for Critical Chain Scheduling

- Use CCS in important projects that can afford a dedicated team.
- Apply CCS in companies that are in a time-to-market race, always striving to shave off their cycle times.
- Support CCS with performance measurements that promote behavior of transferring an activity's output to the succeeding activities as early as possible.
- Deploy CCS where there exists a strong performance culture willing to take on 50 percent probable estimates.

Benefits

Beyond every schedule's purpose of having the project team understand the timetable for project activities and their personal time commitments, the CCS intends to improve the results of the project team. Hence, as experiences with the CCS indicate, the project should see considerable improvements in schedule and cost performance because the CCS:

- Is an important eye-opener. Simply, the CCS recognizes that the interaction between activity durations, dependencies, resource requirements, and resource availabilities has a major impact on project duration.[22]
- Protects a deterministic baseline schedule. This protection helps fight uncertainties by using feeding, resource, and project buffers to set a realistic project deadline. As a result, the CCS offers significant potential for radical acceleration of project completion times. Project managers at 3M and Lucent reported up to a 25 percent reduction in project cycle times when using the CCS approach.[23] For those in a rapid-project-cycle-time business, this may be a tool worth trying.
- Makes a case for truth in activity duration estimation.[24] In contrast to other tools without a mechanism to prevent project managers from building contingency safety into activities, CCS's "no contingency allowed" mechanism eradicates such tendencies.

THE HIERARCHICAL SCHEDULE

The hierarchical schedule is a multilevel schedule with varying amounts of detail at each level (see Figure 6.10). It is an effective tool to have in a project manager's PM Toolbox, particularly when the project manager is applying a rolling-wave scheduling technique

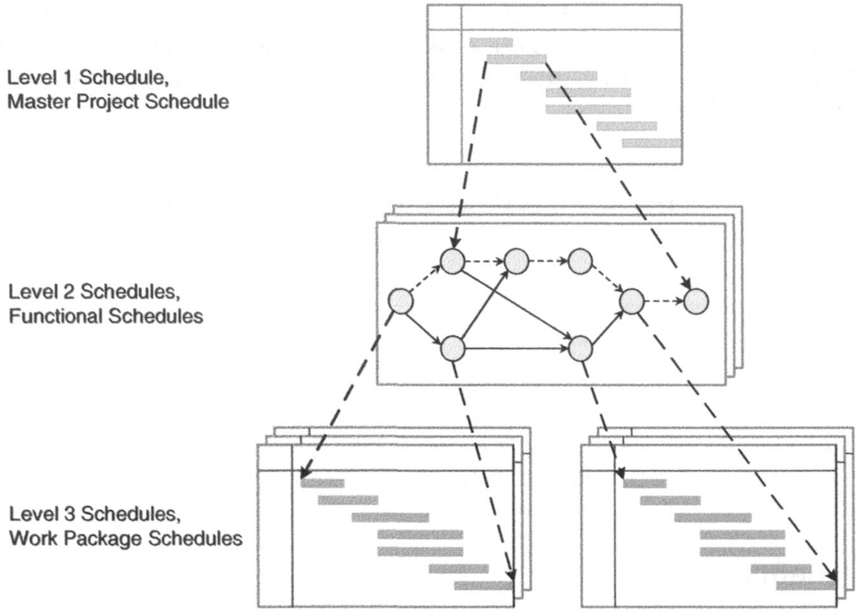

Figure 6.10: An Example Hierarchical Schedule

because it helps integrate the schedules of earlier and later phases. The hierarchical schedule also works well in multitiered scheduling techniques used on larger projects.

A hieararchy is created when activities on a higher-level schedule are disaggregated into several activities and sometimes entire schedules at a lower level. Typically, the schedules from various levels are connected at major milestones or events.

Constructing a Hierarchical Schedule

Developing hierarchical schedules hinges on the size of a project. A very large project can easily use three levels, while a medium-sized project cannot warrant more than two levels. To offer a sense of a more difficult situation, we will use a three-level hierarchy here. In developing the schedules, the rules for building a particular schedule type—whether a Gantt chart, network diagram, or milestone chart—should be applied.

As with all other project schedules and network diagrams, project scope provides information necessary for a good understanding of the project activities to be scheduled. Such scope information will be furnished to or developed by those who are responsible for certain work packages of the project. When scheduling the work packages, they will rely on the information about availability of resources and seek guidance from the particulars from the schedule management system.

Construct the Level 1 (Master Project) Schedule

Level 1 is a summary schedule of the project, which is usually a Gantt or milestone chart format. It is an outline that will be used throughout the project as a tool to report progress to top management. Since it is developed during project planning, it is considered to be an initial plan as well. Included in the schedule are only principal activities and key milestones from Level 1 or 2 of a WBS (the project is level 0). Everything in the schedule is roughly estimated. For example, overall timing of phases, required resources, major dependencies, and major events in the schedule are all roughly estimated. It is important to highlight events that need critical attention, such as material requirements, vital tests, and completion dates. Linking the development of the schedule with the definition of project objectives is a good strategy because this is the time when the purpose and implementation methods for the project are shaped. In this linking, you should be able to easily chart, rechart, and evaluate multiple alternatives of the schedule in order to select the most viable alternative. Because the schedule is rough in its nature, it should not be used for the total integration of all project phases. That is why you need a Level 2 schedule.

Figure 6.10 shows the master schedule for a project named OCI, consisting of eight work elements from Level 1 of the WBS, each with a milestone at its end. It is in a Gantt chart format and is used for progress reporting to the project governance board, an executive group responsible for the project performance.

For example, the first-level schedule (master project schedule) can be built of work elements from Level 1 of the WBS. Work packages from Level 2 of the WBS can be included into the second-level schedules (functional schedules because they are typically owned by functional units). Finally, constituent activities or task of the work packages (Level 3) would be used to develop the third-level schedules (work package schedules).

Construct the Level 2 (Intermediate) Schedule

The Level 2 schedule will explode activities from the master project schedule, scheduling them in more detail. For this, a common choice is a Gantt or network chart, sprinkled with milestones. This schedule is a middle management planning and control tool, generally to assign responsibilities for work packages (Level 3 of the WBS, for example). Clearly, activities in the schedule are not meant to provide daily or even weekly scheduling and directing of project work, except for the most critical activities. Still, it should be scheduled in sufficient detail to include major and minor milestones, crucial human resources, and sequencing and constraints in the project work. This enables you to scrutinize the structure of the project, dissect dependencies between various phases and milestones, and set boundaries within which shifting activities won't impact project completion.

Project OCI had several Level 2 schedules, each drawn as a time-scaled arrow diagram in the cascade format. The largest of them had nearly ten work packages from Level 2 of the WBS. Essentially, each Level 2 schedule was the functional schedule for a certain discipline—marketing, electrical group, optoelectronic group, software group, and so on.

Construct the Level 3 (Detailed) Schedule

This set of detailed schedules is intended to help execution level managers—work package managers, for example—in directing daily and weekly project work. Although it can be in the network format, a more frequent approach is to use a Gantt or milestone chart. Before getting to scheduling, you should size up available information, assess the size and complexity of the project, and weigh in experience and inclination of the involved project members. Then you should decide which of the following approaches to pursue for Level 3 schedules:

- Create a fully integrated schedule for the entire project.
- Build a complete schedule for each activity from the Level 2 schedule.
- Construct a separate detailed schedule for each phase as the project unfolds, and connect them via the Level 2 schedule.
- Ask each project participant to develop detailed schedules for activities in the Level 2 schedule that he or she is responsible for.

Whatever the choice, the schedule must lay down the day-to-day, week-to-week work that an organization needs to successfully execute and control. It goes beyond saying that the schedule needs to be rooted in available resources, established dependencies, and time targets approved by management.

The choice of the OCI project was to use the Gantt chart format to schedule in detail constituent activities of the individual work packages, keeping the number of activities per chart to less than ten (work package schedules). The total number of all activities in Level 3 schedules was slightly below 500. While the OCI project offers one example of how to structure the hierarchical schedule, many other approaches are possible (see "Milestone/CPM-Milestone Schedule Gets the Job Done").

Milestone/CPM-Milestone Schedule Gets the Job Done

In a six-month, $70m project, a semiconductor company identified a few major milestones to be presented to the executive governance board (Level 1 schedule). To direct the work and review progress weekly, the project management team relied on a CPM diagram with carefully weaved 200+ minor milestones (Level 2 schedule). Minor milestones from the diagram were grouped into separate working milestone charts and handed to the teams responsible for certain technical disciplines (Level 3 schedules). Including 40 to 50 milestones, each milestone chart was the key tool for doing work and reporting progress to the project manager. Rid of complex dependencies typical of the CPM diagram, the minor milestone chart provided clear and simple goals to go after.

Using a Hierarchical Schedule

Hierarchical schedules are used to confront challenges in two major project situations:

- *Rolling-wave scheduling*. When starting some projects, we only have information about an early phase, while details about later phases emerge as the project progresses. In response, at the start of the project, we can develop a high-level schedule encompassing the whole project and then build detailed schedules of the project's major phases as details become available. This approach is termed *rolling-wave scheduling* and is implemented via the hierarchical schedule method.

- *Multitiered schedule information for larger projects.* Since different levels of management have different jobs in a project, each level needs a different detail of schedule information. Different levels in hierarchical schedules provide those different details of schedule information.

Benefits

The use of hierarchical schedules equips project managers with the capability to integrate the scheduling of the earlier and later phases. Without the schedules, our scheduling and our attention would be focused on a piece of the project for which we have information, ignoring the whole project. This would be tantamount to a runner who sees terrain just in front of her, having no idea how long the run is (a mile or 26 miles) or, what major milestones lie ahead (e.g., a steep hill). Such a runner would have little chance to pace herself and successfully reach the finish line.

By using a hierarchical schedule, a project team is not forced to develop a schedule for activities for which they have little or no information at the current moment. Rather, they can build a flexible big-picture schedule, focusing on near-term activities first, and then add longer-term detail as the project progresses through the project cycle. However, the hierarchical schedule also has some pitfalls to be aware of, as described in the example titled "Challenges of Hierarchical Schedules."

Challenges of Hierarchical Schedules

Hierachical schedules are complex. Multiple-level scheduling techniques require well-established process, skills, and involvement and coordination of many project participants. This may make hierarchical schedules a bit baffling and cumbersome for participants, raising the resistance level to its use.

Constructing hierarchical schedules is also a time-consuming endeavor. Detailed scheduling of this type demands time, a resource lacking in too many organizations. For this reason, some project participants may resist the use of multilevel scheduling.

LINE OF BALANCE

Line of balance (LOB) is a tool for the scheduling and tracking of project progress designed for projects of a highly repetitive nature (see Figure 6.11). The LOB schedule displays the cumulative number or percentage of components or units that must be completed by a certain point in time for a schedule to be accomplished.

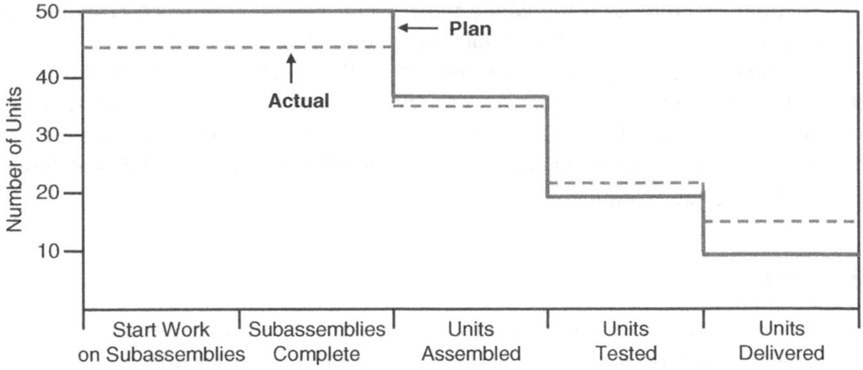

Figure 6.11: An Example LOB Schedule

By presenting the planned completion status at each phase versus actual completion status, LOB illustrates in a very visual manner whether the project is ahead or behind the plan. For executive use, it highlights the potential showstoppers and bottlenecks, urging action to remove the problem areas.

Developing a Line-of-Balance Schedule

As with other scheduling tools, quality information about the scope, responsibilities, available resources, and schedule management system will help users of the LOB to understand the *what* and *who* issues, along with resources and scheduling requirements. What makes the LOB schedule different from other scheduling tools is the need for a bill of materials (for production projects see "A Poor Bill of Materials Can Hurt"), lead times for procurement, and production norms. This information is used to determine the project detail and then the time line for production or construction, as shown in the upcoming discussion on developing an S chart for a multiunit project.

A Poor Bill of Materials (BOM) Can Hurt

Snap-All Corporation, a leader in mechanical fasteners, received a request for special delivery of its Lockmight Hinge System, with conditions that they meet a very aggressive delivery schedule. One of Snap-All's senior executives promised the customer a proposal the same day, asked for an LOB schedule to highlight bottlenecks and assess the likelihood of meeting the delivery date. The first step in preparing the LOB calculation was getting a BOM from someone who understood the parts buildup. Immediately, a manufacturing representative discerned that the BOM was obsolete and did not including all necessary parts. Updating the BOM and checking lead times to acquire the various components took several days. As a result, the executive could not keep his commitment to prepare a fast turnaround on the proposal, and Snap-All lost this business to a more agile competitor.

Set the Objective

The overall objective of the project must be understood in order to begin the LOB development effort. For example, consider a production or construction project aiming at producing a certain quantity of the end deliverables. A production example might include a series of special orders for 50 connection cables per agreed delivery commitments. To make the project more complex, the cables cannot be completed in one batch because of insufficient production capacity. Another project requirement is that the installation of the cable has to occur in 15 houses within the first month. In order to start developing a LOB schedule, the project manager must decide if the main objective of the project is the required production and delivery of the 50 cables, or the installation of the cables in the 15 houses. Understanding the primary objective informs how the project is defined.

Define the Project

The preceding example is a production or construction plan that may be formatted as a network diagram, Gantt chart, or milestone chart (see Figure 6.12). In the plan, we set the control points, which are key points in the production or construction process. Depending on the chosen format, these points are events in the network diagram, or the end of bars in the Gantt chart, or milestones in a milestone chart. They are used to measure the progress of the project.

The example that follows shows such a project for one unit only. Having a project with multiple units to produce or construct requires a project plan for all units. There are two possible scenarios for this. In the first scenario, you can produce all multiple units in one batch, assuming sufficient production capacity. Here, all quantities for one unit are multiplied by the number of units to arrive at a project plan that includes all units. The lead time for every unit would remain unaltered.

The fundamental premise of the second scenario is that the multiple units can be produced in several batches to accommodate for the insufficient production capacity. Considering that the lead time for each batch is equal to the duration of the one unit program, these batches can be fit into the production schedule. As a result, we obtain the plan for producing multiple units.

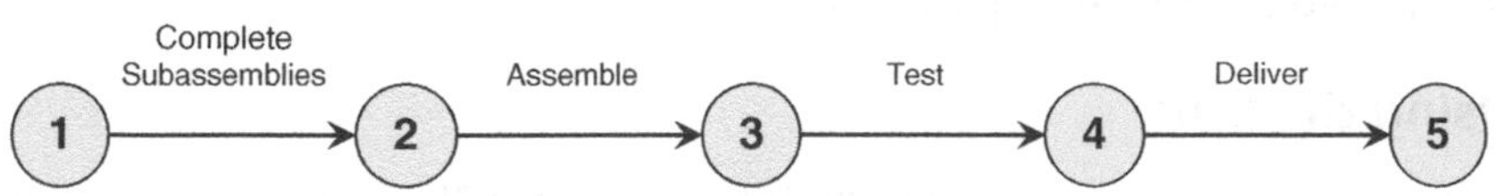

Events in L.O.B.:

1. Start work on subassemblies
2. Subassemblies completed
3. Units assembled
4. Units tested
5. Units delivered

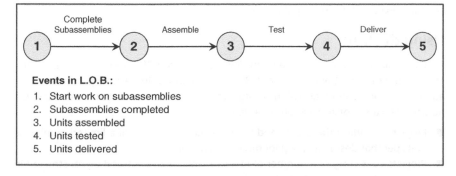

Figure 6.12: Project Plan for a Single Unit in a Multiunit Project

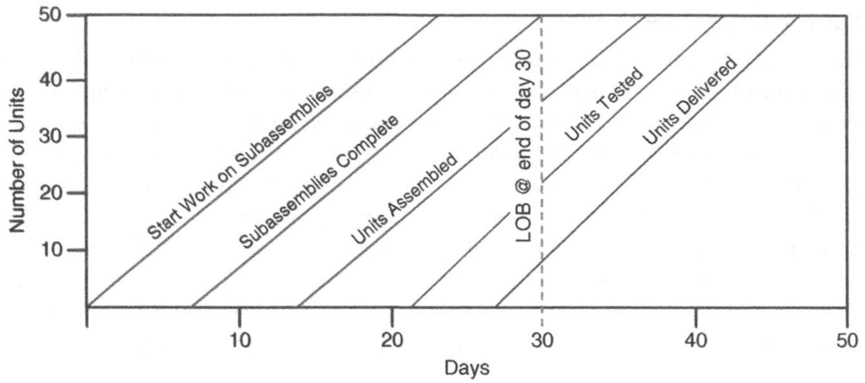

Figure 6.13: S Chart for a Multiunit Project

Develop an S Chart for the Project

In a further step, draw the project plan for multiple units on a graph called an S chart, showing cumulative deliveries against time scale (Figure 6.13) and also cumulative completion schedule for control points. Because of the linear production rate, S curves are straight lines in our example. Essentially, this signifies the planned completion of final units and their intermediate parts or phases. This is for the second scenario—several batches to accommodate for the insufficient production capacity.

Draft and Refine the Line of Balance

Draw a vertical line through the S chart. This is the LOB, a snapshot at a certain point in time (e.g., day 30) that indicates the cumulative number of components or units that are planned to be complete by that date in order to comply with the schedule. To track the progress against the plan, you can draw another line that depicts the cumulative number of actually completed components or units (the dashed line in Figure 6.13). For instance, in the example shown, the first two events are five units behind the plan, the third event is two units behind the plan, the fourth event is two units ahead of the plan, the fifth event is six units ahead of plan, and there is no real bottleneck. All of this is fine and helpful, but most of the time, the first draft may need refinement, including changes necessary to clean up the LOB.

Using the LOB

Although different versions have been developed since LOBs were introduced in the 1950s, more than anything else, the LOB is a tool for low-volume, new-production situations requiring coordination of the design and small-scale manufacturing or construction. Its main areas of application are as follows:

- Projects including the design and then the manufacturing of a limited number of units per that design (e.g., a pilot run for integrated circuits).
- Projects consisting of a number of identical units manufactured or constructed in sequence (e.g., construction of a multiunit housing project).
- Projects of one-off nature (e.g., shipbuilding).

Benefits

By presenting the planned completion status at each phase versus actual completion status at these phases, the LOB shows to the project team in a very visual manner whether the project is ahead or behind the plan.[25] For executive use, it highlights the potential showstoppers and bottlenecks, urging action to remove the problem areas (see Tips for LOB under "Focus on Critical Components").

The LOB also offers value to the project manager in its visual, focused, and concise nature. It provides a visual display of both the planned and actual progress of repetitive activities and provides focus on the project elements that require critical attention. For project stakeholders, the LOB provides the ability to assess project progress at a glance, a quality every reporting tool strives to provide.

If the prerequisites listed previously are prepared and an adequate software program is available, building an LOB diagram with 20+ components can be completed between one and two hours. This time will increase when more people are involved because more time is necessary for their communication.

Focus on Critical Components

In many companies, the prevailing view is that not all components are created equal. From previous experience, some components are known as potential showstoppers in the delivery schedule, while others are not. This is because some components bring with them inherent constraints—for instance, very precise manufacturing tolerances. To respond to different levels of risk related to the constraints, a project manager can use the LOB for monitoring the high-risk components known to cause delay. The point here is this: Keep your eye on the critical few; don't bother with the trivial many.

Tips for LOB

- Focus on critical components or units.
- Use LOB to detect bottlenecks.
- If you have a material requirements planning (MRP) system, use its power to define the project and develop an S chart and LOB.
- Show the LOB to executives when you need their support to solve a problem.

CHOOSING YOUR SCHEDULING TOOLS

Multiple scheduling tools are presented in this chapter that lead to the question, which one or ones are the most appropriate to select and use? Such a decision, of course, depends on your project situation. To help narrow the options, in Table 6.1 we list a set of project situations and indicate how each situation favors the use of the various

Table 6.1: A Summary Comparison of Schedule Development Tools

Situation	Gantt Chart	Mile-stone Chart	Critical Path Method (CPM)	Time-Scaled Arrow Diagram	Critical Chain Schedule	Hierar-chical Schedule	Line of Balance
Small and simple projects	✓			✓			
Short training time	✓	✓					
Focus on highly important events		✓					✓
Increase goal orientation		✓					✓
Large, complex, and cross-functional projects			✓	✓	✓		
Focus on top-priority activities		✓	✓	✓	✓		✓
Strong interface coordination needed			✓	✓	✓		
Need time scale in complex projects				✓	✓		
Very fast schedule			✓		✓		✓
Multitiered schedule for large projects						✓	
Rolling-wave scheduling needed						✓	
Short-term outlook schedule in large projects	✓					✓	
Schedule to support resource planning	✓			✓	✓		
Scheduling, tracking of repetitive projects							✓
Use of templates desired	✓	✓	✓	✓	✓	✓	✓

scheduling tools. Identifying the situations that correspond to your own project is the first step. If the situations described do not describe your particular project situation, brainstorm more situations in addition to those listed, marking how each favors the tools. The tool that has the highest number of marks for identified situations becomes the tool of choice. Note, however, that more than one tool can be used to support any particular project, since some of them complement each other rather than exclude each other. A careful study of the material covered in this chapter will help you determine when this is practical.

References

1. Powers, J. R. "A Structured Approach to Schedule Development and Use." *Project Management Journal* 19 (5): 39–46, 1988.
2. Berkun, Scott. *Making Things Happen: Mastering Project Management* (Sebastopol, CA: O'Reilly Media, 2008).
3. Sipos, A. "Multiproject Scheduling." *Cost Engineering*, 2010. 13–17.
4. Gould, F., and Nancy Joyce. *Construction Project Management* (Upper Saddle River, NJ: Prentice Hall, 2013).
5. Lewis, J. *Mastering Project Management: Applying Advanced Concepts to Systems Thinking, Control & Evaluation, Resource Allocation* (New York, NY: McGraw-Hill, 2007).
6. Segur, P. *Napolean's Russian Campaign* (Alexandria, VA: Time-Life, 1958).
7. Bowen, H. K. *Project Management Manual* (Boston: Harvard Business School Press, 2007).
8. Project Management Institute. *A Guide to the Project Management Body of Knowledge*, 5th ed. (Drexell Hill, PA: Project Management Institute, 2013).
9. Handfield, R. B., Gary L. Ragatz, Kenneth J. Petersen, and Robert M. Monczka. "Involving Suppliers in New Product Development." *California Management Review* 42 (1): 59–82, 2009.
10. Milosevic, D. "Case Study: Integrating the Owner's and the Contractor's Project Organization." *Project Management Journal* 21 (4): 23–32, 2000.
11. Crawford, M. "The Hidden Costs of Accelerated Product Development." *Journal of Product Innovation Management* 9 (3): 188–199, 2002.
12. Miller, R. W. "How to Plan and Control with PERT." *Harvard Business Review* 40 (2): 92–102, 1982.
13. Pinto, Jeffrey K. *Project Management*, 2nd ed. (Upper Saddle River, NJ: Prentice Hall, 2013).
14. Bowen, 2007.
15. Adler, P. S., Avi Mandelbaum, Vien Nguyen, and Elizabeth Schwerer. "Getting the Most out of Your Product Development Process." *Harvard Business Review* 74 (2): 134–152, 1996.
16. Tobis, M. *Managing Multiple Projects* (New York, NY: McGraw-Hill, 2002).
17. Rubenstein, A. M. "Factors Influencing Success at the Project Level." *Resource Management* 16: 15–20, 1979.
18. Goldratt, E. *Critical Chain* (Croton-on-Hudson, NY: North River Press, 1997).

19. Wheelwright, Steven C., and Kim B. Clark. *Revolutionizing Product Development: Quantum Leaps in Speed, Efficiency, and Quality* (New York, NY: Free Press).
20. Leach, L. P. *Critical Chain Project Management* (Norwood, MA: Artech House, 2014).
21. Sipos, 2010.
22. Herroelen, W., and R. Leus. "On the Merits and Pitfalls of Critical Chain Scheduling." *Journal of Operations Management* 19(5): 559–577, 2010.
23. Leach, 2014.
24. Steyn, H. "An Investigation into the Fundamentals of Critical Chain Project Scheduling." *International Journal of Project Management* 19(6): 363–369, 2009.
25. Pinto, Jeffrey K. *Project Management*, 2nd ed. (Upper Saddle River, NJ: Prentice Hall, 2013).

7

COST PLANNING

Foremost on the list of questions in the mind of a project sponsor is the question, "How much will this project cost the organization?" The project sponsor and other senior management stakeholders look to the project manager to answer this question. Furthermore, they look to the project manager to establish a cost baseline to enable the monitoring and control of cost throughout the project life cycle.

Cost planning is the process by which organizational goals are translated into a plan that specifies the allocated resources, the selected estimation methods, and the desired schedule for achieving the goals of the project. The cost-planning process culminates in the establishment of a project budget that represents the organization's (or a client's) investment in the project.

Although we focus attention in this chapter on cost planning at the project level, all levels of the organization are involved in the cost-planning process, as described in Table 7.1. As one can see from the table, creating the detailed cost estimates for the projects is at the center of the organizational cost planning process.

Project cost estimating can be accomplished in various ways as dictated by the needs and policies of an organization, as well as by a particular project situation. Since project managers need to be able to respond to and address varying organizational and project situations, they need to have various cost planning tools in their PM Toolbox. This chapter describes some widely used project cost planning tools.

THE COST-PLANNING MAP

The cost-planning map is a tool for establishing a systematic approach to cost planning in projects (see Figure 7.1). The cost-planning map spells out steps and substeps a project team needs to go through in order to make choices necessary to develop basic definitions, terminology, estimate types, estimating tools, and the process for cost planning. When such choices are seamlessly integrated, the cost-planning map can help establish a culture of cost consciousness within a project.

While any project can find value in using a cost-planning map, organizations with large projects and those with a constant stream of small and medium projects may benefit most. The value of a cost-planning map is in the careful scripting and orchestration of cost-planning tasks that leaves no ambiguity as to what a certain cost estimate type

Table 7.1: Organizational Cost Planning

Sequence	Organization Level	Responsibility
1	Senior Management	Set organizational goals, establish organizational investment budget
2	Functional Management	Establish budget for a functional unit
3	Project Management	Create detailed cost estimate for projects
4	Functional Management	Select projects for functional unit, manage project investment versus functional budget
5	Senior Management	Ratify project selection, balance overall project investment to organizational budget and goals

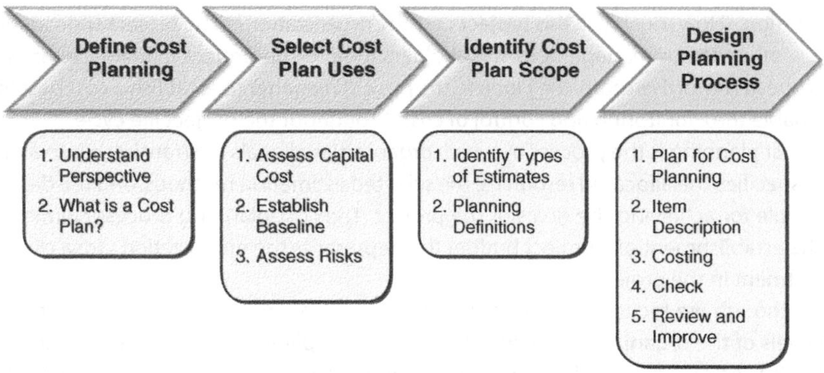

Figure 7.1: An Example Cost-Planning Map

is and how to develop it. This significantly reduces risk of poor cost planning and misuse of a company's resources.

Developing a Cost-Planning Map

Project cost planning is an effort fraught with risks that, if not addressed, may result in serious consequences. To address the risks and avoid the consequences, the deployment of a cost-planning map will establish a well-synchronized and integrated sequence of steps and substeps that are described in continuation.

Prior to deploying a cost-planning map, a project manager should have knowledge about his or her organization's financial policies, financial rate structure, and project staffing policies.

Financial policies dictate the design of the primary elements of the cost-planning model. For example, answers to questions such as which types of cost estimates will be used and with what purpose in the project process depend on the financial policies of the organization. Another example is knowledge of current labor and overhead rates. Similarly, when performing resource planning, you must consider organizational policies regarding staffing and outsourcing, a key element of the costing substep in the cost-planning model.[1]

Define Cost Planning

When configuring a cost-planning model, you should start with two simple questions: "Who is the cost planner?" and "What is a cost plan?" The former really asks about the perspective of the planner, for example, whether the planner is the owner of the project, or perhaps a contractor, and also whether the planner is experienced or just a novice. The approach to developing a cost-planning model will greatly depend on the perspective, experience, and organizational culture in which the planner plans. Think, for example, of a company developing a new product. In developing a manufacturing cost estimate for the product, the company will have to consider its manufacturing process, plant, materials procurement strategy, design for manufacturability approach, and so forth. Each of these will contribute to cost. If manufacturing were outsourced, the contractor is likely to have a different manufacturing process, plant, and materials procurement strategy, all resulting in a different cost planning approach.

Regarding the second question, the project cost plan typically includes a cost estimate and a cost baseline. Take, for example, the estimate. It is an assessment, based on specific facts and assumptions, of the final cost of a project.[2] This assessment and its results will heavily hinge on factors such as accuracy of scope, quality of available estimating data, stage of the project, time allowed for the estimate, estimator's perspective and experience, desired accuracy, available estimating tools, and so on. Consequently, by defining these factors, you will define the makeup of your cost-planning map. The relevance of asking these two primary questions is in getting answers that will help shape the cost-planning map for specific situations.

Select Cost Plan Uses

Once cost planning is defined, determination on how the cost plan will be used must be established (Step 2 in Figure 7.1). Three primary uses are common: to assess the capital cost, to establish a baseline, and to evaluate risks and productivity. As an assessment tool, a cost plan may serve multiple purposes—for instance, substantiating a request for capital appropriations or borrowing funds. In other situations, a cost plan can act as a basis for a proposal, bid, or contract document. Still in other situations, a cost estimate, an element of the cost plan, is compared with other cost plans to validate their accuracy and increase the confidence level. Typically, this type of estimate is called an *independent cost estimate*, or in some companies' parlance, a *shadow estimate*.[3]

The second capability of a cost plan, and also the second substep in this step, is to help establish two baselines—a schedule and a cost baseline. As discussed later in this section, part of developing a cost estimate is identifying necessary resources such as hours of effort that are necessary to complete project activities. This is typically performed hand in hand with scheduling, so that the resource hours of effort may be turned into activity durations, thus setting a schedule baseline. By combining the scheduled activities with their costs, you can develop a cost baseline, also known as a time-phased budget.

The third substep, evaluating risks, aims at establishing a reasonable amount of cost estimate to allow for changes that are likely to occur and drive additional unplanned cost. This amount, called *contingency*, practically reflects project risks in the cost plan.

The motivation for careful analysis of risks and subsequent identification of the contingency amount is usually driven by attempts to lower costs as much as possible. Risks, as we discuss in Chapter 14, may be related to any performance, schedule, and cost uncertainties in work packages. Once cost planning has been defined and the purpose of the cost plan selected, the stage is set for the next step to decide what will be in the cost plan.

Identify the Cost Plan Scope

Vital to this step is the determination of the cost estimate type you want to use in your practice of project cost planning, as well as what exactly the estimates will include. There are many types available, but the three most often used are: (1) order-of-magnitude, (2) budget, and (3) definitive estimates.[4] As Table 7.2 indicates, they differ in many attributes: their purpose, accuracy, cost of preparation, information they require, and type of estimating tool they employ.[5] Each of the estimates can be used as a basis to develop the second element of a cost plan—cost baseline or time-phased budget.

Cost estimates are typically expressed in currency units, such as dollars, enabling an easy comparison across and within projects.[6] In contrast, projects in some industries favor estimates in labor hours, lumping together labor hours of different types of expertise. This is an acceptable practice as long as it doesn't prevent comparison across and within projects. Also, the practice of providing estimates in multiple units of measure when management control requires so is acceptable.

Understanding what exactly an estimate and cost baseline mean calls for concrete definitions of their components and other cost-planning terms. This is why it is advised that the second substep be developing cost-planning definitions, such as those in the "Examples of Cost Estimating Definitions" that follows. They will provide a shared language for all participants to communicate in the cost-planning process that will be designed in the next step.

Table 7.2: Types of Cost Estimates

	Order of Magnitude	Budget Estimates	Definitive Estimates
Use	Feasibility study, project screening, budgeting and forecasting	Budgeting and forecasting, authorization (partial or full funds)	Authorization (full funds), bids and proposals, change orders
Accuracy	−30%, +50% before contingency	−15%, +30% before contingency	−15%, +15% before contingency
Cost of Preparation	0.04% to 0.15% of total project cost	0.15% to 0.60% of total project cost	0.45% to 2% of total project cost
Information Required	Size, capacity, location, completion date, similar projects	Partial design, vendor quotes	Specifications, drawings, execution plan
Estimating Tool	Analogous, parametric	Parametric, bottom-up	Bottom-up, minor parametric
Also Called	Global, conceptual, ballpark, guesstimate, judgment	Scope, sanction, authorization, preliminary, semidetailed	Detailed, control, final

Examples of Cost-Estimating Definitions

Direct cost. An item of cost, or the aggregate of items, that is identified specifically with the project. These costs, such as labor, materials, and travel are charged directly to the project.

Indirect cost. The cost of items such as building usage, utilities, management support labor, services, and general supplies is not easily or readily allocable directly to a project. Indirect costs are accrued and charged to overhead accounts, the sum of which is applied as burden.

Fixed cost. A cost that does not vary with usage. An example may be a database server used for testing on a project.

Variable cost. Expenses that vary according to use, such as the number of hours worked by a person assigned to do tasks on a project.

Most probable cost. This is the cost most likely to occur, which is made up of all the itemized known items and a contingency estimate that together invoke a 50 percent degree of confidence.

Range of accuracy. This is a prediction of the least expected and highest expected cost relative to the most probable cost. Higher quality of estimate, better scope definition, lower project risks, fewer unknowns, and more accurate estimate pricing will lead to a better range of accuracy.

Contingency. An allowance added to an estimate to cover future changes that are likely to occur for unknown causes or unforeseen conditions. Contingency can be determined through statistical analysis of past project costs or from experience in similar projects.

Design the Cost-Planning Process

This is the process that you will use to develop any of the estimates and related cost baselines discussed later in this chapter—analogous, parametric, and bottom-up. Of course, the process for each one of them will differ in terms of level of detail, but the principle steps will be the same. To enable this, you need to design a proper cost-planning process, consisting of several substeps (Figure 7.1). To begin, preplan how you intend to perform cost planning. Although this may sound like an excessive dose of paperwork, in reality it can reduce the total effort for cost planning while minimizing rework. Several specific items are the focus of preplanning. First, thinking through who the end users of estimates are, and with what purpose, may help you select appropriate estimating format and forms. Knowledge of the due date for the estimate and details of the estimate review are crucial to scheduling cost planning work and submitting an estimate of the desired quality. Also, as part of preplanning, you may have to determine the cost of preparing the cost plan and inform the end users.

Developing item descriptions is the second substep, one that may have more impact on the estimate quality than any other factor (other than the contingency estimate). But what exactly are item descriptions? These are descriptions of work task that we want to develop an estimate for. Typically, a complete item description should include several elements. Begin with a quantity and applicable measurement unit, followed by a physical description of the item in as much detail as possible. Continue by stating item scope boundaries that clarify any ambiguities or assumptions, and document any diversions from conventions and standards. Add sources of estimating data (e.g., standard labor rates for each job category).

Computing an estimate for an item is the third subcategory. Within the core of costing is an estimating algorithm or formula that processes project information—for example, item description and a source of cost data for both direct and indirect cost—into costs. Typically, these formulas or algorithms are called *cost-estimating relationships*. Each cost-estimating tool relies on a different cost-estimating relationship. For instance, when number of quantities to produce, unit production rates, and an hourly rate of labor are available, a bottom-up estimate may use the following cost-estimating relationship to calculate labor cost:

$$\text{Labor cost} = \text{Quantity} \times (\text{hrs/unit}) \times (\$/\text{hr})$$

$$\text{Labor cost} = 200\,\text{articles} \times (5\,\text{hrs/article}) \times (\$80/\text{hr})$$

$$\text{Labor cost} = \$80{,}000$$

Again, which cost-estimating relationship for labor cost calculation is used will depend on the type of estimate being prepared.

Labor, materials, and equipment can be estimated by cost estimation relationships using ratios, parameters, cost chunks, or multiplication methods, as illustrated later in this chapter. When costing for individual items is complete, direct costs for all items are often totaled separately from indirect costs, and possibly by categories of work. If requested, this is the time to translate the estimate into a cost baseline.

Costing needs to be checked, which is the fourth substep. This involves validation of calculations, verification of estimating data sources, and peer reviews. With checking done, you can move to the fifth substep—review and improve. Management needs to review the estimate because they are responsible for supervision of the estimate preparation and typically can spot major problems. Then, the estimate can be issued following the principles of sound document management. The cost-planning process does not end here. Rather, it ends when the project is complete. At that time, all actual costs are collected, analyzed, and compared with the cost plan, and historical data is updated. The essence of the cost-planning process is summarized in the example titled "Tips for Cost Planning."

Tips for Cost Planning

Know your user. Ask questions to clarify their needs, item descriptions, and project scope.

Follow the cost planning process. Don't skip process steps. If the process doesn't work, change it.

Go beyond a "number-cruncher" mentality. Understand the big picture and philosophy of the project and its customer.

Document everything. Include assumptions, references, sources, scope exclusions, and so on.

Leave an audit trail. Audits enhance quality of the estimate and demonstrate a process was followed.

Document changes. The estimate you have originally planned for is almost certain to change. Record the change and maintain document revision control.

Create involvement and buy-in. Make experts from each performing functional department part of estimate preparation; after all, they have to live with it during project execution.

Using the Cost-Planning Map

While any project can find value in using it, organizations with large projects and organizations with a constant stream of small and medium projects may benefit most from the cost-planning map. Consistency and discipline in cost planning that the cost-planning map can generate is of vital importance to these users.

Building a cost-planning map is a significant time commitment. In organizations with large projects that are also complex and resource-intensive, heavy involvement of experts from various functions—technical, financial, accounting, for example—is typical, often resulting in hundreds of resource hours required to construct a quality cost-planning map. Developing a cost-planning map in organizations with a constant stream of small and medium projects is less time consuming for lower-complexity projects, although it may still take tens of resource hours.

Benefits

The value of a cost-planning map is in the clarity of the direction that it provides to project teams. Through a careful scripting and orchestration of cost-planning tasks, a cost-planning map leaves no ambiguity as to what a certain cost estimate type and cost baseline are and how to develop them. This significantly increases the consistency of project cost planning and reduces risks of poor cost planning and misuse of company resources.

The cost-planning map also provides structure that increases repeatability from project to project, as well as the quality of each project estimate.

ANALOGOUS ESTIMATE

An analogous estimate is the derivation of a project cost estimate based on the actual cost of a previous project or projects (*analogous* or *source project*) of similar size, complexity, and scope.[7] The estimators may use historical data, or rules of thumb that are modified to account for any differences between the estimated project and analogous project(s). An example of the analogous estimate is illustrated in Table 7.3, while basic features of the estimates are shown in Figure 7.2.

An analogous estimate is generally applied when there is a lack of detailed information about the project. Typically, this is the case early in the project life cycle.

Table 7.3: An Example Analogous Estimate for a Software Project

1	2	3	4	5	6	7
Item	Analogous Size (KLOC)	Analogous Productivity Factor (LOC/Person-Month)	Analogous Effort (Person-Month) 2/3	Target Size (KLOC)	Target Productivity Factor (LOC/Person-Month)	Target Effort (Person-Month) 5/6
1	1	100	10	0.8	80	10.0
2	2	50	40	2.5	40	62.5
3	2	200	10	2.5	160	15.6
4	1	100	10	1.0	80	12.5
5	1	50	20	1.0	40	25.0
Totals	7		90	7.8		125.6

KLOC = thousand lines of code; LOC = lines of code.

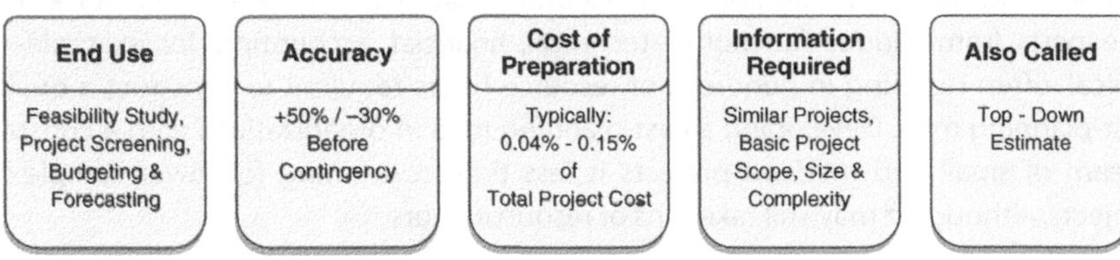

Figure 7.2: Basic Features of Analogous Estimates

Developing an Analogous Estimate

In general, the process of developing an analogous estimate follows the steps previously defined in the cost-planning map. Specifics will vary to reflect the nature of the analogous estimation.

The quality of an analogous estimate is highly dependent on sufficient information about the project scope, historical information about previous projects, resource requirements, and resource rates.

Identifying the target project and analyzing its scope ensures that the project being estimated is understood. For the estimate to develop, however, we need an analogous project, which will be extracted from a historical database of previous projects with similar features. Resource requirements and rates are necessary to express the estimate in appropriate units.

Prepare the Estimate

The starting step involves working through the preplanning specifics such as who are the end users of the estimate, the purpose of the estimate, estimating format, list of contributors and their roles, and available resources for creating the estimate. What follows is a firm understanding of a project's scope, size, and complexity features. In our example in Table 7.2, the scope of the target project is broken down into five major items (column 1), each with the targeted size (column 5). Each item may be a key feature of the software product. Now we can go to the database of previous projects with similar features to search for projects with similar size (scope). The most appropriate project (or projects) is selected as the analog. The mapping of analogous features to the target project is fairly straightforward because the two projects share a common set of items.[8] Our example has chosen one analogous project with the same five items. Analyzing actual data about the analogous project indicates size and productivity (columns 2 and 3, respectively), as well as the effort for the completion of each of its items in column 4, essentially an analogous cost estimating relationship. Then we transfer the information from the analogous project to the target, adjusting it for analogical elements that are not in correspondence with the target project. Specifically, for item 1 in our example, the project team is less experienced and their productivity (column 6) is judged to be 0.8 (judgmental factor) of that for the analogous project team (column 3). Applying a cost estimation relationship that divides an item size value (column 5) by the productivity factor (column 6) yields an item estimate (column 7) expressed in resource hours. To convert to monetary terms, we can multiple the hours by the resource rates.

A sum of all the estimated items is equal to the total project estimate. Crucial in this effort is the ability of the estimators to identify subtle differences in the source and target items and estimate the cost of a target item based on the source item that is analogous.[9] Checking, reviewing, and improving the estimate are the final steps in developing an analogous estimate.

Using an Analogous Estimate

An analogous estimate is a tool of choice when there is a lack of detailed information about a project. Typically, this is the case early in the project cycle. Because other estimating tools have disadvantages of their own as well, an analogous estimate can be used in combination with the bottom-up and parametric estimates described in the following sections.

An analogous estimate operates on the assumption of a limited amount of information about the target project and a very summary type of information about the analogous project. Put together, these two facts mean that just a few hours may be enough for almost any project's analogous estimate.

In analogous estimating, an estimator may choose to estimate only the total target project without breaking it down into items as we have done. He or she may judge, for example, that the target project may take twice the resource hours as the analogous one. This judgmental factor of 2 would then be multiplied by the resources deployed in the analogous project to obtain the estimate for the new target project. Essential for this type of estimation is adequate justification for all judgmental assumptions made.

Benefits

The value an analogous estimate brings to a user is in the little time it takes to develop, while operating with limited available information about the project being estimated.

Also, because the analogous estimate is based on representative past data, the developed estimate can be substantiated and easily justified unless significant judgments in comparison to past projects have been made.

PARAMETRIC ESTIMATE

A parametric estimate uses mathematical models to relate cost to one or more physical or performance characteristics (parameters) of a project that is being estimated.[10] Typically, the models provide a cost estimation relationship that relates cost of the project to its physical or performance parameters (also called cost drivers), such as production capacity, size, volume, weight, power requirements, and so forth. Determining the estimate for a new power plant may be as simple as multiplying two parameters—the number of kilowatts of a new power plant, for example, by the anticipated dollars per kilowatt. Or it may be very complex, such as estimating the cost of a new software development project that requires 32 parameters be comprehended in a cost estimate algorithm. Values of the parameters can be entered into the cost estimate relationship, and the results can be plotted in tablature format or on a graph (see Figure 7.3).

Developing Parametric Estimates

Figure 7.4 shows the basic features of parametric estimates for reference when considering the use of such a method. Prior to developing a parametric estimate, a

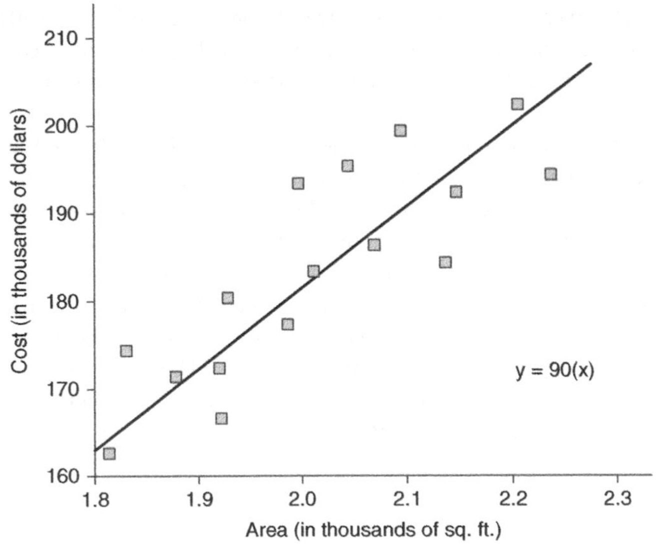

Figure 7.3: Example Cost Estimate Relationship for Parametric Estimates

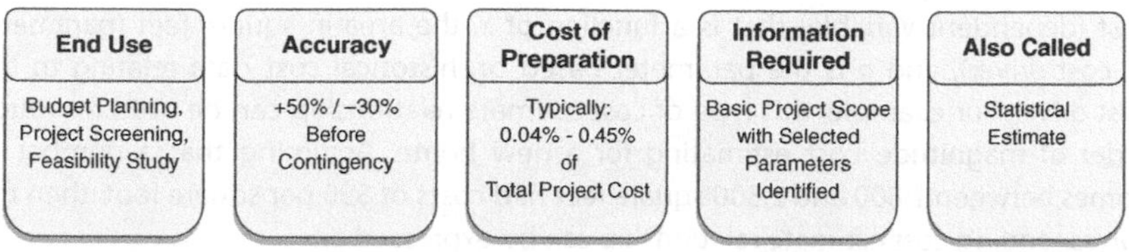

End Use
Budget Planning, Project Screening, Feasibility Study
Accuracy
+50% / −30% Before Contingency
Cost of Preparation
Typically: 0.04% - 0.45% of Total Project Cost
Information Required
Basic Project Scope with Selected Parameters Identified
Also Called
Statistical Estimate

Figure 7.4: Basic Features of Parametric Estimates

project manager should have knowledge about the project scope, the key project parameters (cost drivers) to use for analogous comparison, and historical information from like projects.

Basic project scope description provides understanding of what is being estimated. Its parameters are identified on the basis of the nature of the cost estimation relationship model that will be used to collect and organize historical information, which will be related to the project being estimated.

Prepare the Estimate

As explained in the cost planning map tool, the general process of developing any type of estimate is relatively similar. Per such process the parametric estimating starts with preplanning and continues with developing item descriptions and costing. The last step is unique to parametric estimating. Specifically, it includes database development, model building, and model application.

The purpose of database development is to collect and organize cost data from past projects in such a manner that the data can be used to build a model, which will be employed to estimate the cost of a new project. The first step is to select the framework to organize costs of past projects methodically. That framework is called the *basic work element structure form* and corresponds to that of the desired cost estimate relationship. For example, the structure may include project management cost (e.g., planning, controls), nonrecurring costs (e.g., design and engineering, software, facility), and recurring costs (e.g., production, operations). The necessary cost data from outside and within an organization are collected against the structure and normalized to established ground rules and assumptions. Typically, these databases are maintained in some constant-year price levels and are updated periodically to provide consistency in estimating from project to project.

Building a model aims at determining which equation type would best fit a data sample organized in the database and deriving a mathematical model for the cost estimate relationship that describes the project undergoing estimation.[11] While many mathematical models in the form of equations can be used for the cost estimate relationship, in practice, a lot of cost data can be fit empirically using one of the following forms—linear, power, exponential, or logarithmic curves. When graphed, all of them will resemble either straight lines or well-behaved curves. As shown in Figure 7.3, the simplest cost estimate relationships are as simple as a dollars per square foot relationship, a linear relationship of the form $y = ax$, where y is the estimated project cost (dependent variable) that is a function of *x*, the area in square feet (parameter or cost driver), and *a* is the parameter based on historical cost data relating to the cost driver. For example, this type of cost estimate relationship can be used for rough order of magnitude cost estimating for a new home. Assuming that a number of homes between 1,800 and 2,300 square feet had costs of $90 per square foot, then the corresponding cost estimate relationship can be expressed as:

$$y = 90(x)$$

This simple linear model assumes that there is such a relationship between the independent variable (cost driver) and project cost so that as the independent variable changes by one unit, the cost changes by some relatively constant number. Often, life is not that linear and simple, which leads to the use of nonlinear cost estimate relationships (see the example that follows titled "Parametric Software Estimating"), as well as cost-estimating relationships with multiple independent variables and multiple regression analysis.

How do we determine which equation type would best fit a data sample organized in a database? If we enter all data points from the past projects into a graph, generally the best fit would be the equation type that can be drawn through the data points such that the sum of vertical distances from the cost-estimating relationship curve to the data points above the line are about equal to the sum of vertical distances from the cost-estimating relationship curve to the data points below the line. Or more sophisticatedly, the best mathematical fit is the equation type that minimizes the absolute value of total cost deviation between the data points and the cost estimating relationship curve.

Once the best-fit equation type is determined, model building continues by deriving a mathematical model for the cost estimation relationship. Among many statistical techniques available for this, the method of least squares appears to be the most frequently used curve-fitting method. Although linear in nature, the method can be applied to both linear and nonlinear cost estimation relationship equation types—to the latter only when they are transformed into linear forms.

Parametric Software Estimating

Many parametric software effort models are based on key software parameters as cost drivers. They are usually based on the statistical analysis of the results of previous software development projects.[12] These analyses included key parameters such as system size (e.g., lines of code), complexity (e.g., number of interdependencies), type of application (e.g., real-time operating system), and development productivity (e.g., lines of code developed per hour). One expert suggested 59 parameters (factors) that can impact the outcomes of these cost models.[13] A simple model can take the form of:

$$Z = CY^L$$

where:

Z = estimated project effort (person-months)
Y = Estimated project size (KLOC—thousands of lines of code)
C = Regression coefficient
L = Regression exponent

You can apply this model to estimate the effort for a new software development project by assuming the following values: C = 3.8, L = 1.4, Y = 2.

- $Z = CY^L = 3.8\,(2)^{1.4} = 10.03$ person-months

When database development and model building are complete, you can proceed with the next step—model application, using the cost-estimating relationship derived from past cost experience to estimate the cost of a new project. Clearly, the assumption in applying a cost-estimating relationship is that future projects will be performed as past projects. What if, as is often the case, the future project that is being estimated differs from the past projects in some details? This can be resolved by cost-estimating relationship stratification and cost adjustments. Through stratification our historic database is divided into layers, each layer representing a "family" of data points similar to each other in some respect. Then, a separate curve for each family is fitted. For example, six data points in Figure 7.5 have higher costs than the other nine. A close look reveals that these six data points are for luxurious homes with features such as central vacuum cleaner, surround-sound audio system, stainless steel appliances,

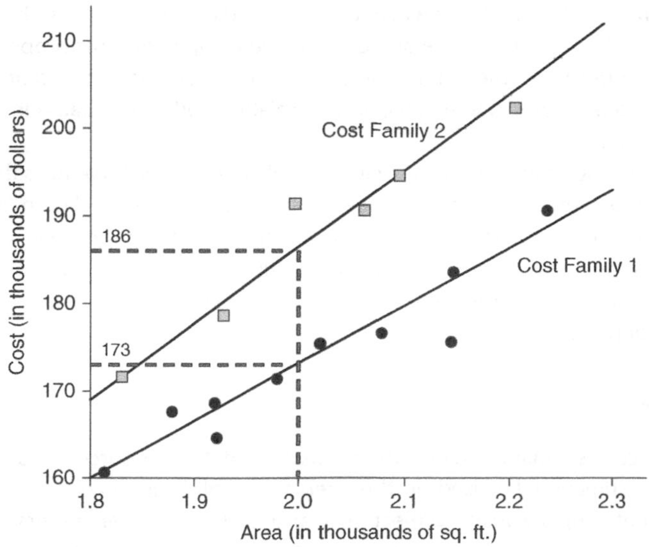

Figure 7.5: Stratified Cost-Estimating Relationships

marble countertops, hardwood floors, stucco work, and so forth, while the remaining nine were ordinary homes with much simpler and less expensive features. Logically, we could stratify our database into two families of homes and fit curves through each of the two subsets of the database, thus obtaining two cost-estimating relationships, as illustrated in Figure 7.5. If we have a square footage of the home being estimated (e.g., 2,000 square feet), we can easily determine that the parametrically estimated cost of either the luxurious ($186 k) or the ordinary ($173 k) home.

Cost adjustments or complexity factors are also used in parametric estimating to adjust the cost-estimating relationship estimated cost to account for differences between the project being estimated and the historical projects used as the basis of estimate. For example, let's assume product development projects from a database typically include an average number of five prototypes. If a project being estimated includes ten prototypes, two methods can help resolve the difference. One is to stratify the database and develop separate cost-estimating relationships for different numbers of prototypes. The other is to separately compute the cost of the prototypes. Whatever the case, when cost adjustment is done, the parametric estimate needs to be checked, reviewed, and improved as discussed in the section on the cost-planning map.

Using Parametric Estimates

Parametric estimates are most often used in the project definition stage as well as in the early design stages when insufficient information is available to develop a bottom-up estimate. Considering that cost estimation relationships typically relate project cost to high-level measurement of capacity or performance, it is exactly this information that is available early in the project cycle. Naturally, such summary information makes parametric estimates very appropriate for calculating comparative cost assessments of alternate project approaches and providing a cross-check to other estimating tools, but not for developing a detailed competitive cost proposal.[14] To be used for such purposes, the parametric estimate must be based on accurate historic information, quantifiable parameters, and a scalable model (applicable in both small and large projects).

The most difficult and time-consuming part of parametric estimating is the methodology development, including database development and formulation of the cost estimating relationship. Depending on the complexity of the database, it may take anywhere from tens of hours to hundreds of hours to develop the database and cost-estimating relationship. Once that's done, actual project estimation can be accomplished in minutes or hours.

Benefits

Parametric cost estimating tends to be faster and less resource consuming than bottom-up estimating. Focused on the need to establish good cost estimation relationships that properly relate project cost and cost-driving parameters, parametric estimates put a focus on cost-driving parameters, disregarding what is less important. This concentration on cost-driving parameters—coupled with greater speed and lower resource consumption—enables parametric estimates to be applied in estimating situations in which detailed, bottom-up estimates are neither practical nor possible. Think, for example, about a cost estimate for a new house. To develop a bottom-up estimate of the house, you need detailed house blueprints, bill of materials, labor rate information, and so forth. A lot of effort and cost is needed to prepare all of this. For estimating the cost of the same house using the dollars-per-square-feet parameter, you only need knowledge of the house size, making it much faster and easier to estimate. Parametric estimates can be produced even though little is known about the project except its physical parameters.

Parameteric estimates are also easy to use and repeatable. The reason for this is that the estimates are based on mathematical formulas, which correlate the present estimate with the past history of resource utilization on similar project types. Still, to enjoy these benefits, you must rely on judgment and experience.

BOTTOM-UP ESTIMATE

A bottom-up estimate relies on estimating the cost of individual work items and then aggregating them to obtain a total project cost.[15] Typically, an in-depth analysis of all project tasks, components, and processes is performed to estimate requirements for the items. The application of labor rates, material prices, and overhead to the requirements turns the estimate into monetary units.[16] Figure 7.6 is a generic version of the bottom-up estimate for simpler projects, but it can be used to estimate both simple and complex projects.

Typically, a bottom-up estimate is developed just before project execution, or even in earlier phases if the required information inputs are available. They are valued for their capacity to produce estimates of good accuracy, which is higher than that of any other estimating tools. Basic features of the bottom-up estimates are summarized in Figure 7.7.

PROJECT BUDGET ESTIMATE

Project Name: Longfellow Estimator: Williams Date: Aug. 5, 2017

1	2	3	4	5	6	7	8	9	10	11
Code	**Item**	**Quan**	**Labor**				**Over-Head (25%)**	**Materials**		**Total $ 7+8+10**
			Unit Hours	Total Hours	Rate $/Hour	Amount (5) X (6)		Unit Price	Amnt	
3210	First Article	10	0.5	5	60	300	75	45	$450	$825
010	Project Total	1	291.5	291.5	65	18947.5	4737		$900	$24,584

Figure 7.6: An Example of a Bottom-Up Estimate

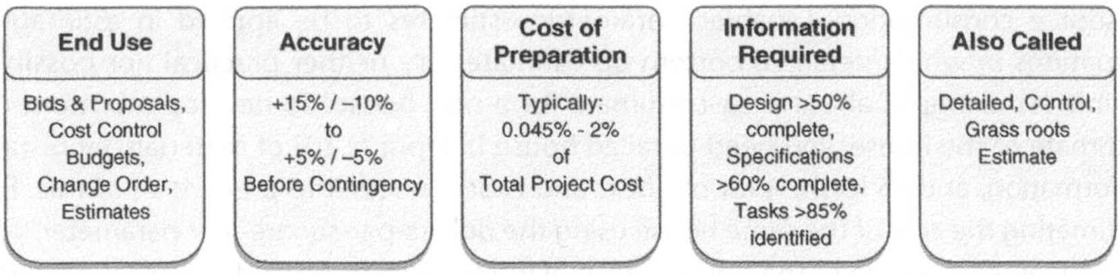

End Use
Bids & Proposals, Cost Control Budgets, Change Order, Estimates
Accuracy
+15% / –10% to +5% / –5% Before Contingency
Cost of Preparation
Typically: 0.045% - 2% of Total Project Cost
Information Required
Design >50% complete, Specifications >60% complete, Tasks >85% identified
Also Called
Detailed, Control, Grass roots Estimate

Figure 7.7: Basic Features of Bottom-up Estimates

Developing Bottom-Up Estimates

To develop a bottom-up estimate, a project manager needs to have knowledge about the project scope, resources needed and their associated labor rates, material cost required, and the project schedule.

Project scope in the form of a work breakdown structure (WBS) provides a framework to organize an estimate and ensure that all work identified in the project is included and estimated.[17] For this to happen, resource requirements that define types and quantities of resources necessary to complete the work are multiplied by resource rates to obtain a cost estimate. Typically, the rates come from historic records of previous project results, commercial databases, or personal knowledge about team members. Considering that some estimates contain an allowance for cost of financing such as interest charges, which are time dependent, the durations of activities as defined in the project schedule are an important input.

Establish the Estimate Format

Once all information inputs are available, work on a bottom-up estimate begins. Normally, the format of the estimate is established in the cost-planning map. A sound practice is to adopt a format that is based on a code of accounts, where a cost code is allocated to each work item. In our bottom-up estimate example, the code in column 1 uses the WBS structure coding. This simplifies the analysis of the project while serving as the basis for cost reporting, cost control, and even information retrieval.[18]

Prepare the Estimate

Once the estimate format is set, there are several steps necessary to prepare the estimate. Since these steps are discussed in general in the section on the cost-planning map, we will shed light on implementation specifics related to bottom-up estimates. First, you need to identify a work item that is being estimated, then determine its quantity, along with the cost of labor (human resources), overhead, and materials. One good way to accomplish this is to proceed area by area or category by category in the project, before adding them up to arrive at the total project estimate. We take that approach in our example cost estimate, where a WBS work package is selected as an item. In our example, the work package is called *first article approval* (column 2). By repeating this process work package by work package, and then aggregating or totaling the individual costs, we obtain the total project cost.

The first-article-approval item requires that the quantity (column 3) of ten prototypes of a high-tech cable be produced with equipment, tooling, fixtures, and materials that will be used later in the course of the regular production. Should the estimated labor cost for a single unit be shown, or a whole project batch of ten items? When project tasks are single and nonrepetitive, the question of quantity is irrelevant. When there are multiple identical items, as in our example, the cost for the whole batch needs to be estimated. Accordingly, our cost estimation relationship will multiply half an hour per unit (column 4) by the ten prototypes (column 3), which is a total of five resource hours (column 5), by a rate of $60 per resource hour (column 6) to obtain the cost of $300 (column 7) per item.

Columns 4 to 7 indicate the labor times and cost for each estimated work package item. While we use monetary units to record cost in columns 6 and 7, we do recognize that project managers will not do so, but instead will only record labor or resource hours in columns 4 and 5. This is an acceptable practice in many industries, where project managers are not expected to manage dollars but resource hours only. Actually, when labor time estimates are used for estimating future projects, the category of resource hours is much more relevant than cost. With the passage of time, the accuracy of cost is eroded by inflation and other factors, while the resource (labor) estimates should remain valid.

Once the direct labor cost is calculated, you can move to the labor overhead (column 8). There are no hard-and-fast rules here, since company policies vary widely. While some companies zealously include labor overhead into the estimate, other companies do not factor overhead labor into the estimate at all. Those who do often have different overhead rates in different parts of the company, and even from one project to another. Very frequently, this rate is based on a cost estimation relationship, calculating it as a percentage of the direct labor cost in column 7. In our example cost estimate, the rate is 25 percent. Typically, the overhead rate relates to the wages and salaries of employees who are not directly connected with the project, such as supervisors, administrators, and support personnel.

So far, the estimate includes direct and overhead labor cost for an item, in our case a work package. Now, we will estimate a net cost of materials required for the item completion (column 10), using a cost estimate relationship that multiplies the cost per unit (column 9) by the number of units (column 3). Material cost typically is comprised of costs of components, raw materials, or services for each item. It can include the cost of larger capital equipment as well, which is left out here for the sake of simplification. While our example for unit prices is based on catalog prices of materials, it is also possible to base it on vendor quotations or standard unit costs for stock items.[19]

With direct and overhead labor cost already available, the materials cost is the last cost piece necessary to figure out the total estimated cost for the item. By adding the costs of all columns, the total cost per item is estimated. Repeating this exercise for each item (i.e., work package) and summing up estimates for all items will lead to a total project cost estimate. If this were a project for an external customer, this would be the time to add profit margin cost to the estimate. The work on a bottom-up estimate ends with checking, reviewing, and improving it.

During the review process, care must be taken to ensure all cost items are included. For an example of the implications of incomplete estimates, see "The Courthouse Disaster."

The Courthouse Disaster

Halfway through its construction, the courthouse project looked like a sure winner for the contractor. The project was on schedule, contract payments were made in a timely manner, and the owner was happy with project performance. Then Greg, the contractor's project manager who also developed the bottom-up project cost estimate that was the basis for the project contract, left the company. A month later, Pete, the new project manager, determined that the entire project budget had been spent, although a lot of work remained. A quick audit commissioned by management revealed the following:

- Greg's project cost estimate was never reviewed by peers or managers.
- A significant monetary loss was to be expected at the end of the project.

When completed a few months later, the courthouse became one of the biggest losers in the company's history, ending $500 k over the estimated cost, almost one-third over the original budget. In the postmortem session, the following improvements were adopted for future cost estimating:

- All major estimates will be reviewed by peers and management
- All major estimates developed under time pressure will be compared to a shadow cost estimate (a cost estimate developed by an independent firm).

Using Bottom-Up Estimates

Both small and large projects, whether simple or complex, are good candidates to apply bottom-up estimates. Typically, the application occurs just before project execution, or even in earlier phases if the required information inputs are available. This generally means that a substantial amount of design work is completed, often exceeding 60 percent.

For their detailed nature, bottom-up estimates are primarily used for cost control budgets, bids/proposals, and change order estimates (see "No Bottom-Up Estimate, No Job!").

The time to develop a bottom-up estimate varies with the size and complexity of a project that is being estimated. A 500-resource hour project without materials and equipment may take an hour or two to bottom-up estimate. In contrast, a team of estimators may spend thousands of resource hours preparing a bottom-up estimate for a $400 m project.

Benefits

The value of the bottom-up estimate lies in their capacity to produce estimates of good accuracy, which is higher than that of any other estimating tools. Subsequently, they are the best basis for cost control.[20]

An added benefit is that a high degree of buy-in can be achieved with this type of estimate because people involved in estimating the project are the people who will be doing the work once the project moves to execution.

No Bottom-Up Estimate, No Job!

"We develop perfect quality software" was an informal motto of the SP Group, a unit of a privately held company. Its clients, divisions of the same company, agreed: The SP Group was doing a great job of developing software applications that had almost no bugs. Happy with the quality, the clients didn't care much about the actual costs of the projects. For a project to be approved and paid for by the client, the SP Group would simply submit an order-of-magnitude estimate ranging from 1,000 to 10,000 resource hours.

Then, the company went public and a focus on profit and demonstrated cost efficiency took over. Unable to respond to the operating principles, all division managers were forced out and new, profit-oriented division executives were brought in.

Project cost estimation also changed. "Sharks," as project managers called the new division managers, flatly refused to look at the order-of-magnitude estimates. Having profit-and-loss responsibility, the sharks wanted to manage their cost and required bottom-up estimates to approve a project. Since the project managers were lacking the expertise to develop such estimates, the large majority of them were eventually forced out. The learning from this real example is that project managers have a responsibility to keep their skills honed not just for current responsibilities, but also for future responsibilities.

THE COST BASELINE

The cost baseline is a time-phased budget used to measure and monitor cost performance on a project.[21] Developed by segmenting estimated costs by time period, the baseline reflects estimated costs and when they are supposed to occur, if executed in a specific way (see Table 7.4). Many projects, mostly large ones, may have multiple cost baselines expressing different facets of cost performance. For example, the baseline may measure expenditures (cash outflows), received payments (cash inflows), or committed costs. In contrast, other projects may have only one cost baseline—an S curve that illustrates how labor hours and material are to be expended over the life cycle of a project.

Typically, the baseline is developed in larger projects as part of initial project planning to forecast its cash flow. The cost baseline offers benefits as a performance measurement baseline, where the project team can gauge efficiency and progress and identify any deviations from planned progress and estimated costs.

Developing a Cost Baseline

To develop a cost baseline, a project manager needs to have knowledge about the project WBS, the project schedule, and the cost estimate for the project.

Table 7.4: An Example Cost Baseline

Work Packages/Tasks	Item Totals $k	Timeline (in thousands of dollars)											
		FEB	MAR	APR	MAY	JUN	JUL	AUG	SEP	OCT	NOV	DEC	JAN
1.01 Select Concept	12	8	4										
1.02 Design Beta PC	8		1	3	3	1							
1.03 Produce Beta PC	8		1	3	3	1							
1.04 Develop Test Plans	2		1	1									
1.05 Test Beta PC	6					3	3						
2.01 Design Production PC	18						3	6	6	3			
2.02 Outsource Mold Design	16						1	7	7	1			
2.03 Design Tooling	3						5	10	10	5			
2.04 Purchase Tool Machines	16									20	140		
2.05 Manufacture Molds	80									10	10	60	
2.06 Test Molds	8									8			
2.07 Certify PC	18											18	
3.01 Ramp Up	30												30
TOTALS	396	8	7	7	6	5	12	23	23	47	150	78	30

A simple definition of cost baselining as the spreading of the cost estimate items over time hints that having a documented cost estimate that includes all cost items is a mandatory starting point. Hopefully, these items can be arranged in alignment with the project WBS. If done so, the knowledge of the project schedule—indicating planned start and expected finish dates for work elements—enables the assignment of the cost to the time period when the cost will be incurred.

Identify Cost Baseline Type and Cost Items

Which types of cost are typically included in a cost baseline? That, of course, depends on the type of baseline being developed. As mentioned earlier, several are available, but the size and nature of the project are major determinants of the baseline type. If the target is to prepare a baseline focused on project expenditures (also called *project spending plan* or *cash outflows* or *project budget*), which is our focus here, consider including a broad menu of cost items, some of which are as follows:

- Salaries and wages of project personnel (in simplest cases this is the only item to include in in-company projects).
- Overhead expenses.
- Payments to contractors.
- Payments of vendors' invoices for purchases of equipment, materials, and services.
- Interest payable on loans, loan repayments, tax payments, shipping fees, duties, and so on.
- Travel expenses.

In case you are establishing a baseline to measure cash inflows, examples of some cost items that may be included are as follows:

- Payments from customers for delivered equipment, materials, and services.
- Loans from financial institutions.
- Tax refunds, grants, and so on.

If the intent is to manage cash flow, you will need items for both cash outflows and cash inflows. Once the cost items to include are identified, it is time to set criteria for cost baselining.

Set Cost Baseline Criteria

The preparation of a cost baseline is essentially an act of establishing the relationship between the cost estimate and time. For this to be possible, there must be clear criteria that determine which project events trigger payments of cost items included in the baseline, and the time intervals between the trigger events and the related payments (see Table 7.5). For payments to vendors, for instance, the trigger events are usually milestones defined in the contractual terms that stipulate how and when the payments are to be made. At other times, such as paying salaries of project team members, labor schedule of their engagement is what triggers their payment at the end of each month. The intervals, whether for payments within or outside the organization, are dictated by the time needed for internal and external communications, approvals and administrative procedures, and company policies bent to take advantage of the time value of money.[22]

Table 7.5: Example Criteria for Cost Baselining in a Product Development Project

Cost Baseline (Time-Phased Budget) Criteria			
Cost or Payment Item	**Schedule Trigger Event or Information**	**Interval between Trigger Event and Payment**	**Comments**
Management and design team	Per labor schedule	1 month	
Vendor subcontracts	Schedule milestones	45 days	This is company policy
Vendor's invoices for equipment and materials purchases	On-site delivery milestones	2 weeks	This is set by the design team to motivate vendors

Performing an appropriate analysis of the criteria and defining them in a written form is highly advisable, for it becomes a crucial foundation for tabulating costs by periods in the process of cost baselining.

Allocate Cost Items to Time Periods

Once the baseline type is chosen, cost items to be included in the baseline are identified, and criteria for baselining are defined, the foundations for allocating cost to time periods are established. Next, one should address coding and arrangement of cost items. Preferably, the project would have its own cost codes (column 1 in Table 7.4) that are consistent with the company's cost coding system or industry standards. If the project is externally funded, the customer may mandate the use of its own cost codes. Items from column 2 may be arranged in different ways. If a cost baseline is being developed on the basis of the bottom-up estimate, the items can be arranged in line with the WBS, as we have done in Table 7.4, using work packages from the WBS for a project. When an analogous or parametric estimate is being used to construct a cost baseline, other methods to arrange the items can be deployed, such as project phases.

Column 3 provides cost estimates for the items, which will now be allocated to certain time periods in a 12-month project. Since reporting is on a monthly basis, time periods are months represented in columns 4 to 15. Item 1.01, Select Concept, will be carried out in months 1 and 2, so part of the estimated $12 k will be expended in month 1 and the remaining part in month 2. How much will be allocated to each month hinges on the following factors:[23]

- The project schedule, indicating the planned start and end dates of the item, along with resource histograms specifying resource requirements by time period.
- The contractual terms.
- Intervals between trigger events and payments.

Similarly, estimates for the remaining items are spread over their months of execution and entered in the appropriate months. The schedule is rarely drawn on the cost baseline, but we included it in Table 7.4 to make the baseline easier to comprehend.

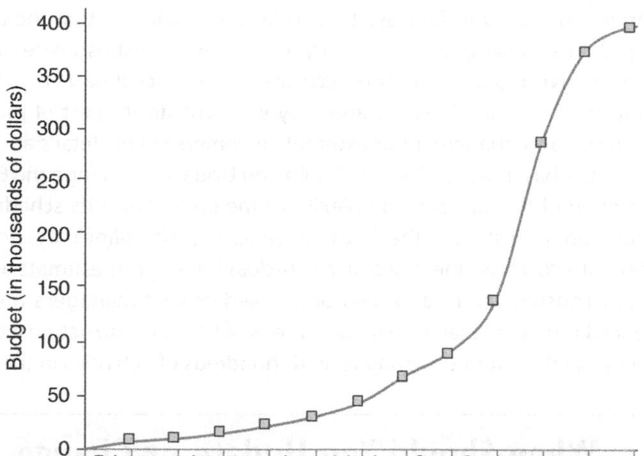

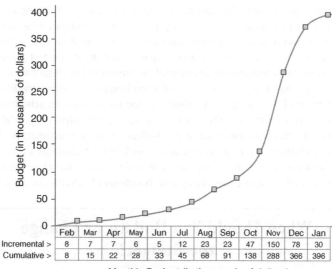

	Feb	Mar	Apr	May	Jun	Jul	Aug	Sep	Oct	Nov	Dec	Jan
Incremental >	8	7	7	6	5	12	23	23	47	150	78	30
Cumulative >	8	15	22	28	33	45	68	91	138	288	366	396

Monthly Budget (in thousands of dollars)

Figure 7.8: A Cost Baseline Displayed as an S Curve

Sum Estimated Cost by Period

Once all item cost estimates are allocated to specific time periods, the next action is to sum estimated cost by periods. This provides information about incremental expenditures by time periods—that is, expenditures for each month—which will be used in the next step to display the cost baseline graphically.

Display the Cost Baseline

The S curve is a popular way of displaying a cost baseline formatted as cumulative expenditures (see Figure 7.8). To calculate cumulative expenditures, add the incremental expenditures for the first period to those of the second period. These are the cumulative expenditures for the first two periods. Add this number to the third period's incremental expenditures to obtain the cumulative expenditures for the first three periods, and continue with this procedure for the remaining periods. When finished, graph the cumulative expenditures (*y*-axis) over time (*x*-axis) to develop a cost baseline in the form of an S curve. As in any type of the cost estimate, this is the time to check and review the cost baseline.

Once the project is completed, there is a lot of value in studying how the initial baseline played out over the life of the project, learning the lessons, and using them to improve future cost baselines.

Using the Cost Baseline

Many experts believe that cost baselining is an unnecessary activity for small projects, because the cost of its preparation may easily outweigh the benefits.[24] In contrast, other projects do have a need for the cost baseline. Typically, the baseline is developed as part of initial project planning to forecast its cash flow. Considering that the cost baseline may be based on an analogous, parametric, or bottom-up cost estimate, sometimes as the estimates evolve and become more accurate, so do the cost baselines. They are reissued at regular or irregular intervals and may even constitute part of project reports submitted to senior management or external customers.[25] For details about updating and changing the baseline, see "When Should You Update or Change the Budget?"

As a function of the size and complexity of the project and its schedule, resource requirements and cost estimate, the time to develop a cost baseline may widely vary. The development of a cost baseline based on a low-detail analogous estimate and summary schedule may consume an hour or two of a skilled project manager's time. However, an experienced project manager may spend tens of hours constructing a cost baseline based on a detailed bottom-up estimate, with hundreds of activities in the schedule.

When Should You Update or Change the Budget?

Dogmatically sticking with the initial cost baseline or time-phased budget when there is a need to alter it serves no purpose and is risky. The need for alteration is triggered by several factors, leading to minor (updates) or major revisions (changes) of the baseline. Updates may occur because of factors such as:[26]

Cost estimate evolvement. As a project progresses, more information becomes available, helping to develop more accurate estimates. Such changes in estimates should lead to the update of the baseline.

Project changes. Management of project changes may require new expenditures, which should be added to the baseline. Changes may be due to unforeseen conditions or from customer-generated changes.

Schedule changes. Changes of time-phasing of project activities during the execution stage are frequent and result in inevitable modification of the baseline.

In addition to these updates (minor revisions), there may be times when a major revision of the baseline is necessary. During project implementation, major unplanned schedule, cost, or technical problems may occur. Or there may be a need to change the project strategy. These typically prompt major revision of the project plan, including a major revision of the cost baseline. Such changes to the cost baseline may happen very rarely, once or twice in the life of a project, if at all. When dealing with updates or changes to the baseline, the key is to manage all modifications and related factors in a proactive rather than a reactive manner when possible to maintain control of the project.

Benefits

The lack of an effective cost baseline, even if a cost estimate and labor requirements are available, poses a major risk to a project—organizing measurement of performance and cash flow is difficult, if not impossible. Therefore, constructing the baseline offers benefits of using it as a performance measurement baseline. In this capacity, the baseline is a basis for comparing actual costs (when they occurred) with planned costs (when they were supposed to occur). This, then, is a way to gauge efficiency and progress, attracting management's attention to any deviations from planned progress and estimated costs.

Cash flow forecasting is another benefit that an effective baseline provides. It informs management or the customer in advance of the funds that must be made available in order to procure resources and use them to sustain project progress. When properly performing this role in the course of project implementation, the cost baseline should be modified to reflect performance and progress to date. Some risky consequences of not managing cash flow in a project, and how to avoid them, are described in the example titled "The Museum Design Company."

The visual power of displaying a cost baseline as an S-curve format is impressive, further strengthening the case of its simplicity and visual nature as a benefit to the project manager and stakeholders.

The Museum Design Company

The Museum Design Company (MDC) found itself in what appeared to be a paradoxical situation: It had multiple contracts, but no positive cash flow. How was this possible? Loaded with top design talent and known for a strong track record of superb technical quality, MDC had no difficulty landing project contracts to design military museum exhibits. But John Riddle Jr., CEO of MDC and an accomplished designer, had to borrow from his bank on a regular basis to make the payroll. Puzzled by this, Riddle asked for professional help.

He was advised to study the cash inflow and outflow S curves for each project within their portfolio. Since S curves were not available, project managers were tasked to develop them for each project. Once developed, the majority of the curves looked like the one illustrated in Figure 7.9a. The S curves clearly showed that the difference between the funds obtained from the customer (cash in) and payments paid for designers' salaries, overhead, and loan interests (cash out) was negative throughout the project. The exception being at the end when it was zero.

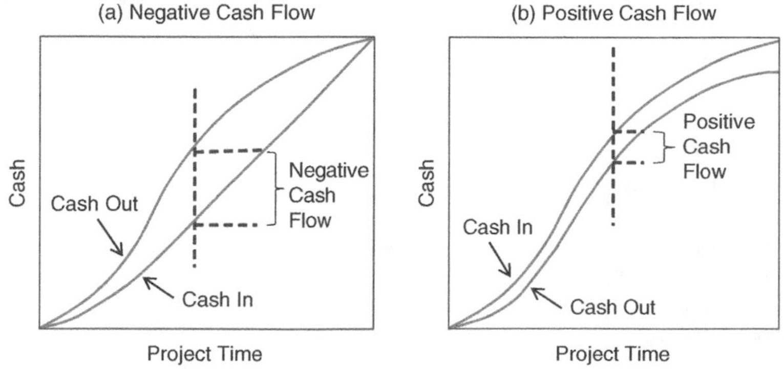

Figure 7.9: Two Possibilities: Negative and Positive Cash flow

(continued)

This was the source of the paradox. MDC was burning through its profits, forcing it to continually borrow from the bank. Continuation of the practice would certainly force MDC out of business. Riddle concluded that the negative cash flow situation had to be avoided at all costs on future projects. The cost baseline for each future project was set up so the difference between cash in and cash out was positive (Figure 7.9b), enabling MDC to eliminate the need for costly loans.

Table 7.6: A Summary Comparison of Cost-Planning Tools

Situation	Cost-Planning Map	Analogous Estimate	Parametric Estimate	Bottom-up Estimate	Cost Baseline
Provide cost-planning methodology	✓				
Show the amount of estimated funds		✓	✓	✓	✓
Show time-phasing of estimated funds					✓
Organizations with stream of small projects	✓	✓	✓	✓	
Organizations with large projects	✓	✓	✓	✓	✓
Based on past experience		✓	✓		
Higher accuracy required				✓	✓
Lower accuracy required		✓	✓		
A few hours to prepare		✓			
Medium time to prepare			✓		
Longer time to prepare	✓		✓	✓	✓
Need estimate for project screening, forecasting			✓	✓	
Need estimate for budget authorization				✓	
Need estimate for cost proposal/change orders				✓	✓
Make decisions very early in project life cycle		✓			
Estimate in project definition/early design			✓		
Before execution, design substantially complete				✓	

CHOOSING A COST-PLANNING TOOL

This chapter features five tools with clearly designed purposes. For two tools, the cost-planning map and cost baseline, the purposes are so distinct that they do not compete with other tools for use in cost planning. While the cost-planning map strives to establish a systematic methodology for cost planning, the cost baseline aims at providing a time-phased budget.

The remaining three tools may be used in combination or a single tool can be chosen for a particular application. That calls for matching the project situation with the tool that favors the situation. Table 7.6 can provide a project manager guidance on tool selection.

References

1. Project Management Institute. *A Guide to the Project Management Body of Knowledge*, 5th ed. (Drexell Hill, PA: Project Management Institute, 2013).
2. Humphreys, K. K. *Project and Cost Engineers' Handbook*, 4th ed. (Boca Raton, FL: CRC Press, 2014).
3. Ibid.
4. Chemuturi, M. K., and Thomas M. Cagley. *Mastering Software Project Management: Best Practices, Tools, and Techniques* (Plantation, FL: J. Ross Publishing, 2010).
5. Ostwald, P. F., and Timothy S. McLaren. *Cost Analysis and Estimating For Engineering and Management* (Upper Saddle River, NJ: Prentice Hall, 2003).
6. Project Management Institute, 2013.
7. Meredith, J. R. *Project Management: A Managerial Approach*, 8th ed. (Hoboken, NJ: John Wiley & Sons, 2011).
8. Chemuturi and Cagley, 2010.
9. Stewart, R. D. *Cost Estimating*, 2nd ed. (New York: John Wiley & Sons, 2001).
10. Humphreys, 2014.
11. Venkataraman, R. R., and Jeffrey K. Pinto. *Cost and Value Management in Projects* (Hoboken, NJ: John Wiley & Sons, 2008).
12. Kile, R. L., and USAFCA Agency. *REVIC Software Cost Estimating Model User's Manual, Version 9.0* (Arlington, VA: Revic Users Group, 2001).
13. Ostwald and McLaren, 2003.
14. Chemuturi and Cagley, 2010.
15. Project Management Institute, 2013.
16. Stewart, 2001.
17. Project Management Institute, 2013.
18. Humphreys, 2014.
19. Chatfield, C., and Timothy Johnson. *Microsoft Project 2010 Step by Step* (Redmond, WA: Microsoft Press, 2010).

20. Ostwald and McLaren, 2003.
21. Project Management Institute, 2013.
22. Verzuh, Eric. *The Fast Forward MBA in Project Management* (Hoboken, NJ: John Wiley & Sons, 2012).
23. Chatfield and Johnson, 2010.
24. Meredith, 2011.
25. Project Management Institute, 2013.
26. Project Management Institute, 2013.

PART IV

Project Implementation Tools

8
SCOPE MANAGEMENT

When it comes to managing a project, the one constant that can be counted upon is change. Changes happen, or attempt to happen, at a constant pace on projects. Each change has the potential to modify the scope of the project, having an impact on the work performed. Scope management is therefore necessary to determine if each change is necessary, how it will impact the work performed, and then ensure that each necessary change is implemented properly.

Without scope management we get uncontrolled scope change, commonly referred to as *scope creep*. We are all guilty of it. Just recently I took on what I estimated to be a two-hour, one-person project to clean out the closet in the master bedroom. I efficiently collected all the items that I determined I no longer needed on a daily basis, boxed them up, and tagged them as either donatable items or keepers for later use and to be stored. The donated items were packed in the car for delivery, while the storable items were brought to the garage.

However, no room was available to store the items in the garage, so the project scope changed to include cleaning out the storage portion of the garage, which in turn led to additional project scope creep to rearrange the storage space in the attic to make more room for items moving from the garage. The two-hour project had expanded uncontrollably to a full two-weekend, two-person job. Sound familiar?

This simple example shows that project management requires focus to get the required work completed, which in turn requires project scope management. The Project Management Body of Knowledge (PMBOK) Guide defines project scope management as "the process to ensure that the project includes all of the work required, and only the work required, to complete the project successfully."[1]

When a project is in the implementation phase, a project manager must be diligent about maintaining control of project performance, especially in comparing actual performance against what has previously been planned. Project control involves establishing checks and balance techniques that compare planned activities (project baseline) against what has actually occurred.

Arguably, the most critical control involves managing changes to the project scope. Changes in scope will affect the tasks that are performed on a project, which has a direct effect on project resources, which in turn directly affect the project cost and schedule. Changes in project scope therefore create a cascading effect as depicted in Figure 8.1.

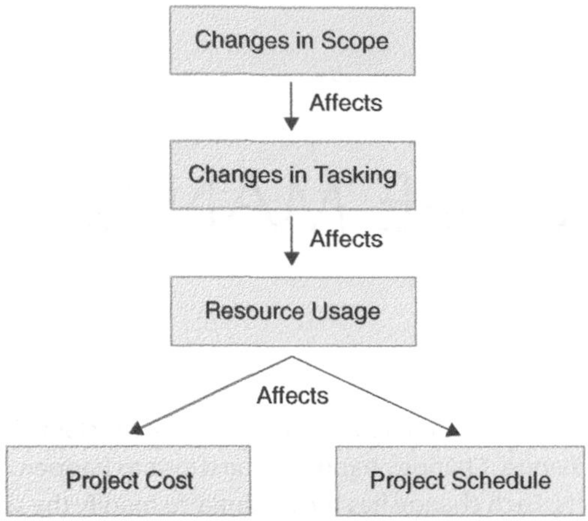

Figure 8.1: The Cascading Effect of Project Scope Change

When one understands the cascading effect of project scope change, it becomes very apparent why this is so. It is a well-documented fact that unmanaged scope change is a leading cause of project failure (see "Scope Change: If You Have to Do It, Do It Early").[2]

Scope Change: If You Have to Do It, Do It Early

Uncontolled project scope changes are known as project killers, because they:

- Cause delay.
- Increase the cost.
- Damage morale and productivity.
- Spoil relationships among project participants.

Why such a far reaching impact? First, more often than not, changes cause work to have to be repeated in the impacted activities. Second, any activities related to those directly impacted by an uncontrolled change will likely need to change as well. This means that the earlier you make the change, the less rework and less damage you inflict on the project. Early in the project, very few activities have been worked on and completed, so therefore there is less rework that will have to take place. In contrast, if a scope change comes in late in the project life cycle, significant and costly change may need to occur, and any work that has to be redone becomes wasted effort and cost.

Consider, for example, if a change comes in late on a product development project. The change may cause redesign and associated redevelopment, the repurchasing of factory tooling, fixtures, and materials, the remaking of prototype systems, and so forth. All this work has significant impact on the project. Even a seemingly insignificant change of a team member during project implementation may set a team back by a number of weeks or months. The lesson to learn here is to think hard and make changes early in the project cycle, and put a very critical assessment eye on changes that come late in the project.

We have made the case for diligent scope management on a project. The remainder of the chapter provides high-value tools that should be a part of every project manager's PM Toolbox for establishing and performing project scope management activities as part of their project management practices.

PROJECT SCOPE CONTROL SYSTEM

Controlling scope on a project cannot be handled in an ad hoc manner, regardless of the size or complexity of the project. Scope management requires the establishment and use of an effective project scope control system.

The project scope control system is the most prominent tool applied to project scope management. Because scope changes are certain to happen on any project, there must be order to process, document, approve, and manage the changes. The project scope control system establishes the scope control protocols, or rules, for a project that defines how changes to project scope will be managed once a project baseline has been established.[3] Scope control protocols will normally include the following:

- Scope change requests must be made in writing.
- The benefits gained from a proposed scope change must be clearly articulated and documented.
- Scope control roles and responsibilities must be established.
- An approval process must be documented.
- A decision maker has to be appointed and anointed.
- Approved scope changes must be incorporated into the project plan.
- Scope changes must be adequately communicated.
- A set of standardized tools must be used.

Many project managers find it useful to create a visual representation of the project scope control system, an example of which is illustrated in Figure 8.2. By incorporating other scope control tools such as the project change request and project change log into the scope control system, organization change control policy can be translated into a practical project work flow for managing and controlling changes to project scope.

Establishing a Project Scope Control System

Establishing a scope control system that harmonizes a process, actions, tools, owners, and their interactions in controlling project scope calls for a thoughtful system that will likely need to be tailored for each organization.

The effectiveness of the project scope control system is dependent on sufficient information about the project scope (scope statement or statement of work), an established project work breakdown structure (WBS), and the appointment of a scope control coordinator.

Identify Scope Control Roles

The first step in establishing a project scope control system is to identify the various roles and responsibilities to be performed by the project actors. Every project manager

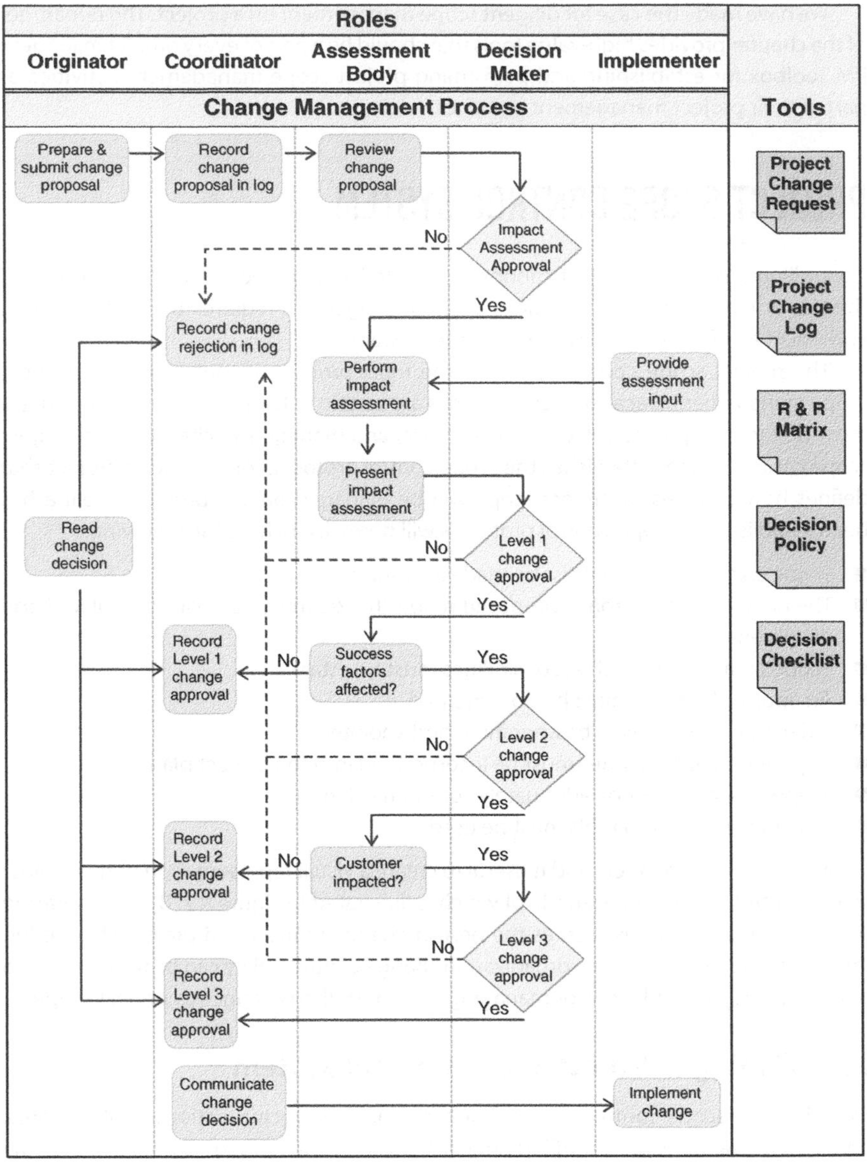

Figure 8.2: Example Scope Control System

(or project office if one exists) is charged with identifying and customizing the scope control roles that are appropriate for the particular organizational and project situation. The following roles are generic in nature and can be used as a starting point for a customized project scope control system:

- *Change originator*. The change originator is the person introducing a scope change proposal to the project. He or she is responsible for clearly articulating and documenting the scope change proposed, the need for the change, and the quantifiable benefit of the change if approved.

- *Scope control coordinator*. The scope control coordinator is the person who shepherds a proposed scope change through the scope control process. The coordinator is the person who is responsible for ensuring that a scope change proposal moves through the evaluation process efficiently, that the decisions are being made effectively, and if approved, that the change in scope is communicated and implemented. The scope control coordinator is also responsible for creating and maintaining the project change log.
- *Change authority*. The scope control change authority is the set of cross-functional subject matter experts on a project who evaluate information associated with a proposed change in scope. The change authority will assess the need for the scope change, the benefit gained, and the impact of each proposed change, and then provide feedback and recommendations to the decision maker based on its assessement.
- *Decision maker*. The scope change decision maker is the person or persons who makes a determination to proceed, or not to proceed, with each proposed scope change. More than one decision maker may be required depending upon the level of impact a proposed scope change will have on a project. This is discussed in more detail in the following sections.
- *Implementers*. Implementers are the various project actors who are responsible for implementing the approved scope changes. Prior to implementation activities, the implementers are also likely responsible for providing input to project baseline plan changes resulting from a proposed scope change.

It should be pointed out that people may serve various roles within the project scope control system. For example, a project manager may serve as the scope control coordinator, a member of the change authority, and a scope change decision maker. Likewise, a project specialist may serve as a scope change originator, a member of the change authority, and an implementer.

Because of the multiple role possibility, it is good to include a responsibility matrix (Chapter 3) as part of the project scope control system.

Document the Scope Control Process

The heart of the project scope control system is a documented and approved project scope management process. This process is commonly referred to as the change control process. Figure 8.2 contains a generic scope control process that can be used as the basis of a customized process for your organization.

The scope control process should clearly identify the primary activities associated with project scope control and the decisions that need to be made in order to effectively control project scope.

It should be noted that the process shown can either be made more robust, or leaned out, to fit a particular organizational or project need.

Establish the Scope Control Decision Policy

The project scope control system and its associated scope control process can break down quickly if a project decision policy is not documented to guide the practices associated with the key scope control decisions that need to be made. The decision policy should describe the decision method that will be employed—consensus, consultative, or authoritive. The decision boundary conditions associated with each decision as well as the various decision levels—project level, executive level, customer or client level—must also be established. Finally, the decision maker for each decision and decision level must be documented.

Gain Senior Management Approval and Commitment

A project scope control system will fail to be effective (or even used) if it is not supported by the members of an organization's senior management team. It is recommended that a senior management sponsor be identified for the newly established project scope control system, and verbal support be clearly evident to the entire organization to establish implementation empowerment.

Beyond senior management sponsorship and approval, a commitment on the part of senior managers to ensure that the organization uses the project scope control system as designed must be established. This involves two primary things. First, senior managers must play their role in the system, most often as a scope control decision maker. Second, they must set expectations and ensure that other project actors play their part and don't try to circumvent the system. For instance, it is not uncommon for a disgruntled scope change originator, who has been told his or her change proposal has been denied, to escalate to a senior manager and get an override to the decision. It is also not uncommon for a senior manager to attempt to mandate a scope change. Both examples demonstrate a lack of commitment on the part of senior management and also destroy the empowerment of the scope control system by changing authority body and the project manager.

Using the Project Scope Control System

A project scope control system should be used on each project that is subject to change, and put into use when the project requirements are locked in.[4] With this, the project scope control system is in place when the first change to project requirements emerges, setting the tone of scope control throughout the project. This is especially vital in larger projects where there is a stronger need for scope control. Because of their lesser scope and complexity, smaller projects can utilize a simplified scope control process—the project manager often may be the only person available to handle scope changes. In circumstances like this, an informal, but well-understood scope control system would be more logical than having a formal system. Project management maturity of an organization is also a factor that affects the formal or informal use of a scope control system (see "Can We Handle the Change?").

Can We Handle the Change?

Project-oriented organizations with a well-established change management process typically have change control policies in place, often accompanied by a template change management plan. Such a plan, then, gets adapted for a specific project by its team, which already may have a change coordinator aboard. In such cases, change coordinators tend to be well versed in change management, spending all or much of their time on coordinating changes.

Other organizations which are less mature in project management offer an example of the opposite experience. Not having change management policies or template plans in place, they may expect a project manager to take on a part-time role of the coordinator, developing a change management plan and coordinating all changes through the change management process.

Using a formal project scope control system involves a number of activities. First, the various roles have to be assigned to individuals involved in managing scope for the project, and the individuals have to begin acting in the capacity defined by the roles. Next, the change authority has to be formed. For many projects, the change authority is referred to the change control board (CCB) and is made up of a cross-section of project stakeholders. Within a short period of time, the various tools that will be used to manage project scope have to be identified and developed. Finally, the change management process needs to be implemented with the emergence of the first scope change request. The following steps (or a customized set of steps) are performed to appropriately manage proposed changes to project scope.

Submit a Scope Change Request

Naturally, anyone involved in a project, inside or outside the team, is welcome to submit a scope change request, assuming the project scope is not frozen. All proposed scope changes have to be submitted in writing using the project change request (detailed later in the chapter). A description of the proposed change, the reasoning for the change, and a benefits justification have to be included with the scope change request.

Record the Change and Distribute the Request

After receiving a scope change request, the coordinator records it in a project change log (detailed later in the chapter) and distributes copies of the project change request. Typically, copies go to the change authority and any specialists whose subject matter expertise is needed to assist in assessment of the request.

Review the Change Request

The originator prepares a briefing and is responsible for presenting the proposed scope change to the change authority. The presenter should be prepared to answer more in-depth questions regarding the proposed change, potential impacts, benefits, and implementation options.

Recognizing that proceeding with a more in-depth evaluation of the change impact and implementation options may divert critical project resources from their normal project work, the change authority acts on behalf of the project to decide if the proposed scope change is approved for further evaluation. One of three decisions can be expected at this stage: (1) the proposed change is rejected; (2) the proposed change is denied but with a request for additional information; or (3) the proposed change is approved for further impact assessment and a formal approval decision.

Assess Impact

When an impact assessment is approved, an owner is assigned to shepherd the proposed change through the evaluation of implementation options and associated impact estimations. Depending on the nature of the proposed change, both a technical assessment and programmatic assessment may be required. It is good practice to have likely implementers involved with the impact assessment.

Make the Scope Change Decision

The change authority's responsibility is to make an approval or rejection decision, which is then recorded in the project control log. When a scope change is rejected, a copy is stored in a master file and a copy is returned to the originator, explaining the decision and the grounds for rejection.

When a scope change is approved, the coordinator makes the change official and sends it to those impacted by the change to carry it out. Also, the coordinator informs the originator and other project stakeholders. Finally, it is not unusual that the change authority finds an impact assessment incomplete, prompting the coordinator to request more information from the originator and assessment team.

It is common that multiple levels of decision authority be required, as defined by the decision policy and the level of authority granted to various decision makers. Level 1 changes include those that do not negatively affect the project success factors, and approval decision resets with the project change authority. Level 2 changes are those that *do* have an impact on the project success factors, and as such, require an additional level of decision approval by the project sponsor or another member of top management. Level 3 changes are those that impact the customer or client directly. These changes require yet another decision approval by the customer or client.

Update the Change Log

The coordinator is responsible for recording the change decision, any actions required, and supporting information in the project change log. The decision is then communicated to the appropriate project stakeholders.

Modify the Project Plan

When a scope change is approved, it normally creates a change in the work that has to be performed on a project, which in turn drives a change in tasks by one or more project team members. Some scope changes are small and relatively insignificant and therefore do not require a change to the project baseline. Other scope changes are moderate and require the project manager to *modify* the project baseline. Still other scope changes are large and affect multiple aspects of the project WBS causing a *reset* to the project baseline.

The significance of the scope change is contained in the impact analysis performed. When changes in tasking and deliverables are required, the change first of all has to be captured in the project WBS. From there, changes to the project budget and schedule have to be performed in order to keep the project in alignment. The project team will then be working under a new project baseline.

Implement the Change

The approved change now has to be broadly communicated to all stakeholders affected. A standard and consistent change communication plan should be developed and managed by the change authority and utilized by the project manager.

Once communicated, the final step in the scope change process is for the various project specialists to implement the change as part of their project execution duties. The final entry in the project control log for all approved changes is an indication that the scope change has been fully implemented.

Benefits

The scope control system's value is in bringing order to the project scope change process. Through a methodical prescription of the sequence and arrangement of roles, responsibilities, tasks, and tools involved in the system, this order significantly diminishes the possibility of problems, including scope creep, budget overruns, and schedule slippage.[5] Making the scope change process known in advance also helps direct behaviors of project participants, eliminating notorious perplexity of people involved in project changes.

When fully documented, the pictorial appearance of a scope control system makes it user friendly and easy to follow, building on a natural ability of humans to better process graphical rather than narrative information. The visual impact is further strengthened by presenting the scope control system as a sequence of steps, a format that adds more transparency and user utility to the process.

PROJECT CHANGE REQUEST

Some project managers despise change requests and have a tendency to say "no" to every request. However, these project managers should broaden their viewpoint to realize that many changes create additional value for a project, and therefore have positive consequence. Most change requests are a result of several factors:

- *Value enhancements*: Changes that will increase the value proposition of the project.
- *External events*: Changes caused by the environment in which a project operates such as new laws and regulations put into place.
- *Errors or omissions*: Changes that result due to forgotten or unrealized features or work during project planning. Hey, we are human!
- *Risk response*: Changes identified as a result of necessary risk mitigation or elimination responses.

However, the impact of changes on a project's schedule, cost, quality, and other matters may easily surpass the awareness or expertise of a single change originator.[6] As a consequence, the project may severely suffer and in certain cases even collapse (see "Scope Creep by Design"). For this reason, it is very important to ensure that the benefit and value of each change is understood, and the impact then evaluated in a disciplined and professional manner before performing a change.

The project change request is a must-have tool to help a project manager and his or her change authority perform a multiperspective evaluation of proposed changes. The process, structure, and content of the project change request help businesses make decisions of a higher quality, keeping project scope, cost, and schedule in check and in alignment with one another.

Scope Creep by Design

Scope creep, or uncontrolled change of scope, is often perceived as a major threat to projects.[7] But one company faced with highly uncertain semiconductor fab projects controls the creep one change at a time. At the time of defining the scope historically prone to many changes, the company identifies a bucket of money equal to 10 percent of the project budget that is called AFC (allowance for change). Its purpose is to pay for the scope items that can't be predicted. To bring control to the process, every time such an item emerges, it is treated as a scope change and the project manager has to formally approve it. This very successful practice has helped the company to proactively manage scope creep from a budget perspective.

Developing a Project Change Request

While the project change request appears as a simple form (Figure 8.3), it does require some thought, which begins with preparing solid information inputs.

Major inputs to the application of the project change request are the scope baseline with the project WBS, the scope statement, and the scope control system.

Project change request forms come in many different styles and formats. When developing a custom project change request for your project or organization, use

Change Request Identification
Change Title:
Change Number:
Originator:
Submittal Date:

Change Request Detail
Description:
Business Benefit:
Implications of No Change:
Priority: ☐ 1 – critical: *"project cannot move forward unless this change is made"* ☐ 2 – high: *"project success criteria impacted without this change"* ☐ 3 – normal: *"project value proposition enhanced with this change"* ☐ 4 – low: *"improvement in ease of use or performance gained by this change"*

Impact Assessment	
Assessment Owner:	
Cost:	
Schedule:	
Resources:	
Deliverables:	
Impact Summary:	

Alternatives
Alternative Approach:

Recommendation and Approvals
Recommendation:
Decision: ☐ Approved ☐ Rejected ☐ Deferred

Figure 8.3: Project Change Request Template

the change management process contained in the scope control system to guide the design and content to be included in the request form. Generally, there are five content sections that we recommend including in the project change request:

1. Change request identification information
2. Change request detail
3. Impact assessment
4. Alternative solutions
5. Recommendations and approval decision

Change Request Identification

This first section of the project change request focuses on the quick identifiers for each requested scope change. Included in this section should be a concise change request title, a unique identification number, the name of the person originating the change request, and the date on which the change request is first submitted to the project change authority for consideration.

Change Request Detail

This section is intended to provide sufficient detail about a requested change so the project change authority can decide if there is sufficient cause for an impact assessment to be performed. The description of the change request must be sufficiently precise to provide clear understanding of which feature, deliverable, or work package will be changed and in what manner.

Additionally, this section must contain the justification for expending resource time and potentially changing the project baseline in order to incorporate the change. Justification should be articulated by describing the business benefits that will result from the change, if it is implemented. From the opposite perspective, we recommend that a well-formed justification also include a statement of the impact if the change is *not approved* and implemented.

Finally, the change request originator should be required to make a determination of the criticality of a requested change. As shown in Figure 8.3, we typically use a 4-point scale of criticality that ranges from low priority to critical priority. One of the major problems with regular, formalities-oriented treatment of project change requests is that they are unnecessarily slow. To overcome this problem, you need a feature in the system that ensures fast responses to urgent change requests, which is marked in the project change request form as "critical." The intent is to inform the change authority that the requested change needs urgent consideration and approval. This may mean that the authority will have to act promptly. Depending on the policy rules, if a critical change is being requested, they may choose to interact in a face-to-face meeting, on the phone, or via an intranet-based program (see "Fast-Tracking the Change Process"). It is of utmost importance that critical changes should not be an excuse to circumvent evaluation of matters related to quality, performance, reliability, safety, or any other aspect. Building safeguards that enforce appropriate consideration of the change may be a useful aid.

Fast-Tracking the Change Process

ODI Incorporated's primary customer made it very clear that they had become frustrated with slow change request responses when their design chief made the following comment: "We are not willing to put up with your long turnaround time of our major change requests!"

> Keen to retain customers, ODI redesigned its change management procedure, adding three vital changes. First, a rule was made that turnaround time for major customer requests will be 48 hours. These changes required a significant evaluation effort, including involvement of design, tooling, and manufacturing engineers, as well as marketing and purchasing experts. In addition, these people and their representatives were not collocated, so communication among them was time consuming and slow. Second, in response, ODI built an intranet site that significantly sped up the communication. Third, instead of using consensus decision making on the change board, typically a slow method, one of the board members was nominated the approver while others were considered reviewers and inputters only. The system redesign led to a drastic improvement, helping ODI to reestablish good relationship with this very important customer.

With the appropriate justification information in hand, and a sense of change request urgency, the project change authority now has what it needs to decide if an impact assessment is warranted, and if a change request needs to be fast-tracked. If so, an assessment owner is assigned and he or she becomes responsible for filling out the next section.

Impact Assessment

Some organizations have a habit of approving major changes to the scope without ever referring back to the original (often called baseline) scope. This practice results in an ever-changing scope, making it a moving target. The risk is that ever-increasing scope may inflate to the point of representing a significantly changed project that needs planning and implementation different from the current baseline. To prevent this, screen all proposed changes by identifying how they will impact the scope.

Should the change have minor, if any, impact, it may be treated as a minor corrective action, not impacting baseline scope, cost, schedule, and resources. In contrast, we may identify a major change to scope of work, funding, or schedule requirements that may warrant a replanning (or rebaselining) effort, including changes to the scope statement, WBS, schedules, budget, and resource allocation. To make all involved fully aware of such consequences, spend adequate effort to assess the impact of the change on scope, quality, cost, and schedule and document it in the project change log.

It is difficult to evaluate the impact of changes on schedules without a good network diagram, simply because a network diagram is where dependencies between activities are shown that help us analyze how a change to one deliverable and its activities will affect dependent activities. Still, more often than not, that is what occurs—a schedule impact assessment is made on the basis of a gut feeling. To safeguard against risks related to such assessments, rely on network schedules to produce reliable estimates, even if you are dealing with a small project.

Requiring a cost estimate for the proposed change is a well-meaning strategy to prevent cost surprises. That humans have a tendency to underestimate the cost is well documented in many books and papers. Several decades ago as well as today, missing an estimate by 20+ percent is not unusual. What is unusual is the failure on management's part to take this tendency into account when evaluating a change request. Asking for a detailed estimate when the change request is proposed is a sound management safeguard against this tendency. If the change is major, it is possible to go further and request an estimate from an independent source to compare with the estimate of the change originator.

With schedule and cost impacts identified, the underlying assumptions about resource impact used to derive the schedule and cost impact should be explicitly stated. Even though a project manager may have budget and schedule contingency to utilize to implement a scope change, availability of resources to do the work may be a constraining factor. For full evaluation of impact, a discussion concerning which resources are needed and for how long is necessary.

Finally, a thorough impact analysis should also include information about changes to project deliverables and the final project outcome. Will any project deliverable be impacted by the change? If so, how? Will the quality of the project outcome be affected and how? These are just a few of the critical questions that need to be asked and answered to complete the analysis.

Alternatives

Some project change request forms include a section requiring the originator and the specialists evaluating the change request to include a statement on alternative solutions to the stated need. We feel this is a best practice and, as such, we have included this field in our example project change request.

This requires the team to spend time in the *possibility space,* thinking about alternative ways to implement the change. In practice, it is sometimes the case where a simpler and lower-impact solution is found than the original path that was being pursued. This is where engaging the project specialists in the evaluation process pays off.

Recommendation and Approval

Armed with an impact analysis and alternative solutions, the final pieces of information needed in a well-designed project change request are the recommended direction and documentation of the final decision. We include three decision alternatives to consider: (1) reject the change; (2) approve the change; and (3) defer the change to a later date or event (such as a software release following delivery of the final project outcome).

Using the Project Change Request

As conventional wisdom goes, a change should be proposed as soon as it is needed. We would like to add a little bit more precision here. The early conceptual stage of scoping a project is the time when there is limited if any use for a project change request as a change management tool. With the scope still conceptual in nature, it is impractical to attempt to control changes. In the later stages of scope definition, however, it is practical to start using the project change request. For example, a new product development project that has not experienced the beginning of design work or has not completed any product specifications would not need to apply formal management of change requests. Rather, it would start using a project change request for changes when they either constitute a departure from the agreed design specification, or affect the work to be performed for product design.[8] The project change request would continue to be used from that point forward.

When do you stop applying the project change request? Although the question may seem less than meaningful, there will come a time when any change may impede the project, potentially leading to costly waste and rework. An effective action in such situations would be to impose a scope freeze, a mandate that no changes will be considered unless an overriding reason exists. An example may be a customer-funded requirement to add a new safety feature to a product once it is in final testing and validation. The scope should be frozen before a project enters final test and verification activities.

The project control log is quite efficient to use; only a few minutes is needed to enter information into the log. But that is only the mechanics, which will likely be preceded by a substantial analysis that is a function of the size and complexity of the proposed change. In a small-scale change, determining its scope, cost, and schedule consequences may take 15 to 30 minutes. At the other end of the spectrum are major changes to projects, where a group of experts may spend a week or two to fully assess the requested change's impact on the project's business purpose and goals.

Benefits

As a result of its process, structure, and content, the project control log brings to the project manager the value of making conscious decisions instead of letting change happen in an ad hoc manner. Naturally, the scope change decisions tend to be of a higher quality. When the decisions are based on or coordinated with other tools necessary to assess the change impact, project scope, cost, and schedule are kept in check and in alignment with one another. Also, the documentation of changes reduces project participants' confusion, and results in better-controlled scope change, lower total cost, and fewer delays.

THE PROJECT CHANGE LOG

Project changes may not come in small numbers; rather, they may proliferate. This creates a need for a tool with the purpose of recording, numbering, and coordinating the flow of project changes (see example project change log in Figure 8.4). Administered by a coordinator, a project change log records each change request and assigns it a unique identifier, making sure the decision about it—whether it has been approved or rejected by the change authority—is recorded as well. When a request is approved and the change implemented, that information also becomes part of a project change log.

The project change log provides good oversight of all requested, rejected, approved, ongoing, and completed changes. Such clarity is bound to decrease the confusion about what changes have or have not occurred that is often present when a project change log does not exist.

Project Change Log

Project Name: Rattlesnake
Project Manager: Unger

Change Summary				Assessment			Approvals		Status
Id	Submit Date	Originator	Description	Cost	Schedule	Other	Assessment Approval	Change Approval	
1	4/22/17	Allenbach	Printer upgrade	$6000	1 week	1 person	Approved	Approved (L1)	In Implementation
2	5/9/17	Westheim	Upgrade GUI Software	$12000	3 weeks	2 people	Approved	Approved (L2)	Completed
87	12/7/17	Reed	Add test probes	$15400	11 days	1 person	Approved		In Assessment
n	...	...	...	...	...	...	...	...	...

Figure 8.4: Example Change Control Log

Developing a Project Change Log

Forms such as project change logs have a major advantage in that they look simple to develop. Their appearance as a simple spreadsheet, however, fails to inform us that it takes some information and energy spent in an orderly sequence of steps to produce a meaningful log.

Development of a project change log begins with gathering information previously developed. The scope control system provides full understanding of the project scope change rules that are necessary for ensuring the right information is contained in the log. The project change request form provides the necessary information for the summary section of the project change log, as well as additional information about the impact assessment. Finally, a change coordinator should be appointed to administer the use of the change log.

If one were to perform a web search for example project change logs, many different formats would emerge—some complicated and overdesigned and some too simplified to provide meaningful information. We believe that a well-designed project change log should provide information in four important areas:

1. A high-level summary of each change requested,
2. The outcome of the impact assessment,
3. The approval decision outcomes, and
4. The current status of the requested change.

Developing a project change log centers on creating a format that, when filled out, provides the necessary information in these four areas.

Change Summary

The first section of the project change log should give the reader a quick overview of each scope change that has been requested. Summary information begins with a unique identifier for each requested change (usually as simple as a chronological numbering system) and a brief description of the change. Also included in the summary should be the name of the person who originated the request and the date that it was submitted to the project change authority. All the summary information should be documented on the project change request form and can be directly transferred to the change log.

Assessment Information

The project change log should provide the reader a concise understanding of the impact to the project if the change is implemented. The assessment fields should be tailored to match the way in which an organization typically measures project impact. In the example shown in Figure 8.4, we chose to show project budget impact, time impact, and resource impact. These fields should be customized to match an organization's particular approach to impact assessment.

Decision Outcomes

Moving a scope change from the point of request to implementation requires at least one, and usually multiple, decisions. All change control logs must serve as the documented reference for all decisions associated with changing the scope of a project in order to create a historical record of approved project scope change. A decision outcome entry must be included in the project change log for each decision point identified in the change management process.

Scope Change Status

Finally, it is good practice to provide a field in the project change log that communicates the current status of each requested change. Current status may include a number of states that a change request may be in. Some example status states include "In Assessment," "In Approval," "In Implementation," "Completed," and "Rejected."

Attempt to keep the project change log focused on the four essential pieces of information described previously, and try to avoid overdesigning the log format. As described in "It Only Takes a Spreadsheet," exotic or enterprise-level tools are not required to develop a highly effective project change log.

It Only Takes a Spreadsheet

Final accounting is a painful part of any contractual project, especially if there were a lot of approved project changes. It is even more painful when there is no project change log, as was the case in a project that approved several hundred changes over a two-year project execution. Because of a lack of scope change policy and several changes in the project manager position, the log was never established. Then, at the end of the project, it took the project team several months and thousands of dollars of their time to track down all requested, approved, and rejected changes to include in a final change control log requested by the client. Eventually, the team learned a very simple lesson the hard way: it only takes a spreadsheet to create a useful change log.

Using the Project Change Log

A project change log should be developed and in use as soon as the project scope goes under change control and the first requested scope change emerges. The change coordinator is responsible for keeping the project change log current with the latest project information.

As soon as a project change request is submitted to the project change authority, the requested change should be entered in the project change log. In turn, as each requested change moves through the assessment, decision, and implementation steps of the change management process, the project change log should be consistently updated to capture the necessary information as required by the change log design.

Special diligence needs to be performed to ensure that each decision is documented in the log. If various levels of decision approval are required by the change management process, each change level must be logged along with the decision date and decision maker's name.

Use of the project change log does not end with the logging of the change decision, unless a change is rejected. For the changes that are approved, the log must remain open and in use until it can be verified that a change has been fully implemented. Although it sounds paradoxical, sloppy change coordination may fail to catch a change that was never requested but implemented nonetheless (see "What Do You Do with an Unrequested Change?") Should this happen, the change coordinator must ensure that the change is entered into the log and fully documented to at least provide a historical record of the change.

What Do You Do with an Unrequested Change?

Alan DeFazio was a computer engineer with no prior experience in contractual project work. When he learned that the project's computer vendor went out of business, he simply ordered better and more expensive equipment from another, more reputable vendor. After all, that's what he had done many times for his own company's internal needs.

Four months later when the equipment was delivered, the project manager was more than upset at DeFazio. The project manager asked the very pointed question, "Why did you change computer specs without going through the change request procedure?" Even worse, the project manager refused to pay the price differential between the original and new equipment, saying "I have no money in the budget and to get it I have to go beg my chief financial officer."

The epilogue? After several months of frustration, the project client approved the change. The moral of the story: Having a change procedure in place means nothing unless you train people to use it.

Benefits

For projects that are fraught with changes, the project change log serves as a repository of change that provides good oversight of all requested, rejected, approved, ongoing, and complete changes. The value of this information is in its potential to prevent both cost losses and project delays, by indicating to the decision makers the impact of major changes on the project. Such clarity is bound to reduce the confusion often seen in situations when a project change log is absent.

The project change log fosters an environment of transparency, offering fundamental but brief information on all changes associated with a project. If a simple design is used, all project stakeholders will be able to gain insight on any and all scope changes in a short amount of time, therefore serving as a communication booster.

The project change log also provides a complete historical record of all changes to project scope, therefore providing documented evidence on why changes to project cost and schedule were required. Additionally, it provides a record of who approved each change and when. This information is many times a required deliverable in contractual-based projects.

THE SCOPE CONTROL DECISION CHECKLIST

Throughout a project's life cycle, the need and desire for change will come from project team members, the project sponsor, senior leaders, customers and clients, and other project stakeholders. All requests for change will require a series of decisions to be made to determine if a change is worth considering, if resources should be directed to assess the impact of a change, if the change should be implemented, and who needs to provide the appropriate resources and approvals.

Decision-making consistency therefore is necessary to ensure that project scope management practices are effective.[9] Decision consistency is crucial for establishing and maintaining buy-in for the scope management process, and to prevent various project players from attempting to circumvent the decision process. A standard set of decision questions, formulated into a scope control decision checklist, is a valuable tool for a project manager to establish the level of decision consistency required on his or her project.

Developing a Scope Control Decision Checklist

A scope control decision checklist will be different for every organization because every organization will likely have a slightly unique scope control system and process, complete with decision-making nuances.

Developing a standard set of decision questions for an organization is a good practice as it drives consistency across the projects being executed within the organization. The questions contained within the checklist can be developed by first understanding the change management process. Questions will center on what decisions have to be

Table 8.1: Sample Scope Control Decision Checklist

Status	Checklist Questions
☑	Does the change strengthen the project objectives?
☑	Does the change negatively affect the project objectives?
☑	Does the change modify any of the original project objectives?
☑	Can the change be implemented within the current budget?
☑	Can the change be implemented within the current schedule?
☑	Do we need additional resources to implement the change?
☑	Does the change affect any project deliverables?
☑	Does the change affect the customer?
☑	Have we considered alternative solutions?
☑	Does the change require senior management approval?
☑	What is the impact of not implementing the change?
☑	What base assumptions have been made concerning the change?
☑	What risks does the change introduce?
☑	What risks does the change reduce or eliminate?
☑	Do we need a contingency approach?
☑	How do we verify correct and complete implementation?
☑	Who is the final decision maker?

made as part of the process, what information is needed to make the various decisions, and what decision outcomes are desired.

Table 8.1 illustrates a sample set of questions that can be used as a reference for developing your own decision checklist, which should then become part of your PM Toolbox.

We have included a status column in the sample checklist that, if adopted, can be used to indicate whether the question has been asked and if so, if appropriate action has been taken. Some project managers add a third column to the checklist labeled "documented" to indicate if the answer to the question has been documented.

Using the Scope Control Decision Checklist

Once the initial scope of a project has been established, change will certainly begin to occur. Before this occurs, the scope control system must be put in place and a documented decision checklist established as a core tool within the system.

The decision checklist can be used to eliminate decision process and decision outcome ambiguity from the very beginning of the scope management process. All project stakeholders will observe and experience consistency in the way decisions are being made on a project.

Obviously, not all answers to the questions contained within the checklist will come quickly and easily. Depending on the complexity associated with a requested change, it may take a number of days, or in some cases, weeks to discover the right information needed to make a decision. Use the checklist throughout project planning and execution as a guide and focusing mechanism to ensure that effective, consistent, and expedient decisions are being made.

Benefits

The primary value that the scope control decision checklist provides a project manager is found in two key areas: decision quality and decision consistency.[10]

High-quality decisions are those that are informed, reasoned, and thorough. Improving the quality of a decision begins with understanding the four primary steps in making a project scope change decision:

1. Description of the proposed change and benefits
2. Identification of the cost of change
3. Identification of solution options
4. Understanding the risks involved with the decision

The questions contained in the scope control decision checklist are designed to improve the information used for decisions in each of the steps listed before.

Equally important to decision quality is decision consistency, which involves making high-quality decision on a regular basis. Consistency in decision making comes with approaching decisions in the same manner over time. The scope control decision checklist provides consistency by documenting a standard set of questions to be asked and explored for each project scope change decision facing the project manager and his or her change authority.

References

1. Project Management Institute. *A Guide to the Project Management Body of Knowledge*, 5th ed. (Drexel Hill, PA: Project Management Institute, 2013).
2. Robertson, S., and James Robertson. *Mastering the Requirements Process: Getting Requirements Right*, 3rd ed. London, England: Addison-Wesley Professional, 2012).
3. Heagney, Joseph. *Fundamentals of Project Management (Worksmart)* (New York, NY: AMACOM, 2011).
4. Kerzner, H. R. *Project Management: A Systems Approach to Planning, Scheduling, and Controlling* (Hoboken, NJ: John Wiley & Sons. 2013).
5. Lock, Dennis. *Project Management* (Farnham, Surrey, England: Gower, 2013).
6. Verzuh, Eric. *The Fast Forward MBA in Project Management* (Hoboken, NJ: John Wiley & Sons, 2012).

7. Robertson and Robertson, 2012
8. Pinto, Jeffrey K. *Project Management*, 2nd ed. (Upper Saddle River, NJ: Prentice Hall, 2013).
9. Andler, Nicolai. *Tools for Project Management, Workshops and Consulting: A Must-Have Compendium of Essential Tools and Techniques* (Erlangen, Germany: Publicis Publishing, 2011).
10. Heath, C., and Dan Heath. *Decisive: How to Make Better Choices in Life and Work* (New York, NY: Crown, 2013).

9

SCHEDULE MANAGEMENT

Schedule management consists of assuring that project work is accomplished according to the planned timeline, assessing changes in the timeline based upon changes to the project scope, and establishing a new baseline schedule when necessary. Generally, there is minimal concern if work is completed earlier than planned, so primary attention is focused on preventing schedule slippage—the primary exception being if early work completion creates a cash flow problem later in the project.

Schedule slippage is common and needs to be a primary focus of project managers during the execution stage of a project. Schedule slips caused by scope changes or major resource adjustment are fairly easy to detect. However, humans are inherently poor at managing their time—both personal and professional (see "Tips for Better Time Management")—and as a result, most slippage occurs one day at a time and is more difficult to detect. Project managers need to be vigilant about preventing slippage from accumulating to a conspicuous and unacceptable level. To do so, they need to have tools that provide early warning of potential and realized schedule slips.

The tools presented in this chapter provide a variety of options for early warning detection of schedule slippage. Some tools are more suited for smaller projects, some for larger projects, others for complex projects, and still others for simple projects. Since the type of project a project manager will be called upon to manage will likely vary over the course of his or her career, it is recommended that each tool become a part of all project managers' PM Toolbox.

Tips for Better Time Management

Better management of our time can help to improve our efficiency in both our personal lives and in our professional lives. The following ten tips for better time management are shared by Shirley McDowell, a seasoned project manager for a large financial institution.

1. *Prioritize*. Since it is impossible to do everything, learn to prioritize the important tasks and let go of the rest. This is done to avoid unwanted

(*continued*)

delay in the important tasks and at the same time avoid any chaos resulting from the delays.

2. *Know your deadlines*. When do you need to finish your tasks? Mark the deadlines clearly so you know when the work needs to be completed.
3. *Target to complete early*. With the deadlines identified for your tasks in the previous steps, plan your activities in a manner that will result in early completion if all goes as intended. This will leave room for the unexpected things that will inevitably occur.
4. *Know the results intended*. Make sure you understand exactly what it is you are trying to accomplish at the end of each task. This will help you know what success looks like before you start. It will also assist in knowing when to *stop* working a task.
5. *Create a daily plan*. Plan your day before it unfolds. The plan gives you a good overview of how the day will play out and helps focus the mind.
6. *Focus*. We are all guilty of excessive multitasking. Focus on one task at a time until it is completed or at the intended state you planned for the day. Focus solely on what needs to be done according to your plan in order to increase your efficiency.
7. *Plan for interruptions*. Plan to be pulled away from what you are doing—it's inevitable. This means not creating a daily plan that won't accommodate unexpected interruptions in your day. Allow for the interruptions instead by making them part of your daily plan.
8. *Learn to say no*. Don't take on more than you can handle. For additional tasks and activities that come your way when you are focusing on your high-priority tasks, give a firm "no" or defer them to a later time.
9. *Block out distractions*. It is important that you avoid any forms of distractions while working on the tasks that you have identified as high priority. When focusing on high-priority tasks turn off the phone, close your browser, close instant messaging, close your e-mail. Don't give people your attention unless it's absolutely crucial in your business to offer an immediate human response. This will improve your concentration.
10. *Don't procrastinate*. There is no benefit of putting off tasks that you find uninteresting, difficult, or undesirable if they are important to accomplishing the work you have been assigned. Instead, try taking on these tasks early in the day when you are fresh and have the energy to focus on completing the tasks as quickly as possible.

Although schedule slips are common, they are not inevitable. Given early detection, there is much a project manager can do to eliminate or limit them. For example, they can subdivide tasks into smaller chunks of work; they can reassign resource responsibilities; they can add additional resources, different skill sets, or different skill levels; they can take advantage of schedule slack by adjusting task start/stop dates; and so on. If schedule slippage is detected early enough to enact control actions, project managers stand an excellent chance of managing to their baseline schedule and meeting their timeline commitments. Beginning with the burn down chart, the following tools are designed to assist project managers perform their schedule management duties effectively.

THE BURN DOWN CHART

The burn down chart is a graphical representation of the amount of work completed over a defined period of time as compared to the amount of work initially planned for completion. The chart is a very effective way for the project team to communicate schedule performance on a periodic basis, particularly in reporting schedule performance to the project manager. One of the most powerful attractions of the burn down chart is that it involves psychology by emotionally tying a metric to schedule completion through a visual representation of a path counting down to zero.

As illustrated in Figure 9.1, burn down charts are represented as standard *x-y* charts, with the *x*-axis representing time and the *y*-axis representing units of work completed.

Project managers should keep in mind that burn down charts are not designed as a precise report on overall project schedule performance. We advocate using the chart to measure schedule performance at the deliverable, milestone, work package, or task level in preparation for discussions aimed at determining status at the lowest levels of the project hierarchy. With this recommended usage, the burn down chart will work in tandem with the other schedule management tools presented in this chapter to determine and communicate project-level schedule performance.

Developing the Burn Down Chart

Developing a burn down chart requires a detailed understanding of the work to be completed by each functional or subproject team working on the project. The work breakdown structure (WBS), or product breakdown structure, and the project baseline schedule provide the information needed to create the chart structure which is represented as work completed over a designated period of time. Additionally, any approved scope changes that have not been incorporated into the baseline schedule will need to be evaluated for impact to the chart.

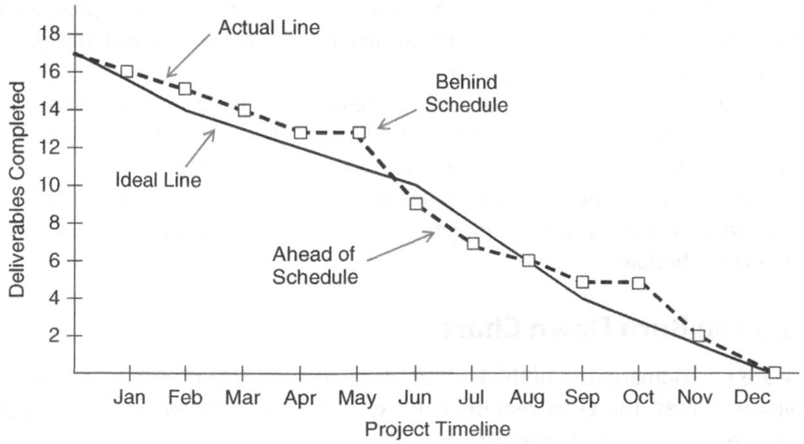

Figure 9.1: An Example Burn Down Chart

Create the Burn Down Template

An effective burn down chart needs to reflect the amount of effort required to complete the work defined over time in a quantifiable manner. Therefore, thought must be put into the most effective way to represent the work effort when creating the template.

As stated earlier, the *x*-axis of the chart represents the project timeline, or a portion of the timeline if a rolling-wave or iterative project execution methodology is employed. Time zero on the *x*-axis represents the point in time when project execution begins, or the point in time when a particular iteration begins.

The *y*-axis represents work completed. In Figure 9.1, we chose to use the number of deliverables completed as the unit for measuring the amount of work completed. The deliverables were derived from the project Gantt chart, segmented for the appropriate owner, and displayed on the *y*-axis of the chart template.

The other measure of work completed we considered using for the *y*-axis was number of hour consumed. Although this is a valid measure of work completed, it is a very weak measure because it does not represent work *output*. As we know, it is possible to consume hours of work and generate no output.

A common measure of work completed used by some project managers is a percentage, as in percent of work completed. We strongly recommend *avoiding* this measure for burn down charts. The problem with using percentage is that is practically impossible to equate work completed with a percentage. How do you distinguish between 30 percent or 40 percent complete without quantifiable units of output to gauge the percentage amounts? If you do have quantifiable units to gauge the percentages, use the quantifiable units instead.

Keep in mind that the finer the unit of measure for work completed, the shorter the delay between updates on schedule performance.[1] This is illustrated in Figure 9.2. The chart on the left shows that three deliverables will be tracked for schedule performance over a 30-day period, the chart on the right shows ten deliverables to be tracked. One can plainly see that a delay in schedule performance would not be reported in the scenario on the left until day 10 (one-third of the performance period expired). Meanwhile, the delay in schedule performance on the right would be reported on day 5 (one-sixth of the performance period expired). Obviously, this scenario provides the greatest opportunity to enact control actions in a timely manner and with the greatest probability of recovering project progress.

The final element of the burn down chart template is the ideal completion line, or ideal line. The ideal line represents the planned completion of work over time and is derived from the baseline project schedule. Many project managers represent this as a line with constant slope. We used a slight iteration of this approach in Figure 9.1 by directly mapping the line to the completion dates of the deliverables as represented in the baseline schedule.

Using the Burn Down Chart

There are two primary uses of the burn down chart: schedule planning and schedule monitoring.[2] First, the chart represent the work that is intended to be completed over a specified period of time with an agreed upon scope of work. During project

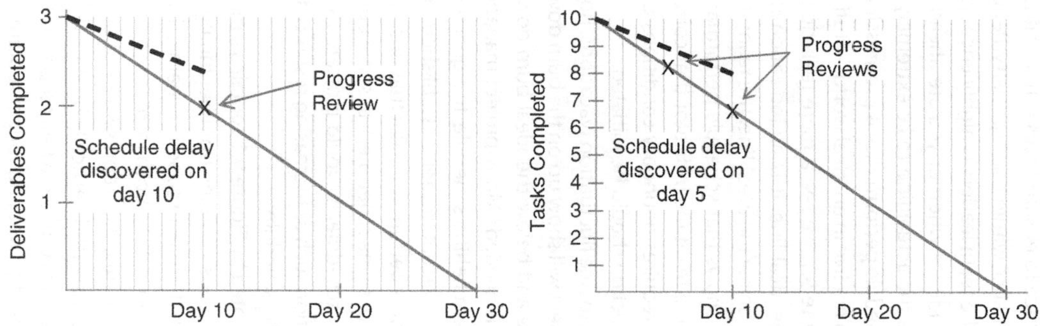

Figure 9.2: Work Completed Cadence versus Information Delay

planning, as the scope is being determined, the burn down chart can be used as an early and low-percentage indication of when the project completion date would likely occur under various scope and resource scenarios. Second, the chart provides a visual representation of the actual work completed during project execution and provides a comparison of planned versus actual schedule performance.

Requiring each functional team involved on a project to create and maintain a burn down chart is an excellent practice to ensure focus is maintained on schedule performance in addition to the performance within their specialty. Periodic and regular review of schedule performance with the team leaders is highly encouraged.

At a glance, the chart can indicate whether you are ahead or behind schedule as planned. Ideally, you would like the actual line to lie extremely close to the ideal line, which would indicate actual schedule performance is progressing as planned. However, this is seldom the case as we do not live in an ideal world—stuff happens and plans are always flawed. Referring to Figure 9.1, we see a more realistic project scenario. When the actual line moves above the ideal line, it is an indication that the team is behind schedule. Conversely, when it is below the ideal line, the team is ahead of schedule.

When the burn down chart shows that schedule performance is behind the baseline plan, the tool can be used as a real-time indicator that corrective action is needed. Using the chart in this manner requires boundary conditions be established to determine exactly when a corrective action should be initiated.

The normal workflow and resource fluctuation (vacations, holidays, multiple project assignments) inherent on a project will show up on the burn down chart as fluctuations in schedule performance above and below the ideal burn down line. So how do you know when a corrective action is needed? Many project managers establish corrective action triggers, in terms of percent behind schedule, to signal the need to initiate a corrective action. For instance, if a 10 percent trigger is established, corrective actions would not be needed for normal schedule performance fluctuations as long as the schedule is less than 10 percent behind the baseline schedule. If at some point the project becomes greater than 10 percent behind schedule, it would indicate that the project manager needs to initiate a corrective action such as increasing resources or resource utilization, decreasing scope, or crashing the schedule.

The project scenario represented by Figure 9.1 contains a major assumption; do you know what it is? It assumes no scope changes from beginning to end. This really is an ideal scenario, as we all know that project scope is dynamic. Figure 9.3 illustrates how the burn down chart is modified to include both increase and decrease in scope.

When an increase in scope occurs, it is represented as an increase in work to be completed, and the ideal line is increased vertically by the additional amount of work required (such as additional tasks). In contrast, when scope is decreased, the ideal line is decreased vertically by the amount of work no longer required (fewer tasks). Once the scope is reset, normal use of the burn down chart again commences.

Benefits

Burn down charts provide value in their simplicity of structure while demonstrating the power of visual project planning and control techniques. They reveal the execution

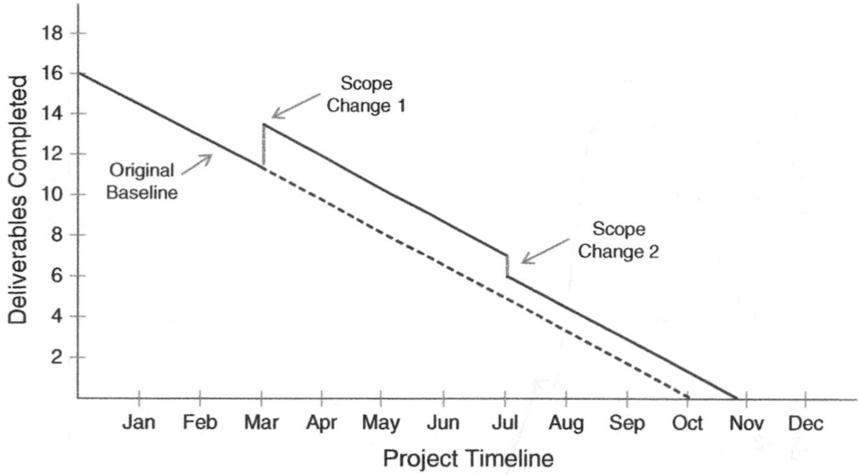

Figure 9.3: Burn Down Chart with Scope Changes

strategy being used, show progress made against the baseline plan, and open the door to discussions about how to best proceed when performance is not occurring as initially predicted.

Although the burn down chart is not designed as a precise indicator of overall project schedule performance, it provides great value for determining schedule performance at the subteam or task owner level. As will be discussed in the tools which follow, in order to determine schedule performance at the project level, it first has to be tracked and understood at the working level where tasks are being completed and deliverables are being created.

THE SLIP CHART

The slip chart tracks progress of the overall project schedule by showing an estimate of how much time the project is ahead of or behind the baseline schedule at the time of reporting (see Figure 9.4). When updated on a regular, periodic basis, the slip chart provides a project manager and project stakeholders a good, overall view of schedule performance in relation to the plan. When consecutive estimates are linked, a trend line is formed. Even though it provides a near-term view of schedule performance and is only moderately valuable as a predictive tool, this has not prevented many project managers from using it in this manner, however. As a result of more practitioners subscribing to proactive project control, the slip chart has taken on the role of helping predict the project completion date and signal the need for corrective actions to fight potential completion date slips. While this does not alter the basic design of this tool, it does require an innovative frame of mind to use it as a predictive tool.

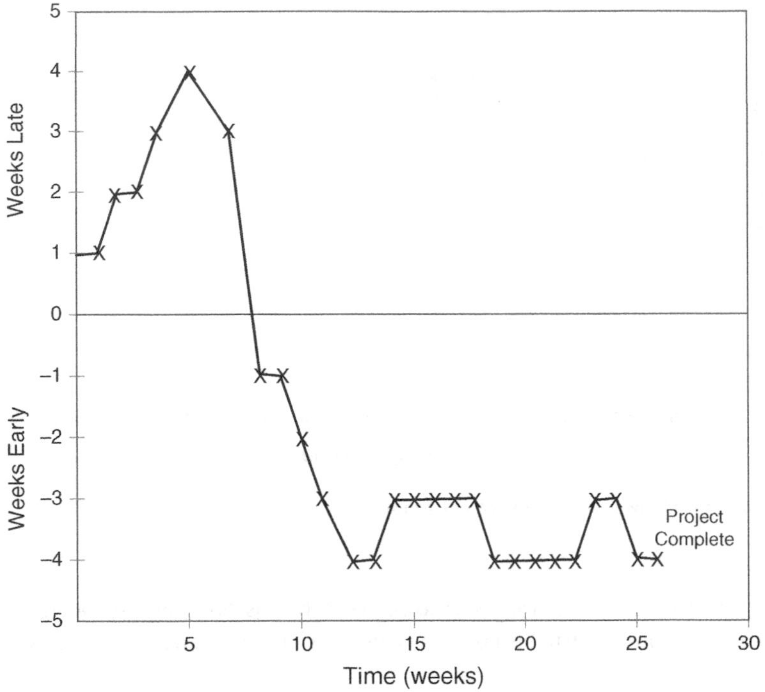

Figure 9.4: An Example Slip Chart

Developing the Slip Chart

Developing a slip chart needs to be precipitated by the gathering of critical information associated with the project. The baseline schedule, preferably a critical path diagram or network diagram, is needed to anchor the status of work completed against the original plan. Current schedule performance status is also needed to compare actual progress to planned progress for each project task. Finally, all approved scope change requests that have not already been incorporated into the current baseline schedule are a necessary input for developing a quality slip chart.

Monitor Project Schedule Performance

Best schedule management practices begin with the regular monitoring of actual execution progress to the planned timeline. For smaller projects, schedule performance can be monitored in periodic team meetings, assuming they are occurring on a regular cadence. For larger, more complex projects, critical path monitoring may have to occur as a separate exercise, outside of the regular team status meetings. In either case, the cadence of the reviews is driven by the length and pace of the project. For example, a waterfall-style project with a two-year timeline would only require weekly or biweekly schedule monitoring. By contrast, a six-month agile-style project may require schedule monitoring every day or every few days.

Review Activity Progress as a Team

With the current status of schedule progress in hand, now is the time to evaluate overall schedule impact with respect to the current progress. The eyes of the project team are primarily on those who own critical activities, and secondarily on near-critical activity owners. With a vivid portrayal of activities' actual progress and any variance from the baseline, one after another, the owners deconstruct activity dependencies to assess their impact on the progress of subsequent activities. Other owners join the impact analysis, receiving and giving more information with the purpose of calculating how slips of individual activities on the critical path combine to establish how much slip exists in the overall schedule.

Review Project Progress

For the first slip chart review meeting, create a template by laying the project timeline along the horizontal axis, "time zero" in the middle of the vertical axis, and early and late increments above and below the "time zero" point. Figure 9.4 illustrates the format of the horizontal and vertical axis, with the unit of time a variable that can be set to days, weeks, months, or quarters depending on the overall length of a particular project.

For subsequent slip chart review meetings, indicate the amount of slip associated with the critical path or critical chain, based on current schedule performance. If it is determined that the schedule end date will be later than planned, go *up* the vertical axis by the appropriate time and mark an "x" along the current date. If the schedule end date is pulled in and will be earlier than planned, go *down* the vertical axis by the amount the project will complete early. Repeat on a regular basis and connect the periodic updates via a line that reveals an overall schedule performance trend.

Predicting Schedule Completion

As stated earlier, the primary utility of the slip chart is in the demonstration of historic and current schedule performance against the baseline plan. However, it can and is used by some project managers to predict the changes to the project end date. Since the slip chart focuses on the project critical path, positive or negative slips indicated on the slip chart can be carried forward to adjust the project completion date.

For example, in Figure 9.4, the slip chart indicates that the project is four weeks behind schedule when the project status is reviewed during week five. If a critical path schedule is the information source, the project manager can then use this information to push the project end date out by four weeks. However, caution must be exercised with this practice. As one can see, the schedule performance on this example project is quite dynamic. The slip is changing consistently and there is an overall wide variation in performance ranging from four weeks behind schedule to four weeks ahead of schedule. If the project manager uses the practice of changing the project end date each time a change in schedule slip is indicated on the slip chart, he or she risks causing frustration and losing credibility on the part of the project sponsor, stakeholders, and partners (see "The Window May Be Closing"). This retrospective view of the project illustrates that using a second schedule management tool with better predictive capabilities in tandem with the slip chart is normally better practice.

The Window May Be Closed

Soon after the start of a website development project, the project slip chart showed a three-week slip. The team added the slip to the project completion date, predicting the project would be three weeks late. Here is an example of how this extrapolation may be a risky practice. One of the project's later critical activities, a one-week-long rapid prototyping, was to be subcontracted to a vendor, who accepted the activity's start date with the comment, "If you come to us a week later, that's fine. If you come later than that, our window will be closed. At that time, add seven extra weeks to the planned delivery date for the prototype. We have already committed to starting another project at that time."

Apparently, the extrapolation is misleading. The predicted completion date is not three but at least seven weeks late. The learning here is to be careful with extrapolations and ensure schedule risk is fully comprehended.

Using the Slip Chart

Small and simple projects can benefit from the slip chart, as can large and complex projects. When applied to track progress in these projects, the chart can work off both the Gantt chart and network diagrams (except the critical chain schedule). Working off the latter is easier and more accurate because the amount of time each activity is behind or ahead of the baseline schedule can be translated into the amount of time the critical path is ahead of or behind the baseline schedule. As we know, the amount of time the critical path is ahead of or behind the schedule is equal to the overall project's time ahead of or behind the baseline. In contrast, with the absence of dependencies between activities, the Gantt chart may pose a challenge when you are converting each activity's amount of time behind or ahead of the baseline schedule into an overall project's time behind or ahead of the schedule. It is this same reason that makes the prediction of the project completion date off the slip chart based on network diagrams easier than based on the Gantt chart.

A smaller, knowledgeable team may be able to develop a slip chart for a 25-activity schedule in about 30 minutes. More project activities will likely increase the time to develop.

The slip chart is best used to monitor project schedule performance, providing a backward-looking view of historic project schedule performance. The information gleaned from the slip chart can be used as input to other schedule management tools with superior predictive capability if a project manager is in need of estimating future schedule performance based on historical and current information. Misusing the chart is bound to reduce its value (see "Three Errors").

Benefits

The slip chart's value is primarily in its ability to record the history of project progress and thus reveal the historic trend.[3] For this value to be further enhanced, an extra step needs to be taken—use the historic trend to forecast future schedule trend and organize actions to deliver the project as planned.

The visual and simple nature of the slip chart adds value to both the project team and executives viewing the chart, making it easy and quick to create and easy to understand the information it conveys.

The slip chart is also very informative. Since the chart focuses on performance relative to the critical path, it immediately indicates when corrective action is needed to attempt to pull the schedule performance back in line with the baseline plan.

Three Errors

The executives of a leading food processing company love slip charts. Consequently, all major projects they oversee are required to submit the chart on a monthly basis. According to Seth Accordino, a senior project manager with the company, "My team gets together on a regular basis and the major emphasis of the meetings is for the activity owners to report their schedule status. I then sit down by myself and prepare the slip chart, and send it to the execs when I'm done. They review it and tell us what corrective actions to take."

There are at least three errors here related to the use of the chart. First, the project manager does not include his team in the preparation of the slip chart. Second, the team is not involved in identifying the corrective actions. Finally, the executives identify the actions without having full understanding of the issues involved, robbing the team of ownership to resolve the issues.

THE BUFFER CHART

The buffer chart plays a very similar role for project managers using the critical chain scheduling as the slip chart does for critical path scheduling (Chapter 6). It measures the status and consumption of buffers established by the critical chain schedule methodology to provide an early warning system to protect the project's end date. First, the buffer chart takes an instantaneous snapshot of buffers' percent consumed relative to the percentage of the work completed on the critical chain (see Figure 9.5). Consecutive snapshots taken at regular periodic intervals are then linked on the chart to obtain a line indicating the trend. For example, in Figure 9.5 the line suggests that the buffer is being consumed at a faster pace than the pace of progress in completing the critical chain activities. In other words, the line answers the question "How are we doing today?," providing information to make a proactive decision to impact the schedule buffer. For the project shown in Figure 9.5, that decision might be to initiate actions to recover the project buffer.

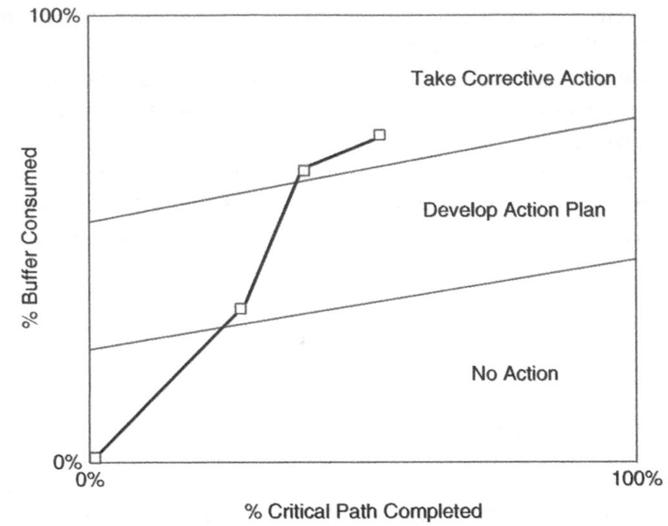

Figure 9.5: An Example Buffer Chart

Developing the Buffer Chart

Developing a buffer chart begins by gathering important information associated with the project that will affect the timeline. The baseline schedule, in the critical chain format only, is needed to anchor the status of work completed against the original plan. Additionally, current schedule performance status is needed to compare actual progress to planned progress for each project task. Finally, all approved scope change requests that have not already been incorporated into the current baseline schedule are a necessary input for developing a quality buffer chart.

Monitor Schedule Performance

As explained earlier, best schedule management practices begin with the regular monitoring of actual execution progress to the planned timeline. For smaller projects, schedule performance can be monitored in periodic team meetings, assuming they are occurring on a regular cadence. For larger, more complex projects, critical chain monitoring may have to occur as a separate exercise, outside of the regular team status meetings. In either case, the cadence of the reviews is driven by the length and pace of the project.

Review Critical Chain Schedule Progress

Developing a buffer chart is most effectively accomplished through involvement of the critical leaders on a project. This normally involves the people responsible for the primary tasks and associated deliverables. The critical chain schedule review meeting is focused on updates from those owning activities being executed on the critical chain, and secondarily on the subordinate merging paths.

With their activity status information prepared in the previous step, the activity owners answer the crucial question, "How many days are remaining on this project activity?" While measuring the project health in this fashion is beneficial for overall good control of the project, it also helps with the next step—monitoring the buffers.

Knowing how many days remain on activities underway indicates the completion date for the activity. The date, then, provides background information to answer the next question, "What percentage of each buffer is consumed?" The emphasis is on the consumption of all buffers, including the project buffer and all critical chain feeding buffers on the merging paths. Typically, the monitoring occurs in a climate without pressure or focus on the estimated activity completion.[4] Rather, there is a realistic expectation that the estimates may vary on a daily basis and may even go beyond their baseline duration estimates. As long as the activity owners continue sticking with critical chain scheduling principles of work and behavior, actual durations of their activities are of no concern.

Monitor the Completion of the Critical Chain

While the consumption of a buffer is of utmost significance, its consequences can only be understood in the context of the performance on the activity chain associated with the buffer. In the progress review meeting the team needs to estimate the percentage of work that is completed on the critical chain and other activity chains. When this information is available, the team can compare the percentage of each buffer consumed with the percent complete of the activity chain associated with the buffer, and be able to establish the project's status or health at any given time. This comparison is conveniently illustrated on the buffer chart (Figure 9.5).

Create and Update the Buffer Chart

Prior to the first review meeting, the buffer chart template must be created. The horizontal axis represents the percent complete of the critical chain, from zero to one hundred percent. The vertical axis represents the amount of schedule buffer consumed, again from zero to 100 percent.

The chart is then sectioned into three parts, representing the action needed relative to the combination of critical chain completed and buffer consumed. Following the guidelines established by Eliyahu Goldratt who pioneered the chart, we divided the buffer chart illustrated in Figure 9.5 into thirds.

Once the buffer chart is created, the progress review meetings focus on periodically updating the chart. To update, begin by marking the critical chain (or activity chain) percent completed on the horizontal axis at the time of the progress meeting. Go vertically upward until reaching a point equal to the percent of buffer consumed, and connect this status point with point zero on the horizontal axis. Repeat drawing status points in each progress meeting, creating a line consisting of connected consecutive status points.

Using the Buffer Chart

Being an integral part of the critical chain schedule methodology, the buffer chart's use is closely linked to how the critical chain schedule is used. The purpose of the chart is to provide an anticipatory tool with clear decision criteria. Buffers, expressed in time units, are used to measure activity chain performance. The crucial point is to establish explicit action levels for decisions expressed in terms of the buffer size, measured in days. Goldratt, the developer of the critical chain scheduling method, proposes the following decision criteria.[5] If a buffer is negative—for example, the latest activity on the chain is late compared to its original completion date—and you penetrated the first third of the buffer, take "No Action" (see the graph). Should you start consuming the middle third, it is time to assess the problem and develop a "Plan" of action. Once you are within the final third, you need to "Act." Note that these hold true for both the project buffer and critical chain feeding buffers.

For the buffer chart to be beneficial, it should be updated as frequently as one-third of the total buffer time. The reason is simple—the decision criteria are based on thirds of the buffer length. For example, whether a buffer is less or more than a third of the total buffer late (less or more than 5 days for a 15-day buffer) determines the type of action we take. In contrast, the chart in Figure 9.5 uses slightly different decision boundaries. See the example titled "You Need to Experiment" for an explanation.

You Need to Experiment

The buffer chart is based on the distinct philosophy of critical chain scheduling. To really comprehend its potential and put it to best possible use, you need to experiment with it and find the decision trigger comfort zone. For example, some companies modified the original criteria for using the buffers. Rather than relying on buffer consumption thirds as decision triggers given by the originator of the tool, they chose decision triggers that change as the consumption of the critical path changes. Decision triggers are borderlines between zones of "No Action," "Plan," and "Act," mandating which action type to use. Look at the chart in Figure 9.5 and notice the subtlety of the slope of the decision trigger lines. The higher the percent of critical chain completed, the higher the decision trigger boundaries as a percentage of buffer consumed. This, of course, makes sense in general—the more work the project team completes, the more buffer consumption the project team can tolerate. But an exact amount of "More" should be picked by the company to fit its business purpose and nature of projects.[6] Experiment to find the decision levels that best fit your company's projects.

Benefits

The value of the buffer chart is in the forward-looking view of current schedule performance it provides. With decision criteria built into the tool, it also forces the necessary conversations between project managers and their sponsors when critical performance trigger points have been reached.

The visual nature of the tool also aids in the presentation of schedule performance status to senior executives and other stakeholders who may not have an intimate understanding of the project details.

Additionally, the buffer chart is partially predictive in nature. The chart is designed to act as an early warning system, prompting the project team to take different actions in situations with different levels of buffer consumption. The fact that the real actions are taken only after a significant portion of the buffer is consumed makes the approach only partially proactive.

THE JOGGING LINE

The jogging line is an excellent tool for showing the project schedule performance in aggregate form. It shows the amount of time each project task is ahead or behind the baseline schedule by drawing a line representing schedule performance in relation to the current date. In that manner, the line provides a snapshot in time indicating the fraction of work completed, and what remains to be completed (see Figure 9.6).

The information gained by assessing the amount of time each task is ahead of or behind the baseline schedule is used to predict the project completion date and create corrective actions necessary to eradicate any potential delay. Since the jogging line focuses on all project tasks and is created as a project team activity, it is best suited for use on smaller and less complex projects.

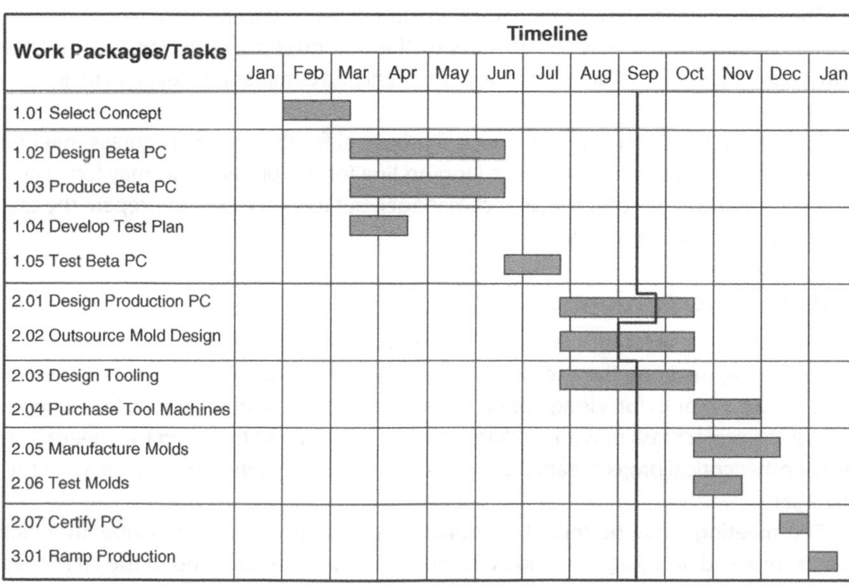

Figure 9.6: An Example Jogging Line

Constructing a Jogging Line

Construction of a jogging line begins with gathering critical information associated with the project timetable. The baseline schedule (either in Gantt or time-scaled arrow diagram [TAD] format) is needed to anchor the status of work completed against the original plan. Current schedule performance status is needed to compare actual progress to planned progress for each project task. Finally, all approved scope change requests that have not already been incorporated into the current baseline schedule are a necessary input for jogging line construction.

The jogging line is most effectively and efficiently constructed as a project team exercise, but before gathering the project team members, it is best to get a sense of the current status of each of the primary work elements of the project.

Review Progress with Tasks' Owners

In spirit of the days when managing-by-walking-around was common, we suggest the project manager have a conversation with each activity (or work package) owner, one-on-one, to gain a good understanding of the current status of the project work scheduled to be taking place at that point in time. As illustrated in Figure 9.6, not all project tasks will be of interest, as some will have already been completed and some are not scheduled to begin yet. For the tasks that are currently being worked, the project manager should inquire about the progress of each task that the person is responsible for completing, whether the current progress differs from the plan, and if so, what issues led to it, and when they anticipate finishing the tasks. If a variance to completion exists, each task owner should be able to explain what can be done to finish the tasks as originally planned.

These questions focus on each of the constituent tasks (work package) of the project. Now is the time to ask how the answers to the task questions will translate into the activity progress—ahead or behind the plan, predicted completion date, and major corrective actions.

Essentially, this initial conversation is a rehearsal for the project progress meeting that will focus on construction of the jogging line for the project. This may look like a time-consuming activity for the project manager, but it is necessary to prepare the task owners for the progress meeting.

Call the Project Progress Meeting

Periodic progress meetings instill the discipline and regularity in reviewing strides that are being made on the execution of a project. They should be called on a regular basis—once a month for a long project or once a week for a short project, for example. Each of the project task or work package owners are required to attend the meeting, as well as other critical project team representatives such as quality assurance or a product manager.

The meetings may be formal, sit-down sessions with an official scribe, in which case, a clear and timed agenda should be communicated beforehand to the attendees.

Or for fast-paced projects that are long on workload and short on time, it is perfectly appropriate to stage short, on-the-go, stand-up meetings. Again, the key is regularity and meeting discipline.

Review Project Progress

With the task owners assembled in the meeting and the information gathered from the initial conversations with the task owners, focus on asking the following five questions to gain a sense of overall project progress:

1. What is the variance between the baseline schedule and actual performance of each task?
2. What are the issues causing the variance?
3. What is the current trend and current prediction of the completion date given the current performance?
4. What new risks have been discovered that may affect the predicted completion date?
5. What actions should be taken to bring performance back in line with the baseline (if a variance exists)?

For those familiar with the work of Deming, one will recognize that the five questions above are an integral part of the "Plan-Do-Study-Act" cycle (see "Project Assessment Questions and Deming's PDSA Cycle").

Where there are interfaces and dependencies between task owners, focusing on them is crucial, since that is where the potential seams may appear on the project. Depending on the culture of the organization, task owners can bring written reports or just verbal information to the meeting.

Project Assessment Questions and Deming's PDSA Cycle

The five questions of project assessment are in harmony with Deming's Plan-Do-Study-Act cycle, a circular approach to project performance improvement. Once the schedule baseline is established in the *Plan* step and project work is being carried out in the *Do* step, project assessment questions one, two, three, and four come into play in the *STUDY* step (see Figure 9.7). Here, the schedule variance is established, its cause determined, and trend forecasted based on current issues and future risks. Then, in the *ACT* step, question five leads to the identification of connective actions, which will be planned for and implemented in the plan-do steps of the next project performance assessment cycle.

(*continued*)

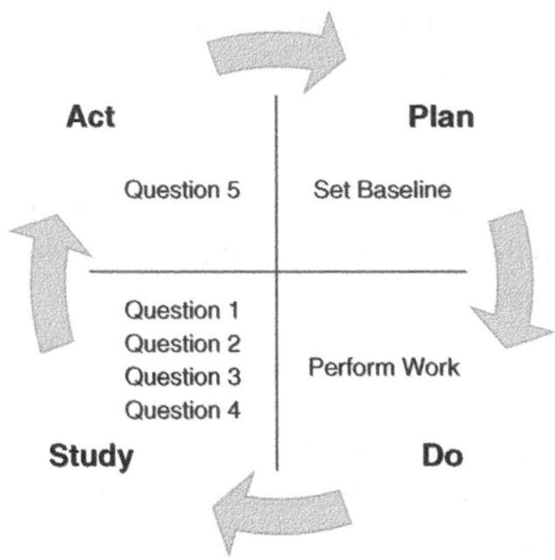

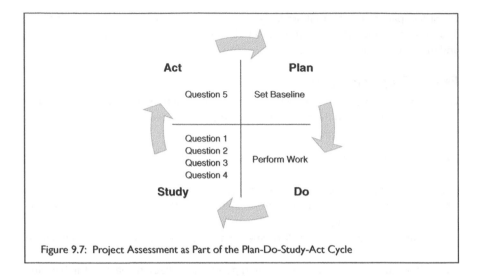

Figure 9.7: Project Assessment as Part of the Plan-Do-Study-Act Cycle

Draw the Jogging Line

Now is the time to assess the current progress of the project by drawing the jogging line based on performance status information from each of the task owners. If the task owners use a burn down chart to track schedule performance, the current performance information can be carried over to the jogging line. Begin the exercise by showing a useful-sized plot of the baseline schedule (preferably one that shows the critical path).

Next, mark on the schedule the date of the progress meeting, whether it's called the data date or, more informally and also more traditionally, the reporting date.[7] From that date, draw a line vertically down until reaching the first task or work package that is currently being worked on. The task owner will tell how many days the work is ahead or behind the baseline. It is of vital importance that their information is reliable. If that's not the case, see what may happen in the example titled "Are You Getting Accurate Information?"

Draw a horizontal line to the left of the data date for as many days the actual schedule is behind. Or, if you are ahead of the baseline schedule, draw the horizontal line to the right of the data date for as many days. At that point, draw a vertical line crossing the first activity. Repeat this exercise for all other tasks being worked on at that time. When vertically crossing the last activity, draw the line horizontally back to the data date and then turn vertically downward. In this way, the jogging line begins from and ends at the data date.

Are You Getting Accurate Information?

Pamela Rice, project manager for a major appliance manufacturer, is struggling with inconsistent progress reporting information from one of her team leaders. According to Pamela, "In a progress meeting, Jim told me his deliverable was three weeks behind schedule. A week later, he reported that he caught up with the deliverable and was right on schedule. Knowing that only one person was working on the activity, I polled several experts and asked how many hours it would take to catch up on the work. The answer was about 140 hours. That means his engineer would have to have worked about 180 hours in the past week."

Pamela is clearly receiving inaccurate status information, which is delivered verbally on a weekly basis. This is a scenario where another tool such as the burn down chart should be used to force more accurate status reporting from within the project team.

Predict Project Completion Date

As shown in Figure 9.6, work package 2.01 is ahead of schedule by approximately a week, while work package 2.02 is behind schedule by about the same amount. If either of these tasks are on the project critical path, an adjustment to the project completion date, as well as the timing of downstream dependent tasks, will have to be made.

Knowing how much each activity is ahead of or behind the schedule prepares you for making an educated forecast of the project completion date. If changes to the baseline schedule result, the next discussion in the progress meeting has to be about actions that can be put into effect to finish as originally planned. It is best to have the initial discussion about corrective actions with the task or work package owners (instead of doing it alone) in order to gain multiple perspectives of options and constraints. The project manager can then assess that information to make a decision on actions to put into play.

Using the Jogging Line

Small and less complex projects are best suited for the jogging line's use. In such situations, when it comes to tracking progress only, the line works well with both the Gantt chart and TAD. If you decide to use the jogging line for larger or more complex projects, it is easier to apply it with a TAD instead of a Gantt chart since the TAD includes dependencies between tasks. This makes it easier and more reliable to translate the amount of time each project task is ahead or behind the baseline schedule into the projected completion date for the project. This is not the case when working on small and simple projects. Here, having the dependencies is not an issue, and the Gantt chart will suffice.

When constructing a jogging line, the following guidelines should be used:

- The line should be drawn on an appropriate baseline schedule in the form of a TAD or Gantt chart.
- The line should be continuous.
- It should start at the data date.
- It should cross each task to indicate its time variance.
- It should end at the data date.

Keep in mind that the jogging line only provides a snapshot in time of project performance, and must be periodically updated. A reasonably skilled and prepared project team can prepare a jogging line for a 25-task Gantt chart or TAD in 15 to 30 minutes. As the number of tasks or work packages grows, so does the necessary time to construct and update the tool.

Variations

There are several Gantt chart–based tools for schedule management that project teams usually consider before discovering the value of the jogging line (see Figure 9.8):

1. Shaded Bar Method
2. Plan vs. Actual Bar Method
3. Percent (%) Complete with Plan vs. Actual Bar Method

The Shaded Bar Method uses shading to indicate the portion of the activity that has been completed. While very visual, the shaded portion cannot show when the implementation of the activity started, how much of the activity scope was really completed, and how much the actual implementation is behind or ahead of schedule.

Only one of these shortcomings is resolved by the Plan vs. Actual Bar Method, which visually compares the plan bar with the actual bar—we see when the implementation of the activity started.

Percent (%) Complete with Plan vs. Actual Bar Method also relies on the plan and actual bar but adds percent complete for each bar. The strongest of these three methods, % Complete with Plan vs. Actual Bar Method, suffers from an inability to reveal how much the actual implementation of an activity is behind or ahead of the schedule. By adding actual bars, it also increases the number of bars on the Gantt chart, making it more complex. Faced with this increased complexity, many would see the jogging line as more effective.

Benefits

The value of the jogging line is in both its historic and forecasting power. The former means that it accurately tells a task owner and the project team the history of work progress. Using the history, then, of course, they can forecast future schedule trend and strategize actions to deliver the project as planned if a variance is determined.

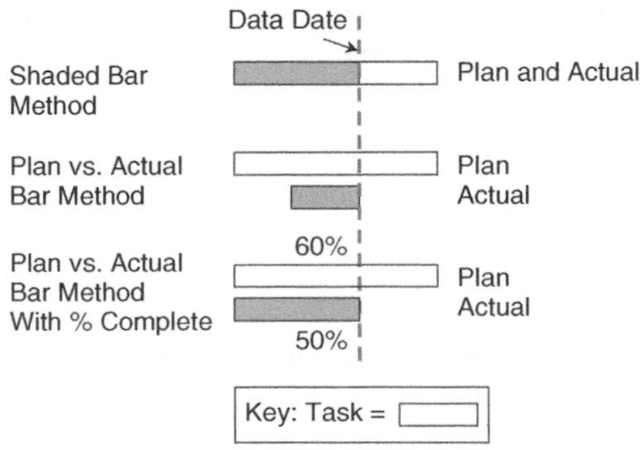

Figure 9.8: Less Effective Alternatives to the Jogging Line

The jogging line provides additional value due to its visual and simplistic structure and presentation. It visually indicates the work done, supplying the project team with an invaluable means of communication. Additionally, in a matter of minutes, almost any project participant can read and draw the jogging line, and then use it to predict schedule performance trends. When used for predicting trends, the jogging line helps build an anticipatory mind-set, equipping the project team to act in advance to combat expected difficulties.

THE MILESTONE PREDICTION CHART

Like other schedule management tools, the milestone prediction chart anticipates the expected rate of future project progress by focusing on major project events: milestones, major deliverables, and project completion. Figure 9.9 illustrates an example of a milestone prediction chart. Note that the vertical axis shows the team's predicted completion date for a specific milestone or deliverable, while the horizontal axis shows the date the prediction was made.

Obviously, the beginning point on the horizontal axis is the time when the schedule baseline is prepared, and its milestone dates are marked on the vertical axis. Once the project work is kicked off, the team reviews progress regularly and makes milestone predictions. By connecting all predictions for a particular milestone into a line, we can obtain

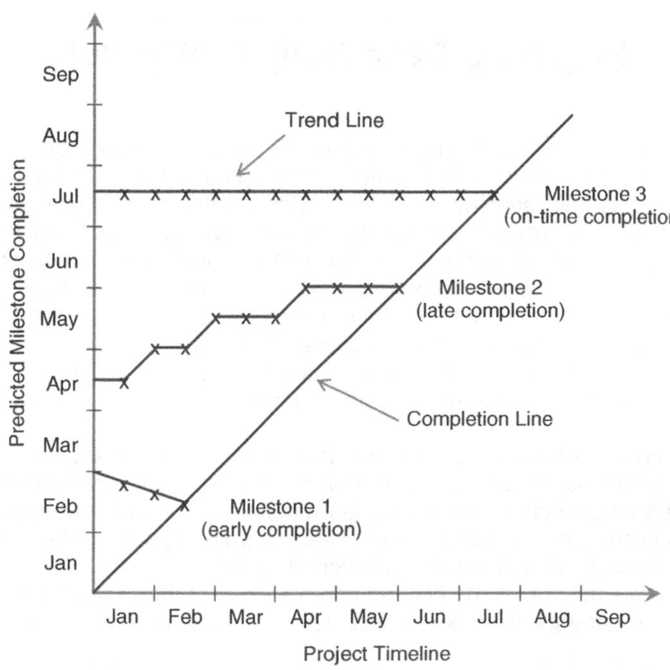

Figure 9.9: Example Milestone Prediction Chart

the milestone trend line. If the line approaches the completion line moving upward, the trend would indicate a slip in the milestone or deliverable completion date. Delivering the milestone right on time would produce a line approaching the completion line horizontally. If we estimate an early milestone completion, the completion line would be approached with a downward trend. Although it is effective in predicting milestone progress, the chart is even more effective if used to develop actions required to eliminate any potential deviation from the baseline milestone schedule.

Constructing the Milestone Prediction Chart

Construction of a milestone prediction chart begins with gathering critical information associated with the project schedule and meeting the major milestones or deliverables. The baseline schedule, preferably in the form of a milestone chart (see Chapter 6), is needed to anchor the status of work completed against the original plan. Current schedule performance status is needed to compare actual progress to planned progress for each milestone. Finally, all approved scope change requests that have not already been incorporated into the current baseline schedule are a necessary input for milestone prediction chart construction. This background information is needed to create the intent of the chart and to predict future milestone completion dates (see "Forgetting Trend Analysis: Déjà Vu?").

The chart is most effectively and efficiently constructed as a project team exercise, but before gathering the project team members, it is best to get a sense of the current status of each of the primary milestones and deliverables of the project.

Forgetting Trend Analysis: Déjà Vu?

Experience shows that the majority of projects place an emphasis on the evaluation of historical data to determine current schedule status as the core of their schedule management practices, leaving out the trend forecast to help predict *future* performance. As experts have been writing for decades, the primary purpose of project control is to prevent any sudden shocks to the project team, project sponsor, and key project stakeholders. Predictive analysis allows for corrective action before issues mount and recovery options evaporate.[8] In other words, we predict via trends because it is much better if we can predict ahead of time what is going to happen, "rather than just watch it happen." This makes trend analysis the single most important piece of information in project control.

Managers, especially those in industries where time to market is a competitive advantage, hate to suddenly hear that the schedule is going to slip. To them, it is much more meaningful to learn that ahead of time. Having a trend projection provides the managers with early warning signals so they can act while it is still possible to reverse unfavorable trends.[9]

Does this mean that the historic information is of little value? Not really. What we are saying is that history, of course, can't be influenced and the future can, so it is more important for the project team to use the historic information to forecast a future schedule trend and strategize actions to deliver the project as planned. It is for this reason that trend analysis should be the central piece of schedule management.

Gather Milestone Information

Good preparation for progress meetings that are chartered with developing the milestone prediction charts has a tremendous impact on how well you control a project. We suggest the project manager have a conversation with each milestone or deliverable owner, one-on-one, to gain a good understanding of the current status at that point in time. As illustrated in Figure 9.9, not all milestones will be of immediate interest, as some will have already been achieved and some are too far in the future that gathering status is really meaningless. For the milestones in play, the project manager should inquire about the progress of each milestone or deliverable that the person is responsible for completing, whether the current progress differs from the plan, and if so, what causes and issues led to it, and when they anticipate finishing the milestone or deliverable. If a variance to completion exists, inquire as to what the milestone owner thinks can be done to finish the milestones as planned.

Essentially, this initial conversation is a rehearsal for the project progress meeting that will focus on constructing or updating the milestone prediction chart. This may look like a time-consuming activity for the project manager, but it is necessary to prepare the milestone owners for the progress meeting.

Review Project Progress

Periodic progress meetings instill the discipline and regularity in reviewing strides that are being made on the execution of a project. They should be called on a regular basis—once a month for a long project or once a week for a short project, for example. Each of the project milestone owners are required to attend the meeting, as well as other critical project team representatives such as quality assurance or a product manager.

Formally or informally, based on verbal or written information, milestone owners need to provide information on the status toward achieving the milestones that are of interest in the review. Since cracks in the project usually appear during interfaces between milestones or their constituent activities, the project team has the benefit of scanning and dissecting the interfaces and their related dependencies. In this process, the understanding of the interfaces and the impact they may have on milestone progress that the team wants to predict will be enhanced. Face-to-face, enriched exchange of information between milestone owners and others involved has no par.

Develop the Milestone Prediction Updates

During the initial milestone review meeting, the prediction chart has to be created. It can then be used in future review meetings. To begin, mark milestones from the baseline milestone schedule on the vertical axis of the chart; these are the "as-planned" milestones. Next, draw the completion line. Because the vertical and horizontal scales use the same project schedule as the basis, the line has a 45-degree angle relative to both scales.

With the initial predictive chart created, the milestone review can commence; the owner of the first milestone crisply describes its actual progress and, potential variance from the baseline and current issues causing the variance, and gives a preliminary prediction of the milestone completion. Relying on a schedule indicating the dependencies between milestones and activities, the owner opines how the actual status of the milestone can impact the progress of other dependent milestones. A pointed discourse will normally unfold.

Owners of dependent milestones ask for more information and share their opinion about their actual progress, variance, and current issues. They also give a preliminary prediction of their milestone completion, analyze their actual impact on dependent milestones, and review possible future risks that may further affect the milestones.

At this point, preliminary predictions of milestone completions are made; however, the job is not yet complete. Measures must be taken to prevent possible slips. Everybody goes back to the drawing board and analyzes the dependencies between milestones. Corrective actions are formulated if needed and final predicted milestone completion dates are established. Those are marked on the milestone prediction chart at the date of the prediction. A final analysis should be performed to determine if the project end date is affected by any slips in the milestone completion dates.

This exercise should be repeated every time schedule progress is reviewed. Once a milestone is completed, mark it on the project completion line. It should be noted that the milestone prediction chart provides the project team a near-term view of milestone progress and is most effective when analyzing project progress within a one- to three-month window. Progress can only be gauged on milestones which are currently being worked on, or have dependencies on the milestones being worked on.

Using the Milestone Prediction Chart

The milestone prediction chart is primarily designed to predict the completion date for major milestones.[10] It was originally designed to focus on six to seven milestones at the highest level of the project, small or large. Despite this, many project managers apply it with minor milestones numbering in the tens.

The chart is best used in a rolling wave manner where detailed milestone progress is analyzed within a one- to three-month window, which repeatedly moves through time with the progression of the project.

To move beyond the near-term window, an analysis of the dependencies between the milestones has to be performed. Geared with such understanding, the project manager can use the chart to formulate control strategies and report project progress to his or her stakeholders.

Benefits

The base value of the milestone prediction chart is in its ability to create a sense of predictability in achieving major project events, or milestones. Through an environment of disciplined progress reviews, the chart helps predict completion dates for major milestones on a regular basis, helping identify trends and leading to actions to correct possible negative trends.

Additionally, the graphical nature of the chart accomplished through the directionality of the line connecting milestone predictions is unmatched and is likely a major reason for the affinity that both top-level managers and project managers have for this tool (see "The Tool of Choice"). As one expert commented, "These graphic representations are nearly infallible in improving schedule predictions over trying to use Program Evaluation and Review Technique (PERT) or Gantt charts alone."[11]

The Tool of Choice

When a group of senior project managers from a leading sportswear manufacturer engaged in a presentation titled "A Few Good Schedule Management Tools," it took them all of about ten minutes to unanimously choose the tool they would prefer to use on their next project to manage their project timeline.

The overview presentation included details on the slip chart, jogging line, baseline-current-future (B-C-F) analysis, earned value analysis, milestone analysis, and milestone prediction chart. When the presentation ended, the project managers were asked to pick one tool that they believed provided the greatest utility for providing a summary level view of project schedule progress. The milestone prediction chart was chosen as the tool of choice.

B-C-F ANALYSIS

The baseline-current-future (B-C-F) analysis compares the baseline project schedule with two predictive schedules—the first one based on the current schedule performance and the second one derived from the worst-case future scenario (see Figure 9.10). As a result, schedule performance trend is detected, or in other words, where we predict the schedule performance is headed. Most importantly, if the trend is unfavorable, it forces a project team to design actions to prevent it, which is the ultimate purpose of all proactive schedule management tools.

The B-C-F analysis enables the project team to visualize the future of their project schedule and devise actions necessary to get there. The analysis is more applicable in smaller and medium-size projects than in large and complex projects.

The three schedules can be either in the format of the Gantt chart or a time-scaled activity diagram. Whichever the format, the B-C-F schedule is no more than an application of well established scheduling tools with a new twist of anticipatory prediction. Simply, this tool is much more about a novel mind set and approach to schedule management than about a novel tool design.

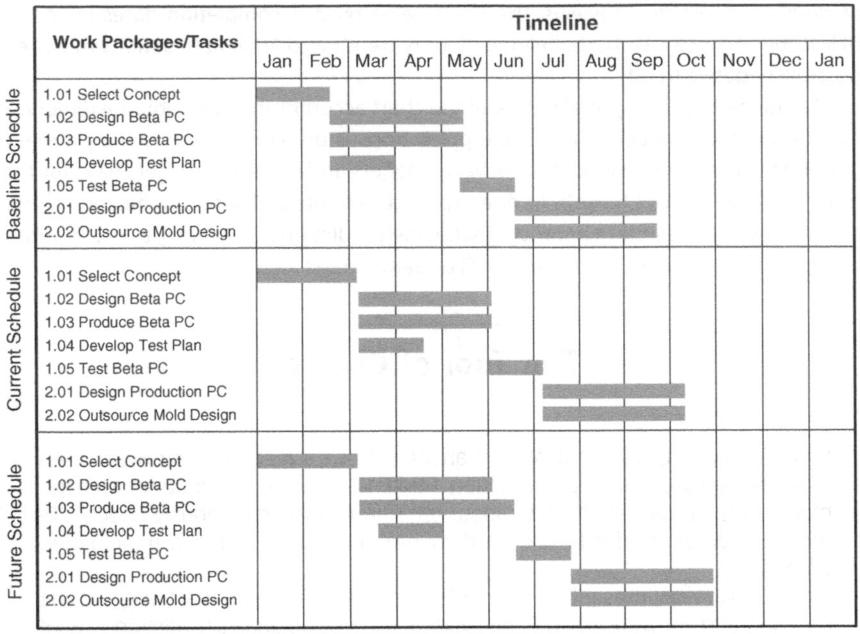

Figure 9.10: Example B-C-F Analysis

Performing the B-C-F Analysis

Performing a B-C-F Analysis begins with gathering critical information associated with the project timeline. The baseline schedule is needed to anchor the status of work completed against the original plan. Current schedule performance status is needed to compare actual progress to planned progress for each project task. Finally, all approved scope change requests that have not already been incorporated into the current baseline schedule are a necessary input for performing a B-C-F analysis.

Prepare the Current Schedule

With a jogging line added to the baseline schedule (see previous section) the project manager is armed with the knowledge of how much each project task is ahead of or behind the baseline schedule, what the current issues are, and what remains to be completed. The project manager and task owners can now forecast the amount of change to the baseline schedule (if any). This will produce a new duration for each task, when it was started, and when it will be finished. Using this information, a new baseline schedule will be drawn. This is the *current schedule*, reflecting how the project is expected to unfold in the future given the task performance to date.

Develop a Future Schedule

To develop the future schedule, the project manager should ask each work package owner the following question: "What is the worst thing that can realistically happen based upon where we are now?" This is a worst-case scenario planning exercise. Visualizing threats, dangers, and risks that the project may encounter helps in developing the future schedule. If the doomsday approach doesn't work, asking seasoned project managers about their worst experiences with the same type of project is another alternative. For example, one may respond that her primary vendor went out of business during her project and set her back by six months. Brainstorming with team members to identify such risks is helpful as well, especially if the organization has an issue risk (see "Issue Database: A Checklist for Future Schedules").[12]

Once the worst-case scenarios have been identified, the next step is to figure out how they can impact the baseline schedule. For this, the project team needs to look closely at the dependencies between tasks in the current schedule. Will the risks impact tasks on the critical path? If the impacted tasks are on a noncritical path, will the total float be consumed? How much could it push out the completion of impacted tasks? If the critical path is impacted, how much will the project completion be pushed out? When the team develops answers to these questions, take the current schedule and extend task durations accordingly to obtain the *future schedule*.

Issue Database: A Repository for Future Schedules

When developing the future schedule in the B-C-F analysis, a project team can go to its issue database as a repository of issues encountered on previous projects. It is possible that issues which occurred in the past may again affect current projects. Therefore, they are potential risks to the current project.

For each issue in the database, the project team asks: "Could this issue become a risk on our project? Would its impact on our project be different from the impact it had on the past projects? Would actions taken on the past projects work on our project, or would we have to use different actions to defend against such risks?"

The database is built over time to include crucial lessons learned about major issues and their impact in past projects. So, it is both experiential and realistic. Using this tactic helps build future schedules of better quality and uses past experience in mitigating risks.

Take Action

The primary purpose of the preceding steps is to equip the project manager with an early warning signal—a signal that says, "Take action to resolve the issues causing the current schedule variance and mitigate risks to the future schedule." If the issues cannot be eliminated, there is a need to rechart the future schedule and find alternatives that would lead the project team to deliver as expected per the baseline schedule. One option is to try to fast-track the project.

To fast-track a project, perform the following activities and evaluate their affects on the overall project schedule:

1. Go back to the future schedule and focus on hard and soft dependencies between activities.
2. Turn any of the sequential activities with hard dependencies into overlapping activities as much as possible, while still observing the hard dependencies.
3. Reexamine all soft dependencies in order to overlap as many activities as much as possible.
4. Also, given the soft dependencies, pick activities that can be performed out of the sequence established in the schedule.

Fast-tracking however, may substantially increase the number of critical activities and paths, putting more pressure on project time management.

Using the B-C-F Analysis

The B-C-F analysis is very beneficial to those using Gantt charts or time-scaled activity diagrams. In projects with Gantt charts, applying a B-C-F analysis requires that the dependencies between tasks be well known, even if they are not shown on the Gantt. When dependencies are formally identified via time-scaled activity diagram, the stage is well set for the B-C-F analysis.

In both applications, the bottom line is to be proactive and insist on applying the B-C-F analysis consistently in progress reviews, both formal and informal.

The analysis is more applicable to smaller and medium-size projects than to large and complex projects. For larger projects with many dependencies, the application of the B-C-F analysis may be too cumbersome and time consuming.

A reasonably skilled and prepared project team of smaller size can prepare a B-C-F analysis for a 25-activity Gantt chart or TAD in 45 to 60 minutes. The necessary time will expand as the project size increases.

When using a B-C-F Analysis, the following guidelines should be used to ensure maximum effectiveness:

- Work with the work package owners to help them give accurate project schedule performance information.
- Strive to develop good enough, not perfect, current and future schedules.
- Insist on maximum interaction among task owners in progress meetings to help them understand how they impact each other and the project.
- Observe which task owners tend to be too optimistic or pessimistic when forecasting their completion times. They may need personal coaching to overcome these tendencies.

Keep in mind that like the jogging line, the B-C-F analysis provides a snapshot in time of project performance and must be performed on a periodic basis.

Benefits

The B-C-F analysis provides project managers a deeper level of schedule forecasting capability, namely the ability to foresee the estimated worst case scenario when schedule slippage occurs and risks begin to mount. Even though there is a perception of "overkill" when a project is running smoothly, one will fully come to appreciate the power of the B-C-F analysis if a project gets in trouble.

For project managers who have been assigned the responsibility of turning a failing project around, there is no better schedule management tool available to help analyze the current situation and forecast future scenarios.

The visual nature and basic structure of the B-C-F analysis provides additional value to a project manager and his or her stakeholders. It vividly indicates current schedule performance on a project, directly overlaid with forecasted future performance.

SCHEDULE CRASHING

Schedule crashing is a method of shortening the total project duration without changing the project logic, which means that the sequence of dependencies between project activities remains the same.[13] To compress the duration, the project usually deploys more resources in performing activities. As a consequence, the total project cost grows as the duration decreases.

Performing Schedule Crashing

Schedule crashing requires a process of disciplined and patient steps outlined here, which we will follow to demonstrate how a schedule is crashed from seven to four days.

A number of inputs are needed in order to effectively crash a schedule. First, the baseline schedule is needed. A network diagram provides the task logic and baseline timeline, which is called the *normal schedule*. Second, the current schedule performance as compared to the baseline is needed to identify the schedule gap that needs to be eliminated by the crashing exercise. Third, any scope changes that have not already been incorporated into the baseline need to be identified and factored into the performance gap. Finally, resource loading and availability, along with the associated labor rates and cost information will be needed to develop the cost impact of the crashed schedule when completed.

Develop a Normal, Cost-Loaded Schedule

This is the baseline schedule developed during project planning. Here resources are assigned to project activities, and their costs are calculated. Without this resource and cost information, schedule crashing the way we define it here is not possible. In Figure 9.11, we give an example of a normal schedule (starting position) with duration and cost for each activity in the table.

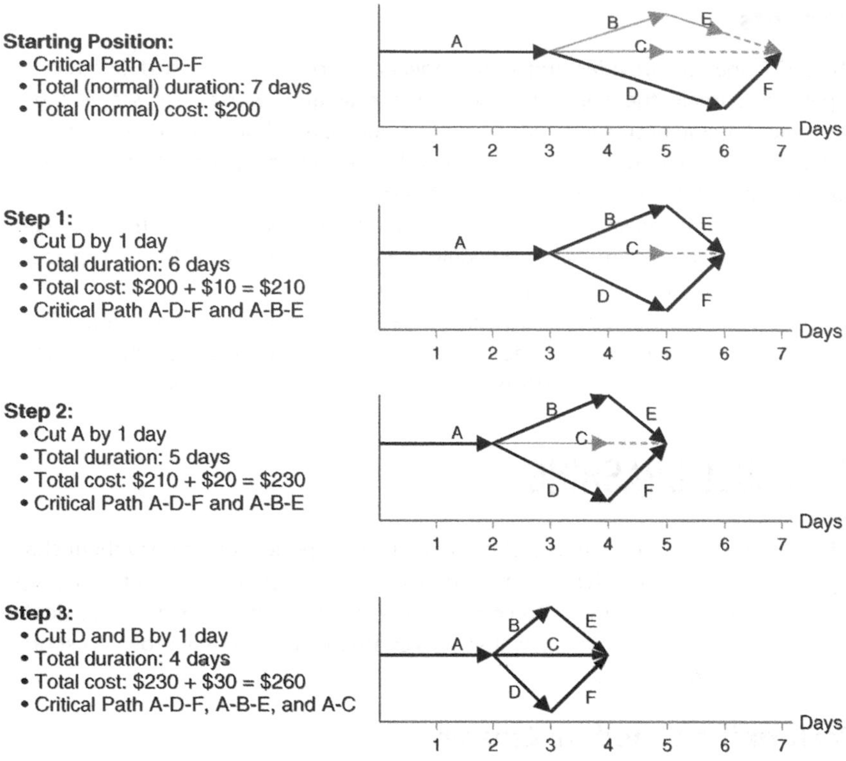

Starting Position:
• Critical Path A-D-F
• Total (normal) duration: 7 days
• Total (normal) cost: $200
Step 1:
• Cut D by 1 day
• Total duration: 6 days
• Total cost: $200 + $10 = $210
• Critical Path A-D-F and A-B-E
Step 2:
• Cut A by 1 day
• Total duration: 5 days
• Total cost: $210 + $20 = $230
• Critical Path A-D-F and A-B-E
Step 3:
• Cut D and B by 1 day
• Total duration: 4 days
• Total cost: $230 + $30 = $260
• Critical Path A-D-F, A-B-E, and A-C
A
B
C
D
E
F
Days
1
2
3
4
5
6
7

Figure 9.11: Schedule Crashing Example

Develop a Crashed, Cost-Loaded Schedule

While preserving the sequence of dependencies between project activities, perform the following steps:

1. Estimate for each activity the shortest possible time to complete it (unconstrained estimate).
2. Ask the activity owners and team the following questions: "What resources do you need to complete each activity in this amount of time?" and "How much does it cost?" This is sometimes a painstaking exercise, requiring a lot of quick and good information. Also, it may take multiple iterations to develop these estimates, called crash durations. In the process, various challenges will be encountered. For example, some activities cannot be completed in shorter time than the normal schedule shows, or some activities will require additional human and nonhuman resources. Not surprisingly, some resources that the project needs may not be available even though there is budget for them. Generally, these estimates will be developed following the rules of time and cost estimating, which are heavily dependent on the knowledge of project technology and productivity of resources.

3. Reestimate the cost of resources and adjust the baseline project budget to reflect the crashed timeline. Adjustments may then be needed to match the timeline to available project budget. This may require multiple iterations of schedule crashing and budget alignment.

Once crash durations and costs are prepared and aligned, the crashed schedule is ready. Table 9.1 shows the duration and cost for each activity in the crashed schedule for our example.

Compute the Cost/Time Slope

All activities are not created equal. Some of them are more costly to shorten than the others. Calculating the cost/time slope will show the cost of reducing duration of each activity by one day. Use the following formula to compute the slope for each activity (see Table 9.1):

$$\textit{Cost/time slope} = (\textit{crash cost} - \textit{normal cost})/(\textit{normal time} - \textit{crash time})$$

This creates the basis to identify the most cost effective activities to crash. But don't start crashing yet. First, the sequence of crashing activities needs to be determined.

Focus on the Critical Path Only

The critical path is the longest path in the network schedule, composed of activities whose float is zero. The duration of the critical path is the minimum time to complete all project activities. Therefore, the duration of the critical path is equal to the total project duration. The only way to shorten the project duration, then, is to shorten duration of activities on the critical path (see the example titled "Wasting Money by Crashing Non-critical Activities"). Simply, crashing the critical path duration by a certain number of days will translate into reducing the total project duration by the same number of days. Now look at the network diagram and identify the critical path (in Figure 9.11 we show the

Table 9.1: Crashed Schedule Duration and Cost Amounts

	Duration (Days)		Cost ($)		
Activity	**Normal**	**Crash**	**Normal**	**Crash**	**Cost/Time Slope***
A	3	2	30	50	20
B	2	1	40	60	20
C	2	1	20	80	60
D	3	1	30	50	10
E	1	1	40	40	0
F	1	1	40	40	0
	Total: 7 days		Total: $200		

*Cost/Time Slope = (Crash cost − Normal cost) / (Normal duration − Crash duration).

critical path for each schedule as a bold line). Working off a time-scaled diagram is the easiest way to crash the schedule because it shows activities with and without float visually, which is why we use it in our example.

Wasting Money by Crashing Noncritical Activities

How often does a project schedule slip? In our experience, project delays are widespread and many of those projects take actions to accelerate and catch up with the baseline schedule. A frequent action of this type is to throw in additional resources indiscriminately to shorten duration of project activities. Since all too often the schedule does not show dependencies, and accordingly the critical path, both critical and noncritical activities are many times crashed. Crashing noncritical activities increases the total project cost without compressing the schedule. This is an unnecessary waste of project budget. The only way to reduce a project's duration without changing the project logic is to crash critical activities.

Crash the Most Cost-Effective Activities

When crashing, we want to do it with the minimum cost increase. For this reason, we don't choose to crash just any activity on the critical path. Rather, we focus on the most cost-effective one to shorten first, by selecting a critical activity with the least cost/time slope. In our example in Table 9.1 activity D has the least cost/time slope on the critical path, $10/day (activity F cannot be shortened). Cut it by one day. Now the schedule is one day shorter, or six days long, and its cost equals the normal cost of the schedule plus the cost/time slope of the activity that was cut, which makes $210 (see Step 1 in Figure 9.11). Continue with cutting critical activities one day at a time, first those with the least cost/time slope—see Steps 2 and 3—until reaching the desired schedule duration of four days and its cost ($260).

Crash Multiple Critical Paths

It is a rare privilege to have a single critical path. More often than not, as we crash activities on the original critical path (activities A-D-F in Table 9.1), new critical paths appear. After cutting D in the first step, there are two critical paths—A-D-F and A-B-E. When this happens, to shorten the total project duration, shorten duration of all, in our case both, critical paths at the same time. This is why in Step 2, we cut activity A, which is on both critical paths, and in Step 3, we cut D on one critical path and B on another critical path.

Failing to shorten one of them will leave the total project duration unchanged; simply, the longest path(s) determines the total duration. As multiple critical paths are crashed simultaneously, follow the rule of first crashing the least slope activity on each of the paths. Having multiple paths is why we enforce the rule of "crash one day at a time."

Using Schedule Crashing

Schedule crashing is primarily a method involving two project scenarios. In the first, the project is in the planning stage, the execution has not started yet, and the project team proposes a schedule for management approval. Management finds the schedule too slow and demands it be shortened. To accomplish this, the team goes back to the drawing board, employing schedule crashing.

The second scenario occurs when the project is under way and the schedule slips.[14] To catch up, the team may use the schedule crashing method. While for both scenarios the team can apply schedule crashing alone, many teams combine it with fast-tracking. Remember, fast-tracking changes the project logic, altering the dependencies between project activities.

Crashing a 250-activity network diagram in a half-day to one-day time is realistic for a skilled project manager. Expect this time to increase with the increase of the team size, since larger teams need more time for internal communications.

Benefits

The value of schedule crashing lies in its capability to provide a way to correct for negative schedule performance variances. Step by step, it shows which activities to crash, what resources it takes, and how much it costs. To all organizations cherishing time-to-market speed, or more generally, fast cycle times, this capability is a significant benefit (see "Five Golden Rules of Schedule Crashing").

Five Golden Rules of Schedule Crashing

1. Crash only activities on the critical path.
2. Crash by one time unit of the schedule at a time (one day at a time for example).
3. When there are multiple critical paths, crash all of them simultaneously.
4. First crash critical activities that are the least costly to crash (the least cost/time slope).
5. Don't crash noncritical activities.

CHOOSING YOUR SCHEDULE MANAGEMENT TOOLS

The seven tools presented in this chapter are designed for different project situations. Most project managers find it useful to employ one to three schedule management tools, and face the problem of eliminating the other tool options. To help in this effort, Table 9.2 lists various project situations and identifies which tools are geared for each situation. Consider this table as a starting point, and create your own custom project situation analysis and tools of choice to fit your particular project management style.

Table 9.2: A Summary Comparison of Schedule Control Tools

Situation	Jogging Line	B-C-F Method	Milestone Prediction Chart	Burn Down Chart	Slip Chart	Buffer Chart	Schedule Crashing
Small and simple projects	✓	✓	✓	✓			
Task-level progress reviews	✓	✓	✓	✓	✓	✓	
Low-level detail needed	✓			✓	✓	✓	✓
Short time to train how to use the tool	✓	✓	✓	✓	✓		
Focus on highly important events			✓	✓			
Large, complex, and cross-functional projects	✓		✓		✓	✓	✓
Fast projects				✓		✓	✓
Projects of strategic importance			✓	✓		✓	
Focus on top-priority activities			✓		✓	✓	✓
Summary detail needed		✓	✓	✓	✓		
Display trend		✓	✓	✓	✓	✓	
Provide predictive analysis		✓	✓			✓	
Little time available for schedule management	✓			✓			
Analyze failing project		✓		✓	✓	✓	
Correct project delays							✓

References

1. Cockburn, Alistair. *Crystal Clear* (Boston, MA: Addison-Wesley, 2004).
2. Taylor, James. *Project Scheduling and Cost Control: Planning, Monitoring and Controlling the Baseline* (Plantation, FL: J. Ross Publishing, 2007).
3. Kahn, K. B., ed. *The PDMA Handbook of New Product Development* (Hoboken, NJ: John Wiley & Sons, 2012).
4. Leach, L. P. 1999. "Critical Chain Project Management Improves Project Performance." *Project Management Journal* 30 (2): 39–51.

5. Goldratt, E. M. *Critical Chain* (Great Barrington, MA: North River Press, 1997).
6. Kania, E. "Measurements for Product Development Organizations: A Perspective from Theory of Constraints." *Visions* 24 (2): 17–20, 2010.
7. Project Management Institute. *A Guide to Program Management Body of Knowledge*, 5th ed. (Newtown Square, PA: Project Management Institute, 2013).
8. Meredith, J. R., Samuel J. Mantel Jr., and Scott M. Shafer. Project Management in Practice. (Hoboken, NJ: John Wiley & Sons, 2013).
9. Kharbanda, O. P., E. A. Stalworthy, and L. F. Williams. *Project Cost Control in Action* (Farnborough, England: Gower, 1981).
10. Pinto, Jeffrey K. *Project Management*, 2nd ed. (Upper Saddle River, NJ: Prentice Hall, 2013).
11. Silverberg, E. C. "Predicting Project Completion." *Research-Technology Management* 34 (3): 46–47, 2011.
12. Lientz, B. P., and K. P. Rea. *Breakthrough Technology Project Management*, 2nd ed. (San Diego, CA: Academic Press, 2011).
13. Kerzner, H. *Project Management: A Systems Approach to Planning, Scheduling, and Controlling*, 11th ed. (Hoboken, NJ: John Wiley & Sons, 2013).
14. Meredith, J. R., and S. J. Mantel. *Project Management: A Managerial Approach*, 8th ed. (Hoboken, NJ: John Wiley & Sons, 2011).

10

COST MANAGEMENT

Project cost management involves management of the processes required to ensure that project work is accomplished within an approved budget, assessing changes to the budget based upon changes to the project scope, and establishing a new budget baseline when necessary. Generally, there is minimal concern if work is completed under budget, so primary attention is focused on preventing budget overruns. The primary exception being if a budget underrun indicates work is not being performed as planned.

Budget variances are common and need to be a primary focus of project managers during the execution stage of a project (see "Common Reasons for Cost Overruns"). Cost variances caused by scope changes or major resource adjustments are fairly easy to detect. However, humans are inherently poor at managing their work and as a result, most budget under and overruns occur one day at a time and are more difficult to detect. Project managers need to be vigilant about preventing budget variances from accumulating to a conspicuous and unacceptable level. To do so, they need to have tools that provide early warning of potential and realized budget slips.

The tools presented in this chapter provide a variety of options for early warning detection of budget variances. Some tools are more suited for smaller projects, some for larger projects, others for complex projects, and still others for simple projects. Since the type of project a project manager will be called on to manage will likely vary over the course of his or her career, it is recommended that each tool become a part of each project manager's PM Toolbox.

Common Reasons for Cost Overruns

It is a well-known fact that a large number of projects complete later than scheduled and exceed cost estimates. No single reason explains all of the cost overruns on projects. However, there are a number of reasons that show up consistently in project cost management studies.

1. *Insufficient funds*. One of the main reasons for project budget overrun is underfunding. This refers to the act of not allocating an adequate amount

(*continued*)

of budget to a project to begin with. Without adequate funding, project success becomes a matter of wishful thinking.

2. *Inaccurate cost estimates*. Accurate cost estimation is crucial to effective cost management. If the cost of the project is underestimated during the planning stage, a budget overrun will eventually occur unless the scope of the project is reduced to match the estimated cost. This is especially true in environments where top managers of a firm set cost targets or not-to-exceed thresholds that artificially limit the amount of budget that is estimated. Likewise, if the cost of a project is overestimated, a budget underrun will occur.
3. *Scope increases*. Increases in scope on projects frequently cause cost overruns. These changes map directly to project requirements. Normally, scope increases are a result of missed requirements during project definition and planning or new requirements that are introduced during project execution. New requirements demand additional work and additional work demands additional project cost.
4. *Extended project schedules*. If project activities take longer to complete than originally planned, the project schedule usually has to be extended to reflect the correct amount of time required. Schedule extensions mean additional costs caused by additional unplanned resource hours. This is especially true as the complexity of a project increases. The higher the project complexity, the higher the number of interdependencies between team members and activities; therefore the higher the probability that interconnected tasks and activities will take longer than planned.
5. *Lack of risk management contingency*. Failure to establish risk management contingency budget for events likely to occur or with severe impact on project completion accounts for many cost overruns. Failure to plan risk management budget contingency creates the base assumption that the best-case execution outcome will occur, and no unexpected impacts to the project will occur.
6. *Poor cost management*. Failure to effectively manage the cost aspects of a project have resulted in many project cost overruns. They can include the failure to clearly identify who is responsible for managing project cost, who has the authority to approve changes to the project budget, how cost performance will be measured and reported, and the identification of triggers to signal the need for cost management corrective actions.

Even though changes to the project budget are common, they are not a foregone conclusion if effective project cost management practices are followed. Cost management practices involve cost performance measurement, forecasted budget at project completion, change impact analysis, recommended corrective actions, and making updates to the project management plan and budget baseline.

Beginning with the cost management plan, the following tools are designed to assist project managers perform their project cost management practices effectively.

THE COST MANAGEMENT PLAN

The cost management plan is a tool that describes how the budget and costs will be managed on a project. It sets the process for how project costs are measured, reported, and controlled by the project manager and his or her team.

Developing the Cost Management Plan

The cost management plan identifies who is responsible for managing a project's costs, who has the authority to approve changes to the project budget baseline, how project cost will be measured and reported on, and how cost variances will be controlled or corrected.

The following elements of a cost management plan can be used as a suggested template that can be modified to formulate a customized plan that meets the needs of your organization.

Purpose

The cost management plan should begin with a statement of purpose for diligent cost management practices on the project, as well as a statement of purpose for the plan. It is also good to include a statement of who is responsible for managing and reporting the project's cost throughout the duration of the project in this opening section of the plan. In most cases, this is the project manager.

Cost Management Approach

This section of the cost management plan explains the approach that will be used to manage the costs on the project. It identifies the processes, procedures, and tools that will be used to manage cost during project execution.

Begin by describing how the baseline budget was derived and where the budget funding is coming from. Also included in this section is the identification of who is involved in managing costs on the project and their specific roles.

Tracking Budget Expenditures

In this section, focus is put on how the expenditures will be tracked on the project while it is in the execution stage. Discuss how project expenditures are tracked, what tools will be employed, where the expenditure data comes from, how often the expenditure data should be refreshed, what expenditure reports should be received for the project and by whom, and what should happen if discrepancies in the data are discovered.

Metrics and Reports

This section of the plan describes the standard metrics that will be used to track and report project budget expenditures. Also covered is a description of expenditure report format, required content, and cadence of delivery to the project manager.

Analyzing Cost Variances

Describe how variances (either positive or negative) between actual performance and the planned budget expenditure are derived and analyzed. Define the control thresholds for the project (see Chapter 12), and what actions should be taken if project performance triggers a control threshold.

Changes to the Project Budget

This section describes the process used to review and approve a change to the project budget, the approvals required to make a budget change, and how a new budget baseline is established.

Project Budget

This final section details the project budget. It is best to present the budget in its various categories:

- Fixed costs
- Material costs
- Other direct costs
- Burden and overhead
- Total project budget

In addition, if there is any budget contingency or management reserve amounts that are being held separately, it is good practice to explicitly state those amounts as part of the budget breakout.

Using the Cost Management Plan

Project costs cannot be managed just by thinking about them on a regular basis. Cost management must be handled and performed as a primary project management process, one that requires forethought and planning. By developing a cost management plan, a project manager establishes the expectations of how cost will be managed on a project and by whom. This in turn establishes the necessary roles and responsibilities for effective cost management among the project team members.

The cost management plan is established during the project planning stage of a project and then put into practice during the project execution and closure stages. It is equally important for small projects and large projects, as well as simple or complex projects.

THE BUDGET CONSUMPTION CHART

The budget consumption chart is a graphical representation of the project expenditures as they occur over the project timeline. The chart is a very effective tool for project managers to communicate current budget consumption status. Like the schedule consumption chart presented in Chapter 9, one of the most powerful attractions of this chart

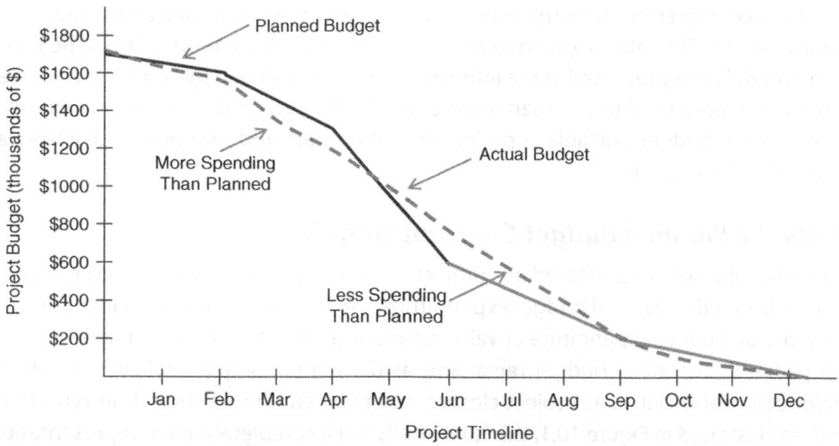

Figure 10.1: Example Budget Consumption Chart

is the psychological benefit of representing progress by a path that counts down to zero project budget available.

As illustrated in Figure 10.1, budget consumption charts are represented as standard *x-y* charts, with the *x*-axis representing the project timeline and the *y*-axis representing the project budget.

Project managers should keep in mind that the budget consumption chart is not designed as a comprehensive tool to measure overall project cost performance, especially as it directly relates to schedule performance. Rather, it is only focused on measurement of project budget spending and serves as a good tool for preparing the project manager for a broader analysis of project cost management status.

Developing the Budget Consumption Chart

Developing the budget consumption chart requires a detailed understanding of the project scope, the resources needed to complete the work defined by the scope, the cost of the resources, and the overall budget for a project. This information is normally integrated into a detailed project budget (see Chapter 7). Additionally, the incremental cost of any approved scope changes that have not been incorporated into the baseline project budget will need to be evaluated for impact to the project budget and incorporated into the chart.

Create the Budget Consumption Template

The budget consumption chart simply represents the amount of project budget available to the project manager at any point in time. Begin by laying the project timeline across the *x*-axis of the chart. If a rolling-wave or iterative project execution methodology is being employed, show only the portion of the project timeline that is pertinent. Time zero on the *x*-axis represents the point in time when project execution begins, or the point in time when a particular execution iteration begins.

The *y*-axis represents the project budget, and should be distributed across the axis in equal amounts. The total project budget is represented at the highest point of the *y*-axis with zero dollars represented at the intersection of the *y*-axis and the *x*-axis. The budget should be represented in this manner in order to illustrate budget consumption over time, with full budget available at project execution start, and zero budget available at planned project closure.

Draw the Planned Budget Consumption Line

Using the information contained in a project-scheduling tool (the Gantt chart is recommended), plot the planned budget expenditure over the project timeline. Next, connect the periodic budget expenditure or value points with a contiguous line that represents the planned amount of budget remaining at the normal project review intervals, or major deliverable dates. At project closure, the planned budget should be zero. Note that, as illustrated in Figure 10.1, this is normally not a completely linear representation due to the variations of resources, materials, and other direct costs; the exception being projects that have constant level of effort resource loading. The chart will normally look like an inverted S chart.

Using the Budget Consumption Chart

The budget consumption chart is used for both the planning and execution stages of a project.[1] First, the chart represents the amount of project budget available over a specified period of time. During project planning, the chart can be used as an early indication of project spending under various scope, resource, and timeline scenarios. This is especially useful in situations where budget is constrained, or available at various points in time—such as projects that rely on government funding from multiple sources or multiple fiscal years. Second, the chart provides a visual representation of the amount of budget spent and amount of budget still available at any point on the project timeline.

At a glance, the chart can indicate whether a project manager is spending more or less budget than planned. Ideally, you would like the actual line to lie extremely close to the planned line. However, we do not live in an ideal world and project spending seldom occurs as planned. Referring to Figure 10.1, we see a more realistic scenario. When the actual line moves above the planned line, it is an indication that the team is spending less than was planned, and therefore more budget is available than planned. Conversely, when it is below the ideal line, more project budget is being consumed than planned.

When variations above and below the planned line show up, it may indicate the need for corrective actions. However, how do you know when a corrective action is needed? Good cost management practices involve the use of corrective action triggers. One such trigger might be percent above or below planned budget consumption. For example, if current budget consumption is 10 percent more or 10 percent less than planned at any point in which the project status is reviewed, it would indicate that a corrective action analysis is needed.

Care should be taken to restrain yourself from reacting too quickly, however. Keep in mind that you are only looking at the project from a single perspective (budget consumption), and that you are missing the project schedule, resource, and risk perspectives. Recall that earlier we stated that the budget consumption chart is best used to prepare the project manager for a broader discussion on project performance. Here's why, it only provides a single perspective of performance and taking corrective action on a single perspective is a risky proposition. Instead, we recommend that the budget consumption chart be used in tandem with other cost management tools presented in this chapter that provide a more integrated view of project performance.

Variations

The project scenario represented in Figure 10.1 contains a major assumption. It assumes that the entire project budget is available when project execution begins. Many times this is not the case, such as in the world of government procurement projects as mentioned earlier. Figure 10.2 illustrates how the budget consumption chart can be modified to include changes in project funding.

When an increase in project funding occurs, the project budget increases and the planned budget line on the chart is increased vertically by the amount of incremental budget received. In contrast, when project funding is lost, perhaps due to descope of work, the planned budget line is decreased vertically by the amount of budget no longer available to the project manager. Once the budget baseline is reset, normal use of the budget consumption chart resumes.

Benefits

Since the budget is often a primary constraint on a project, a quick and easy method to track its consumption and change over time is necessary for any project, large or small. The budget consumption chart provides value to the project manager by conveying project budget status in a simple and powerful manner through visual graphical analysis.

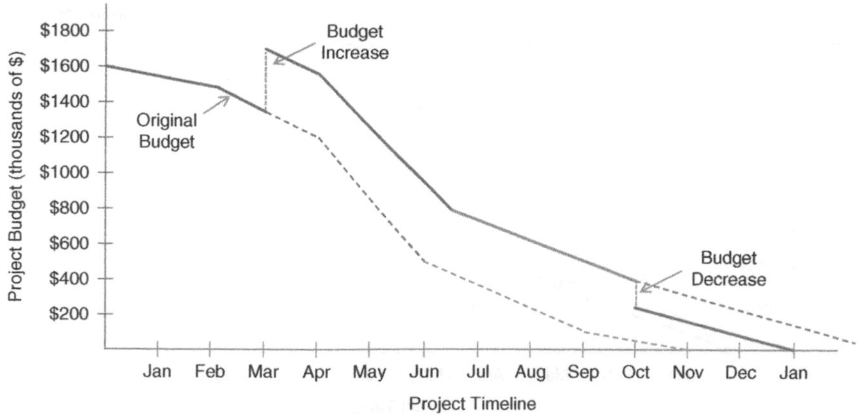

Figure 10.2: Budget Consumption Chart with Funding Changes

Even though the chart is not designed as a precise indication of overall project cost performance, it conveys progress made against the baseline and can be used to provide necessary detail when there is a need to ascertain overall project performance to plan.

Used in tandem with other project cost management tools, the budget consumption chart is a good asset for determining how the project budget is being utilized and helping to trigger corrective actions when variances exist.

EARNED VALUE ANALYSIS

The best method today for precisely determining the overall project performance status generally, and cost status specifically, is earned value analysis (EVA). EVA is a project performance measurement technique that integrates scope, time, and cost data. It periodically records the past performance of a project in order to forecast its future performance (see Figure 10.3).

During project progress evaluation, EVA measures a project's schedule and cost performance to determine whether the project is ahead or behind the plan (schedule and cost variances) and why. Then, final project costs (estimate at completion) and completion date (schedule at completion) are predicted based upon the current performance. While the practical elegance of such an approach comes from EVA's integration of project scope, cost, and time, its special value is in the proactive, predictive approach it enables. In particular, predictions warn us of possible problems, creating opportunities to fix them in a timely manner, and keep the project on its planned course. In summary, EVA strives to establish the accurate measurement of physical performance against

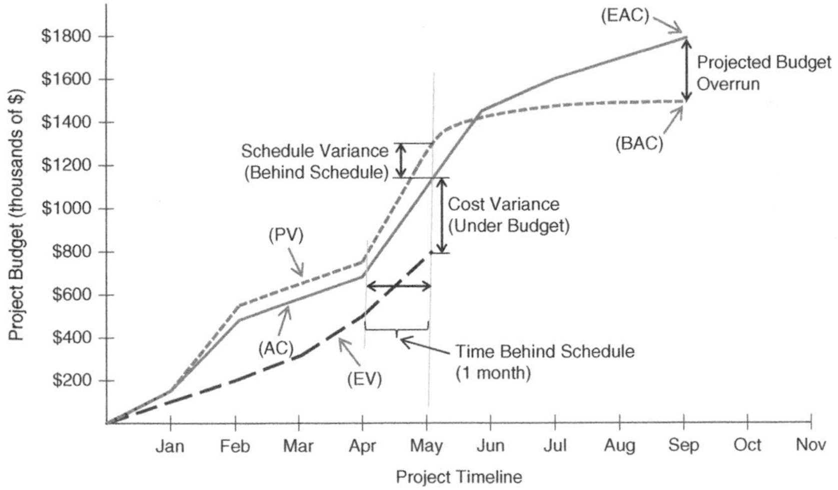

Figure 10.3: Example Earned Value Analysis Chart

Table 10.1: Fundamentals of Major Earned Value Measurement Methods

Term	Acronymn	Description
Planned value	PV	Also known as the budget, the PV is the total cost estimate planned to be spent on an activity (or project in totality) during a given period of time. PV was formally called budgeted cost of work scheduled (BCWS).
Actual cost	AC	The cost actually incurred for the work completed by the report date. AC was formally called actual cost of work performed (ACWP).
Earned value	EV	The total of direct and indirect cost incurred in accomplishing work on an activity by the report date. EV was formally called the budgeted cost of work performed (BCWP).
Cost variance	CV	The difference between the amount budgeted and the amount actually spent for the work performed.
Schedule variance	SV	The difference between the amounts budgeted for the work completed and the amount that was planned.
Cost performance index	CPI	The index reflects the relative value of work done compared to the amount paid for it.
Schedule performance index	SPI	The index reflects the relative amount the project is ahead or behind schedule.
Budget at complete	BAC	The original estimate of the total cost to complete a task or to complete the project in its entirety.
Estimate at complete	EAC	The current estimate of the total cost to complete a task or the project in its entirety.
Estimate to complete	ETC	The current estimate of the amount of funds required to complete the project from a particular point in time.
Report date		The data date or point in time at which the EVA is performed.

a plan to enable the reliable forecast of final project costs and completion date.[2] We describe the deliberate sequence of steps to explain the conceptual simplicity of performing an EVA. To begin, refer to Table 10.1, which defines the basic EVA terminology, and Table 10.2, which defines the key formulas.[3]

Performing Earned Value Analysis

First conceptualized by the industrial engineers of the late nineteenth century, EVA grew to its comprehensive and dominant form under the auspices of government project management practices.[4] In the process of growth, the original simple terminology of the engineers yielded to a confusing terminology. Although EVA has become an effective tool, primarily in large government projects, it has not been able to attract a large following in the private sector—the exception being the construction industry. Private business generally performs work through small and medium-size projects, often managed by companies that have only recently adopted formal project management methods.[5]

Table 10.2: Key Earned Value Analysis Formulas

Term	Formula
Cost variance	CV = EV − AV
Schedule variance	SV = EV − PV
Cost performance index	CPI = EV / AV
Schedule performance index	SPI = EV / PV
Estimate at complete	EAC = (BAC − EV) + AC
Estimate to complete	ETC = EAC − AC

These companies need EVA in a simpler form, built on a terminology as simple as that of the industrial engineers and often based on resource hours as much as on dollars. Stellar, although rare, examples of private organizations pursuing such EVA can be found in industry.[6] Fully valuing both approaches, for large government and smaller private projects, we will first focus on a comprehensive, but simplified approach using current terminology.

Performing an EVA needs to be precipitated by the gathering of critical information associated with the project. The baseline schedule is needed to anchor the status of work completed against the original plan and to compare actual progress to planned progress for each project task. A time-phase project budget is needed to evaluate the value of work completed against that which was planned. Finally, the project scope has to be fully defined and documented.

Fully defining the project scope is not an easy task, of course, especially when you are dealing with a new project fraught with unknowns. Among the available tools for scope definition, it is our belief that the logic of systematic decomposition of project work into successive levels of manageable chunks of work, as in the work breakdown structure (WBS), provides a sufficient degree of confidence that the scope will be fully defined, including all work to do in the project (see Chapter 5). This is why one of the golden rules of WBS structuring is to show all project work. That this is critical becomes clear when we know that EVA may require an estimate of the percent of work completed. If the estimate is 20 percent complete and the project scope is not fully defined, the estimate is inaccurate because it does not include the full scope of work. A disciplined and appropriate application of the WBS can help create a reasonable representation of a fully defined project scope.

The WBS provides the basis for scheduling the project scope of work. Each task will be carefully analyzed to determine when in the project time line will be executed. Details about beginning and ending points of work tasks, as well as their durations, will be determined in the schedule. Such information, along with the approved budgets for tasks, essentially defines scheduled work or *planned value*. As the project implementation unfolds, physically completed work is evaluated and *earned value* determined. Both the *planned value* and *earned value* are derived from the project schedule information and are critical for successful EVA. For that reason, the project schedule is a critical input to EVA.

A fully defined project scope that is scheduled for implementation must be based on careful resource estimates. Specifically, resources to complete each WBS element need to be identified and allocated for certain time periods on the schedule. This creates a time-phased budget of resources. These resource estimates along with the scheduled work help generate the planned value. Also, these estimates combined with the completed work constitute the earned value. The time-phased budget is a critical input to EVA. In summary, EVA requires a fully defined project scope integrated with allocated resources, all translated into a sound project schedule for performance. Often, these three inputs are termed a *bottom-up project baseline plan*.[7]

Set Up a Performance Measurement Baseline

A performance measurement baseline (PMB) should first be established in order to determine how much of the planned work the project team has accomplished at any point in time. Establishing a PMB involves three tasks: (1) determining points of management control and who is responsible for them, (2) selecting a method for measurement of earned value, and (3) setting up the baseline.

The foundation for the tasks is the project baseline plan, which fully defines project scope, integrating it with allocated resources and translating them into a project schedule for performance, all within the framework of the WBS. Given that the WBS has elements on multiple levels, you have to decide which elements (on which level) will be management control points. These points are called *control account plans* (CAPs). Although at first sight this might seem like a confusing term, in actuality its concept is simple—a CAP is a point at which we measure and monitor performance. The makeup of a CAP is shown in the example titled "Key Components of a Control Account Plan."

Key Components of a Control Account Plan

- Narrative scope definition.
- Location in WBS (e.g., Level 1 in a WBS with levels 0 for project, 1 for CAP, and 2 for work packages).
- Constituent work elements (e.g., Level 2 for work packages).
- Timeline (e.g., begin/end dates of each work package).
- Budget (resource hours, dollars, or units for each work package).
- Owner; the person responsible for the CAP (e.g., software project manager).
- Type of effort (e.g., nonrecurring or recurring).
- Methods to measure EVA performance (e.g., weighted milestones).

CAPs may be located on a selected level of a WBS—at Level 1, 2, or 3 (the project is level 0 of the WBS), or all the way down to whatever is chosen as a lowest level to exercise management control. The essence is that a CAP is a homogeneous grouping of work elements that is manageable, which brings us to the issue of its size. How large or how small should a CAP be? According to the current trends in private industry, the size of CAPs is on the increase.[8] One reason is that project managers want to concentrate on CAPs that include larger work elements, which are typically on higher levels of the WBS. Also, they include into a CAP all organizational units responsible for its constituent work elements. The desired result of these trends is to enable project managers to focus their attention on fewer but more vital control points of their projects, making EVA significantly easier to use and much more time efficient.

Such a CAP, then, has a clearly defined narrative scope, location in the WBS, constituent work elements, timeline, and budget. Although the budgets are often expressed in dollars, they can be expressed in other forms—from resource hours, to units, to standards. Because so many project managers manage only resource hour budgets, we will use hours in our examples. To ensure accountability for the budgets, each CAP should be assigned to a person responsible for its performance.

Figure 5.5 in Chapter 5 illustrates how a WBS culminates in a set of work packages that represent tasks or groups of tasks that have definable outcomes. The work packages are then aligned to the various organizational functions that will be responsible for execution of the work packages and delivery of the outcomes. Figure 10.4 extends this view of work organization one step further by illustrating how a Level 1 WBS element can be organized into a CAP.

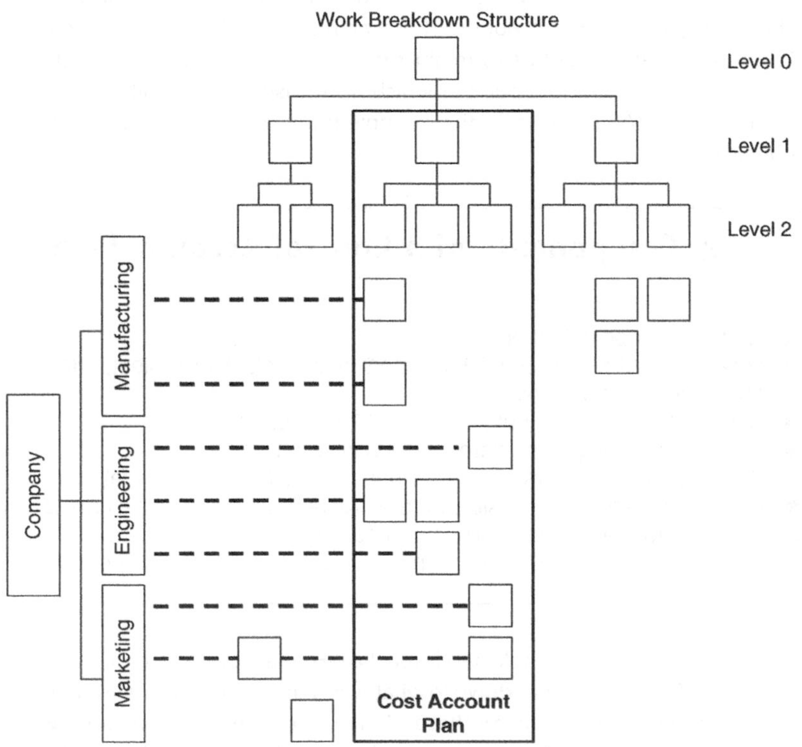

Figure 10.4: Organizing a Control Account Plan

Measurement of a CAP's performance, the cornerstone of an EVA, calls for well-defined methods of measurement. While we review several such methods (see Table 10.3), hard-and-fast rules for selecting the appropriate one do not exist. Rather, the choice you have to make is a personal one, often arbitrary, and may vary on a project-by-project basis. In the selection process, the project team and CAP managers should focus on the ease and accuracy of measurements that can be consistently applied to appropriately support their specific project needs.

The *percent complete* method uses a periodic (monthly or weekly) estimate of the percentage of completion of a work package, expressed as a cumulative value (e.g., 65 percent) against the full 100 percent value of the work package. Hailed as a simple and fast method, which perhaps explains its wide popularity, the method has also been viewed as being overly subjective. Defining work packages' scope well and checking on accuracy of the estimates helps make the subjectivity reasonable.

Fixed formula by work package includes various options: 25/75, 50/50, 75/25, and so on. For example, a 25/75 formula means that when a work package is started, 25 percent of the package's budget is earned, while the completion of the package earns another 75 percent. Any combination that adds up to 100 percent is possible. This is a quick way of estimating, applicable in situations where work packages are short-span and performed in a cascade type of time frame.

Table 10.3: Fundamentals of Major Earned Value Measurement Methods

Type of Method (1)	When to Use (2)	Major Advantage (3)	Major Disadvantage (4)
% complete	Well-defined work packages; management reviews in place; nonrecurring tasks	The easiest method to administer	Made purely on a subjective basis
Fixed formula	Work packages are detailed and short-span; nonrecurring tasks	Easy to understand	Rather subjective
Weighted milestones	Work packages run two or more performance periods; nonrecurring tasks	Perhaps the most objective method	Difficult to plan and administer
% complete with milestone gates	Works in any industry, on any type of project; nonrecurring tasks	Both easy and objective	Requires time and energy to define meaningful milestones
Earned standards	Preestablished standards of performance; nonrecurring or recurring tasks	Perhaps most sophisticated of all methods	Requires the most discipline
Equivalent units	Long performance periods; nonrecurring or recurring tasks	Simple and effective	Requires a detailed bottom-up estimate

Weighted milestones is a method of dividing a long-span work package into a several milestones, each one assigned a specific budgeted value, which is earned when the milestone is accomplished. As objective as it is, the method's success hinges heavily on the ability to define meaningful milestones that are clearly tangible, budgeted, and scheduled.

Percent complete with milestone gates strives to balance the ease of percent complete estimates with the accuracy of tangible milestones. A work package of, say, 600 hours is broken down into three sequential milestones, each budgeted at 200 hours and placed as a performance gate. You are allowed to estimate the first milestone's earned value by percent complete up to 200 hours. To go beyond the point of 200, you need to meet predefined completion criteria for the first milestone. This procedure is repeated for subsequent milestones.[9]

Earned standards is a method often applied by industrial engineers to establish planned standards for performance of work packages, which are then used as the basis for budgeting the packages and subsequently measuring their earned value. For example, the planned standard for producing a cup of lemonade at $0.20/cup is used to budget the work package including the production of 1,000 cups for $200. When 500 cups are produced, regardless of the actual cost the earned value is 500 cups × $0.20/cup = $100. Widely applied in repetitive types of project work, the method's foundations are the planned standards developed from historical cost data, time, and motion studies.[10]

In *equivalent completed units*, a planned work package is earned when it is fully completed. Similarly, a planned portion of it is earned when completed. For example, a work package to build five miles (five units) of freeway is estimated at $3m/mile for a total of $15m. It is fully earned when all five miles are finished. Also, the completion of half a mile will earn $1.5m. Based on detailed bottom-up estimates, the method is favored by the construction industry without ever having been called its real name—the earned value.

After this short review of the six methods, two things need to be mentioned in closing comments about measuring EVA performance. First, note that the work package is the place where the measurement is taken, while measurement for a CAP is a summation of work packages' measurements. Second, there is no single best way to measure earned value for any type of project task. This contingency rule means that different types of tasks will use different methods, and perhaps the most appropriate method is to combine multiple methods, relying on CAP managers to collectively estimate the earned value of individual work packages. For an example of a project where multiple methods are used, see Figure 10.5a. A design project consists of three CAPs, essentially three phases on Level 1 of the WBS. Each of them applies a different method of EVA measurement—percent-complete, weighted milestones, and earned standards. Since each CAP consists of multiple work packages, this means that all work packages within the CAP are measured with the same method. As we remind you that the EVA measurement method is the last on the list of components of a CAP (see the previous example titled "Key Components of a Control Account Plan"), we move to establish the PMB.

The PMB is a time-phased sum of detailed and individually measurable CAPs. What is included into CAPs depends on how companies define cost management responsibilities of their project managers. Many companies allow their managers of internal projects

a) Data on multiple methods of earned value measurement for a project

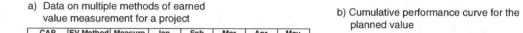

CAP	EV Method	Measure	Jan	Feb	Mar	Apr	May
Conceptual Design	% Complete Estimate	Planned	45	55			
		Earned	20	30	50		
		Actual	35	45	50		
Detailed Design	Weighted Milestones	Planned		100	100	50	
		Earned		100			
		Actual		115			
Prototype	Earned Standards	Planned				100	50
		Earned			25		
		Actual			25		
Total Project	Planned	Inc.	45	155	100	150	50
		Cum	45	200	300	450	500
	Earned	Inc.	20	130	75		
		Cum	20	150	225		
	Actual	Inc.	35	160	75		
		Cum	35	195	270		

b) Cumulative performance curve for the planned value

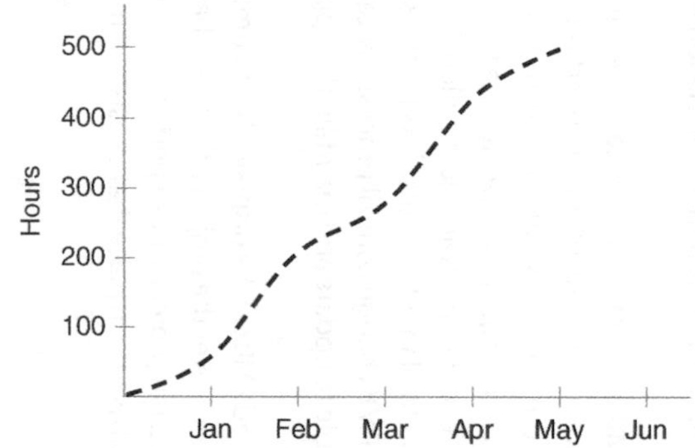

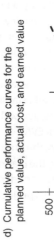

c) Cumulative performance curves for the planned value and actual cost

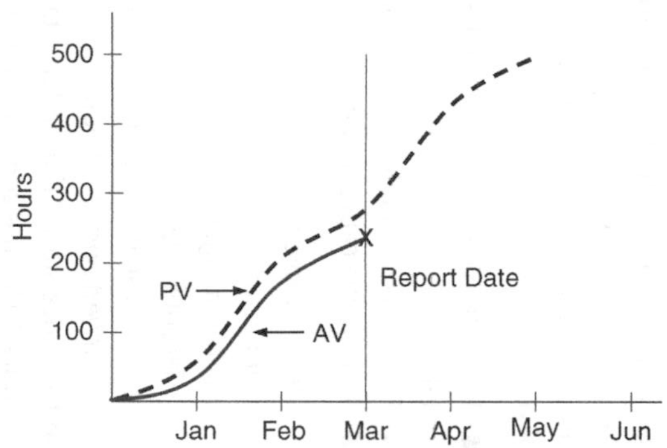

d) Cumulative performance curves for the planned value, actual cost, and earned value

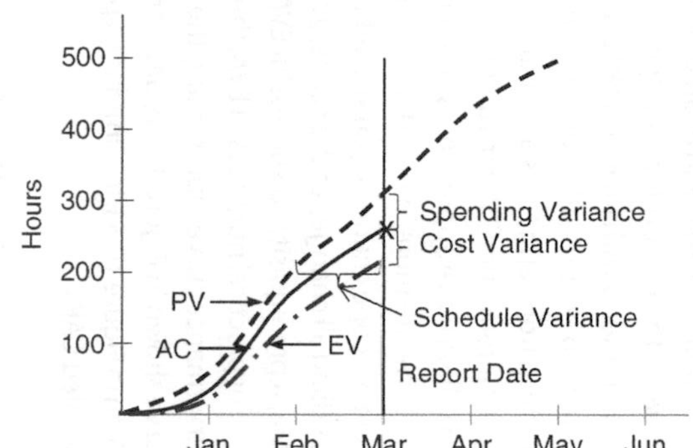

Figure 10.5: Performing Earned Value Analysis

to manage only direct labor hours, which is our focus here. In that case, their CAPs and PMB will include only direct labor hours. On the other end of the spectrum are project managers, whose job is to manage all project costs, as well as management reserves and profit. Accordingly, their PMB will reflect this situation. Still other companies may select that their PMB and related cost responsibilities of project managers is somewhere between the two ends of the spectrum.

In projects with lower uncertainty, a firm PMB with detailed CAPs can be established before project execution begins. What if you have to start executing an uncertain project in which front-end CAPs are detailed out while the later ones cannot be planned for the lack of information? What if a CAP's scope starts changing? The answer for the first issue is the rolling wave approach; as you progress in executing the available detailed CAPs, you will generate more information that enables you to plan other CAPs.[11] As for the second issue of scope changes, you should establish a PMB change control process. By carefully handling all changes to the scope, you will be able to update and maintain the approved PMB, a prerequisite to successful EVA.

For practical purposes of EVA, the time-phased PMB can be displayed as a cumulative performance curve representing the planned value over the project schedule. That is the curve shown in Figure 10.5b, developed for a design project whose performance data is given in Figure 10.5a. In summary, a project's PMB is now in place that consists of multiple detailed CAPs.

Evaluate Project Results

This step compares the actual results of performing the project with its plan (PMB). While the very measurement of performance occurs within individual CAPs, you may monitor and periodically (e.g., weekly or monthly) evaluate performance results at three levels: within individual CAPs, at some intermediary summary level (either a WBS element beyond the CAP or a PBS), and at the project level. This step includes the following:

- Focus on schedule: Evaluate schedule variance (SV) and schedule performance index (SPI).
- Focus on cost: Evaluate cost variance (CV) and cost performance index (CPI).
- Identify the cause of the variances, if any.

Our example in Figure 10.5c shows the comparison, the cumulative performance curve for actual values against the performance curve for planned values. The majority of project managers favor this traditional cost management approach despite its potentially deceptive results. At the end of March in our example, the difference between the two curves, called the *spending variance*, only reflects whether the project stays within the approved budgeted hours. It does not in any way determine the project's true cost performance status. If used for establishing the project's true cost performance status in Figure 10.5c, the comparison would mislead us by indicating that the actual performance is under budget (300 hours – 270 hours = 30 hours), a positive development. This couldn't be further from the truth; the project is in cost trouble, as we will soon see, and that can't be discerned using this planned-versus-actual approach. The reason for this false finding is in comparing points on the planned and actual curves that include different scopes of work. In short, apples are compared with oranges. Another issue with this two-dimensional traditional approach from Figure 10.5c is in that it relates to cost only. To get an insight into the project schedule performance, we need a separate planned versus actual schedule chart, and that one would not match the cost chart. The remedy for these problems is a chart that integrates true cost and schedule performance. This is where the earned value performance curve comes in, as illustrated in Figure 10.5d.

A comparison of the earned and planned value at the end of March indicates the following:

$$\text{Schedule Variance } (SV) = EV - PV$$

$$(SV) = 225 \text{ hours} - 300 \text{ hours}$$

$$(SV) = -75 \text{ hours}$$

This negative SV means that the project has fallen behind its planned work. A look back at Figure 10.5d reveals two manifestations of the same SVs—one drawn vertically is expressed in budget units, the other horizontal in time units. Not surprisingly, you may prefer the one expressed in time units (days, weeks, months). It is generally easier to identify time units of delay by means of the schedule performance index. Before we get there, it is worth mentioning that any time SV is negative, the project is behind the planned work, and any time SV is positive, the project is ahead of the planned work.

Another task in evaluating the schedule position is calculating SPI. SPI quantifies how much actual earned value was accomplished against the originally planned value. In other words, it represents how much of the originally scheduled work has been accomplished at a certain point of time. An SPI equal to 1 means perfect schedule performance to its plan. Any SPI greater than 1 implies an ahead-of-schedule position to the original plan of work. SPI running below 1 reflects a behind-schedule position to the originally scheduled work.

At the end of March in Figure 10.5d, the situation is as follows:

$$SPI = EV/PV$$

$$SPI = 225/300$$

$$SPI = 0.75$$

The SPI of 0.75 indicates that 75 percent of the originally planned work is accomplished. This means our project is behind, more precisely, 25 percent (1 − 0.75 = 0.25) behind the baseline plan of work. Since our reporting date at the end of March is the 90th day of the project, we can tell that our project is 22.5 days (25 percent of 90 days) behind the original work planned.

Schedule analysis in EVA deserves a word of caution. Specifically, anytime you find a schedule delay condition that includes negative SV and SPI that is less than 1, you should know that EVA schedule variance is not based on the critical path information and may be deceptive. Poor schedule performance of some work packages or tasks may be balanced by schedule performance of other work packages or tasks. Therefore, use your critical path schedule and risk analysis in conjunction with EVA schedule analysis.[12] If the late work packages/tasks are on the critical path or are highly risky to the project, complete the work packages/tasks at the earliest possible date.[13]

Now we can move to our second area of interest in this step, calculate CV and CPI. At the end of March in Figure 10.5d, CV is as follows:

$$\text{Cost Variance } (CV) = EV - AC$$

$$CV = 225 \text{ hours} - 275 \text{ hours}$$

$$CV = -45 \text{ hours}$$

and

$$CPI = EV/AC$$

$$CPI = 225/270$$

$$CPI = 0.83$$

The purpose of CV is to indicate the differential between the earned value for the physically accomplished work and the actual cost to accomplish the work. Therefore, the positive CV means that the project is running under budget, while negative CV signals the project is spending more than planned, overrunning the budget. In essence, we are experiencing the latter, consuming 45 more hours than we have allocated for the amount of accomplished work.

CPI is a cost efficiency factor. By relating the physically accomplished work to the actual cost to accomplish the work, CPI establishes the cumulative cost performance position. When CPI is equal to 1, it indicates perfect cost performance to the original budget. Values of CPI exceeding 1 indicate under original budget position, while those less than 1 indicate over original budget position. In our example, the CPI reading of 0.83 tells us that the earned value for the physically accomplished work is only 83 percent of the actual cost to accomplish the work. Putting it differently, for the amount of budget spent, we accomplished only 83 percent of the work that was planned for that amount of allocated budget.

Both CPI and SPI cumulative curves enable a very effective tracking of a project, as illustrated in Figure 10.6. Note that both rate and trend of the indices are crucial here. Key to it is using the cumulative data, rather than incremental data (weekly or monthly). Unlike the incremental data, which is prone to fluctuations, the cumulative data tends to smooth out the fluctuations and is very effective in forecasting the final project results, the focus of our next step.

A quick forecast of the completion date for our example in Figure 10.5d at the end of March is as follows:

$$\text{Schedule at Completion } (SAC) = \text{original schedule}/SPI$$

$$SAC = 150 \text{ days}/0.75$$

$$SAC = 200 \text{ days}$$

This quick method may be risky. As mentioned earlier, an EVA schedule delay condition as in Figure 10.5d has a negative schedule variance, and SPI that is less than 1 is not based on the critical path information and may be deceptive. Therefore, a better

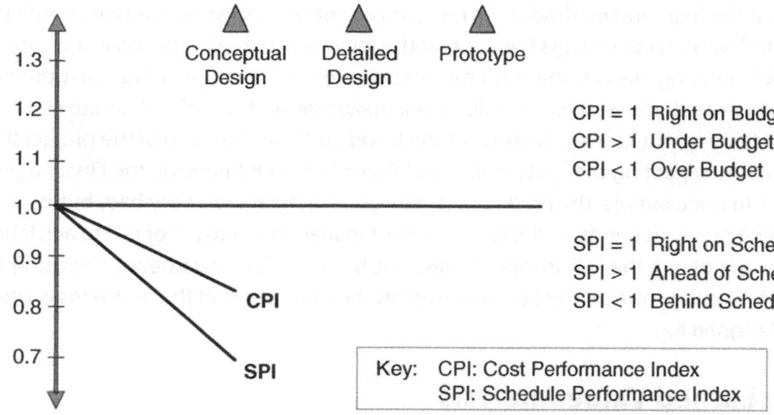

Figure 10.6: Tracking Cumulative Schedule Performance Index (SPI) and Cost Performance Index (CPI)

solution is to predict the completion date based on results of the critical path analysis in combination with EVA schedule variance.

Out of 20+ available formulas to estimate a project's cost at completion of the project, we will only look at two that are frequently used.[14] Here is the low-end formula and forecast:

$$\textit{Estimate at Completion (EAC)} = \textit{budget at complete}/\textit{CPI}$$

$$EAC = 500\ \textit{hours}/0.83$$

$$EAC = 600\ \textit{hours}$$

This means that at the end of the project we would need 600 hours to get this project done, a whole 100 hours variance at completion (VAC) over the original budget. Clearly, this method—called *constant cost efficiency rate*—relies on to-date cost overrun and projects it to the end of the project. Figure 10.3 includes the prediction of the final results developed by means of the quick forecast for SAC and low-end formula for estimate at completion (EAC).

A more rigorous method, based on our forecast on both the cost overrun and schedule slippage to date, is called *constant cost and schedule efficiency rate*:

$$\textit{Estimate at Completion (EAC)} = (\textit{budget at complete})/(SPI \times CPI)$$

$$EAC = 500\ \textit{hours}/0.625$$

$$EAC = 800\ \textit{hours}$$

With this method, we get variance at completion of 800 hours. Some researchers found that the low-end forecast is a reliable measure of the "minimum" hours, while the high-end method produces a forecast of the "maximum" hours we may need.[15] Their claim that the high-end method is the most appropriate forecasting method should be contrasted with a recent study's finding that the low-end method is the most accurate.[16] With such differing views using both methods to develop a range of final cost projections, in our case between 600 and 800 hours, makes sense. This is the absolute essence of prediction—produce a sanity check of the trend and final direction of the project (for major factors impacting the final results, see "Three Factors Influencing the Final Project Results"). In our example, the prediction is not good; actually, it is very bad, but its ultimate value purely depends on the willingness of management to act or not to act. If the option is to not act, the EVA is meaningless—it has no value whatsoever. Choosing to act by developing corrective actions focused on the root causes of the problems is what EVA is designed for.

Using Earned Value Analysis

If there is one single, most compelling reason to use EVA, it is for its predictive ability, the ability to reasonably forecast the final project results during project execution, most of the time. We say most of the time because it is only somewhere at the 15 percent completion point and beyond that a sound, statistically reliable forecast becomes feasible by completing the following tasks:

- Forecast the project's completion date.
- Forecast the project's cost at completion of the project.
- Take corrective actions, if necessary.

EVA is an option for any project, regardless of the industry and size. With the amount of resources at stake in large projects, a full-scale EVA can be easily justified. Simplified versions of EVA such as cost and achievement analysis (see the Variations section) are a good fit for smaller projects. In either case, a good measure of customization is recommended.

Three Factors Influencing the Final Project Results

1. *Sound project baseline*. Only when the scope is well defined, the schedule is realistic, and the budget is accurate can we expect a realistic forecast of the final project results.
2. *Actual status of the project*. The actual status of the project, as quantified by SPI and CPI, will be a vital factor in determining what final results the project will end up. Better SPI and CPI rates and trends indicate better final results.
3. *Corrective actions*. What will management do if the forecast is poor? Not believe it and do nothing? Or believe it and aggressively pursue corrective actions to alter the forecast? This is the moment of truth for management that will critically influence the final results.

Variations

There are several cost control tools that are conceptually founded on earned value, although they are not, at least explicitly, referred to as what they really are—a simplification of EVA. Two such tools that enjoy a high level of popularity are milestone analysis and cost and achievement analysis.[17] Overall, their appeal is in that they use simple terminology and straightforward process, which is perhaps why they are so time efficient. Because of a perception in the PM community that the milestone analysis is a tool of its own, it is described as a separate tool in the next section in this chapter. The cost and achievement analysis is briefly covered in the following example and is illustrated in Figure 10.7.

Based on the scope and schedule for a task, its budget (same as the planned value in EVA) of resource hours is defined. Multiplying the budget by the percent complete will produce the achieved value (equivalent to the earned value in EVA). Actual consumed hours (equivalent to actual cost in EVA) to complete the scope defined by the achieved value are recorded as well. Values for the budget, achieved, and actual cost are then used to predict the final cost for a task. Doing this on a regular basis for each task, in cumulative terms, allows you to sum them to produce budget, achieved, and actual values for the whole project and predict the project's final cost. The approach offers a great way to be proactive in smaller projects.

Benefits

A disciplined use of EVA that is based on a clear understanding of what a company wants to accomplish with it offers multiple benefits. We begin with how EVA can help handle a

COST AND ACHIEVEMENT ANALYSIS

Project Name: ____________ Estimate Date: ____________

1	2	3	4	5	6	7
Task No.	Task Description	Budget (hours)	Percent Complete to Date	Achieved to Date (3) X (4)	Actual Cost to Date	Predicted Final Cost (hours) $(6)+\frac{(3)-(5)}{(5)+(6)}$
12	Prepare Bill of Materials and Routing	8.0	40%	3.2	5.0	12.5
Total		312.0	36.4%	113.6	118.0	206.1

Figure 10.7: Example Cost and Achievement Analysis

fundamental question in today's project business: Is the project on schedule, behind, or ahead of schedule? Using the schedule variance and SPI in conjunction with the critical path method can reliably answer this question. In a similar manner, the CV and CPI play the crucial role in establishing the true cost position of the project by finding whether it is on, over, or under budget.

Each true schedule and cost position may be viewed as a significant step, telling what happened in the project at a specific point of time of its history. Knowing that they cannot change the history, proactive project managers use the position to look into the future and impact it. In particular, research of the past use of EVA indicates that cumulative CPI for larger projects becomes very stable at the 15 percent completion point in the project.[18] Simply, this means that early in the project the CPI exhibits a consistent pattern, enabling reliable forecasts of the project cost at completion. Similarly, the schedule performance index combined with the critical path method can be used for predicting the final completion date. Hence, you can periodically ask, "Given my current performance, what will be my final costs and completion date?" The answer offers trend performance and, if the trend differs from the baseline plan, an early-warning signal. This may be the greatest benefit of EVA, providing project managers with the early-warning signal about possible problems in the future and giving them an opportunity to devise and take needed corrective actions while there is still time to fix the problems.

The true credibility and, eventually, the value of EVA is rooted in its integration of project scope, schedule, and cost. The WBS provides the vehicle for the integration of all project work through its hierarchical tree of deliverables called *work elements*. For each element with its scope of work, resources are allocated, schedule determined, and cost estimated. By measuring current and predicting future performance for each work element, and aggregating them up the WBS hierarchy, we can arrive at a determination of the current project performance and prediction of the future performance for the total project. This integrated and consistent manner of performance measurement and prediction is a vital improvement compared to the traditional separate schedule performance and cost performance charts, a fairly typical approach in the industry.[19]

MILESTONE ANALYSIS

Milestone analysis compares the planned and actual cost performance for milestones to establish cost and schedule variances as measures of the project's progress (see Figure 10.8). A milestone's cost is planned and tracked on the *y*-axis, and its schedule on the *x*-axis. The gap between the milestone's planned and actual cost provides the cost variance. Similarly, the schedule variance is obtained through the differential between the planned and actual schedule for the milestone. Both the planned and actual values are portrayed by cumulative curves. These two curves—as opposed to EVA's three curves of plan, actual, and earned values—are made possible by using milestones as a platform for the integration of scope, schedule, and budget. Although effective in tracking project progress, milestone analysis is far more effective when used predictively to estimate the final project cost and completion date.

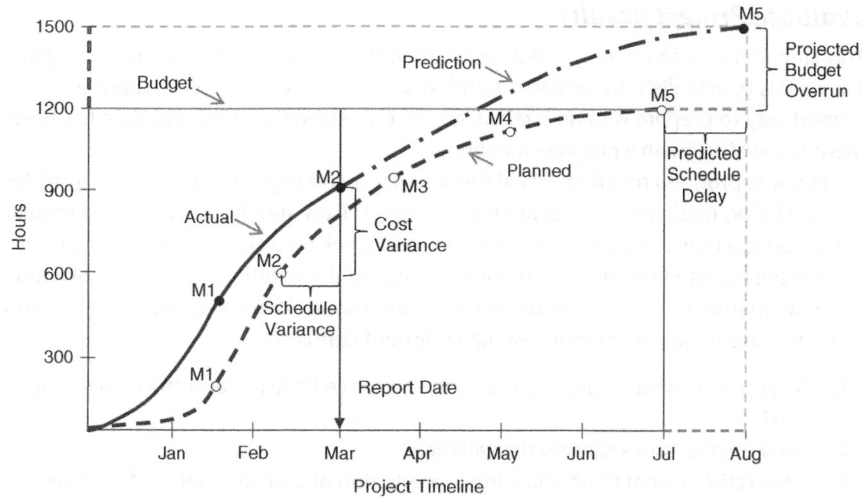

Figure 10.8: Example Milestone Analysis

Performing a Milestone Analysis

The milestone analysis technique integrates project scope, schedule, and cost to determine overall project performance. The quality of the analysis output is completely dependent on the quality of the input. Therefore, a fully defined project scope, detailed project schedule, and a time-phased project budget are all necessary inputs for forming a solid foundation for effective milestone analysis. Additionally, any changes in the project scope that have not been fully comprehended in the baseline plan have to be included before an analysis is performed to create an accurate cost performance determination.

Set Up and Track Milestones

Using the cost baseline (time-phased budget), draw a planned cost performance curve that will be the baseline, annotating the milestones (Figure 10.8). This is a cumulative curve, typically expressed in resource hours or monetary units. A certain number of hours are budgeted for each milestone, and because of the cumulative nature of the curve, when a milestone is reached, the cumulative number of hours for the milestone and all preceding milestones should be consumed.

As the project unfolds, actual cost data are collected and used to draw a cumulative actual cost curve, but what really matters is when a milestone is accomplished and marked on the actual curve. Hence, all performance is measured on the milestone level, following a fixed formula of 0/100 EVA measurement method. When the work on a milestone is started, 0 percent of the milestone's budget is earned, while the completion of the milestone earns a full 100 percent. Because of the cumulative nature of curves, the milestone acts as the culmination point of all previous project work, making its performance equate with project performance at that point.

Evaluate Project Results

This step involves comparing actual results of performing the project against its plan. The goal is to establish the SV and CV, and to identify the cause of the variances, if any. A good way to prepare is to do a rehearsal with milestone owners, and then to implement the evaluation in a progress meeting.

Periodic progress meetings instill the discipline and regularity in reviewing strides that are being made on the execution of a project. They should be called on a regular basis—once a month for a long project or once a week for a short project, for example. Each of the milestone owners are required to attend the meeting and report milestone cost management progress. The project manager should focus on asking the following five questions to gain a sense of cost management status:

1. What is the variance between cost performance baseline and the actual project cost?
2. What are the issues causing the variance?
3. What is the current trend and current prediction of cost completion if we continue with our current performance?
4. What new risks have been discovered and may affect the predicted cost at completion?
5. What actions should be taken to bring cost performance back in line with the baseline?

In our example from Figure 10.8 the variances are as follows:

$$\textit{Schedule Variance} = \textit{Planned} - \textit{Actual}$$

$$SV = 2\ \textit{months} - 3\ \textit{months}$$

$$SV = -1\ \textit{month}$$

and

$$\textit{Cost Variance} = \textit{Planned} - \textit{Actual}$$

$$CV = 600\ \textit{hours} - 900\ \textit{hours}$$

$$CV = -300\ \textit{hours}$$

While the negative variance indicates that the project has fallen behind its plan, a positive variance means the project is ahead of its plan. No variance implies the performance is right on plan. Therefore, in our example, the project is one month late and 300 hours over the budget.

Predict Final Results

Certainly, the most important and also the toughest step is the prediction of the final results. It is important because it enables a predictive look at the direction and trend of the project—where are our final cost and completion date going to end up? The absence of formulas for prediction such as those used in EVA makes the prediction an intuitive, challenging assignment, typically performed in the progress meeting. Such an exercise is very similar to one described in detail in the Milestone Prediction Chart section in Chapter 9. In particular, as the owner of the milestone describes its actual progress, potential variance from the baseline, and issues causing the variance, owners of dependent milestones evaluate the ripple effect of the milestone on subsequent milestones. The ripple effect is analyzed in the context of the critical path schedule, indicating the dependencies between milestones and related tasks. As a result of the analysis, predictions of milestones' cost and completion dates are made, all the way to the end of the project. If the final results are not favorable, corrective actions are developed to alter the trend and set the project back on track.

Using the Milestone Analysis

Milestone analysis is a good candidate for both smaller and larger projects. With its visual power and little time to develop, the analysis serves well the needs of projects with smaller budgets. In larger projects, its primary rationale for use is its ability to supply a summary view of the project status to high-level managers, focusing on major project milestones.

With a bottom-up project plan already in place, a well-versed project team should take no longer than 30 to 45 minutes to perform a milestone analysis that includes five or six milestones. As the number of milestones increases, so will the necessary time for the analysis.

Benefits

Crucial to the benefits of the milestone analysis is an understanding that it is a simplification of EVA. Using a milestone as a precisely defined scope of work, the analysis integrates cost and schedule with the scope, eliminating the need for the earned value curve. As a result, the milestone analysis includes only two curves, as opposed to EVA's three. This makes it more attractive and easier to use than EVA, while providing some of the values offered by EVA. In particular, milestone analysis establishes cost and schedule position, indicates performance trend and detects early-warning signals; integrates scope, cost, and schedule; and facilitates management by exception. Most often, these benefits are confined to smaller projects and larger projects that use the milestone analysis for schedule performance analysis.

Additionally, the graphical appeal and simplicity of the tool makes it easy to visually discern the schedule and cost variances, along with the predicted line of future milestone performance. This is especially appealing to top-level managers, who do not have the time to understand the potentially overwhelming amount of cost management details that can exist on a project.

CHOOSING YOUR COST MANAGEMENT TOOLS

The tools presented in this chapter are designed for different project situations. Most project managers find it useful to employ one to three cost management tools, and face

Table 10.4: A Summary Comparison of Cost Control Tools

Situation	Cost Management Plan	Budget Consumption Chart	Earned Value Analysis	Milestone Analysis
Small and simple projects	✓	✓		✓
Large and complex projects	✓	✓	✓	✓
Formal progress reviews			✓	✓
Informal progress reviews		✓	✓ (simplified version)	✓
Short time to train how to use the tool	✓	✓		✓
Focuses on exceptions			✓	
Provides early-warning indication			✓	✓
Integrates scope, cost, and schedule			✓	✓
Provides a single control system for all management levels in all projects	✓		✓	
Takes little time to apply	✓	✓	✓ (simplified version)	✓
Uses dollars or hours		✓	✓	✓
Summary detail needed	✓	✓	✓	✓
Uses two curves		✓		✓
Uses three curves			✓	
Displays trend			✓	✓
Provides built-in predictive approach			✓	
Little time available for cost management	✓	✓		✓

the problem of eliminating the other tool options. Project managers living under time pressures often ask, "Given my project situation, which are more appropriate to use?" To decide, refer to Table 10.4, which lists various project situations and identifies which tools are geared for each situation. Consider this table as a starting point, and create your own custom project situation analysis and tools of choice to fit your particular project management style.

References

1. Layer, Alexander, Erik Ten Brinke, Fred Van Houten, Hubert Kals, and Siegmar Haasis. "Recent and future trends in cost estimation". *International Journal of Computer Integrated Manufacturing*, 15 (6): 499–510, 2002.

2. Fleming, Q. W., and J. M. Koppelman. *Earned Value Project Management*, 4th ed. (Newton Square, PA: Project Management Institute, 2010).
3. Portney, Stanley E. *Project Management For Dummies*, 3rd ed. (Indianapolis, IN: John Wiley & Sons, 2010).
4. Fleming, Q. W., and J. M. Koppelman. "Earned Value for the Masses." *PM Network* 16 (7): 29–32, 2001.
5. Hatfield, M. A. "The Case for Earned Value." *PM Network* 10 (12): 25–27, 2006.
6. Ten Brinke, Erik Lutters, Tom Streppel, and Hubert Kals. "Cost Estimation Architecture of Integrated Cost Control Based on Information Management." *International Journal of Computer Integrated Manufacturing* 17 (6): 534–545, 2004.
7. Hayes, B., and Jim Miller. "Using Earned Value Analysis for Better Project Management." *Construction Management & Economics* 16 (1): 2002.
8. Fleming and Koppelman, 2010.
9. Zwikael, O., S. Globerson, and T. Raz. "Evaluation of Models for Forecasting the Final Cost of a Project." *Project Management Journal* 31 (1): 53–57, 2000.
10. Hatfield, 2006.
11. Kerzner, H. R. *Project Management: A Systems Approach to Planning, Scheduling, and Controlling*, 11th ed. (Hoboken, NJ: John Wiley & Sons, 2013).
12. Singletary, N. "What's the Value of Earned Value?" *PM Network* 10 (12), 28–30, 1996.
13. Fleming and Koppelman, 2010.
14. Portney, 2010.
15. Hatfield, 2006.
16. Zwikael, Globerson, and Raz, 2000.
17. Lock, D. *Project Planner* (Hunts, England: Gower Publishing, 1990).
18. Johnson, Cheryl. "Implementing an ANSI/EIA-748-Compliant Earned Value Management System." *Contract Management* 46 (4): 29–34, 2006.
19. Christensen, D. S., and S. R. Heise. "Cost Performance Index Stability." *National Contract Management Association Journal*. 25:17–22, 2011.

11

AGILE PROJECT EXECUTION

Contributed by

Peerasit Patanakul

James Henry

Jeffrey A. Leach

Agile is a form of adaptive project execution, heavily used in software development as an alternative to traditional approaches that emphasize the sequential or linear process starting from requirements gathering, planning, designing, software code writing, testing, and implementation. Alternative to a sequential process, agile methodologies emphasize an iterative workflow and incremental delivery of project outcomes in short iterations (see "Value Proposition of Agile Methodologies"). Several popular agile methodologies include extreme programming (XP), Dynamic Systems Development Method (DSDM), feature-driven development (FDD), crystal, lean software development, agile unified process (AUP), and scrum.

Extreme programming, for example, was created to effectively address projects that involve requirement changes. Instead of assembling large, separately developed modules, XP relies on continual integration. XP projects are divided into iterations, approximately one to three weeks long to produce a project outcome (usually in the form of a software release) that is fully tested. These iterations represent the development, integration, and thorough testing of a small amount of new capability. A planning meeting is held at the beginning of each iteration to determine the content of the iteration and to facilitate the incorporation of the changing requirements.

Four values guiding XP are communication, simplicity, feedback, and courage. These four values drive 12 practices that dictate how programmers should carry out their daily jobs. For instance, some of the practices are using user stories in the planning process, working on small releases, implementing test-driven development, focusing on design simplicity, using pair programming, having collective code ownership, and ensuring continuous integration.[1]

Similar to XP, the main features of DSDM include user involvement, iterative and incremental development, increased delivery frequency, integrated delivery frequency, integrated test at each phase, and fulfilling requirements.

FDD is also an iterative and incremental software development process focusing on feature (client-valued functionality).

Scrum is another agile methodology that has been widely used to manage project execution activities. This chapter focuses on tools and techniques used in scrum framework.

Value Proposition of Agile Methodologies

Many practitioners have false expectations of agile methodologies and this can lead to tension between stakeholders. It is *incorrect* to equate agile methodologies with simply being "faster and cheaper," as it leads to misconceptions. Many customers think that because an agile methodology is used, they can ask for anything and the project team can deliver it in two to four weeks—*this is not particularly true*. Customers also seem to believe that their projects should be completed faster and sooner than they would have been completed under a waterfall methodology—*this is not particularly true either*.

Being "faster" is not the value of agile methodologies. The true value is the delivery of customer-usable features earlier in the project life cycle. The focus of any agile methodology is on shorter iterations of work effort to produce project outcomes that:

- Deliver the realization of business value earlier in the project cycle than waterfall approaches may. This could give the customers the opportunity to monetize projects earlier.
- Give the customers the opportunity to prioritize work. There are many cases where an agile project is 80 or 90 percent complete, and the customers decide that the product is good enough as is and drop the last 10 percent of the requirements. Sometimes, the customers still want the last 10 percent of the requirements but decide to wait to complete the original scope of work at a later date as there are other higher priority projects.
- In some cases, allow customers to modify requirements with less impact on cost and schedule.
- Reduce project risks by allowing the customers to see the tangible progress of the project. This enables the customers to fully understand what is being delivered and confirm that what is being built is the same as what is needed.

SCRUM BASICS

The scrum framework provides a structure of roles, meetings, rules, and artifacts that project teams utilize during software development.[2] Within this framework, a project is broken down into self-organizing teams (scrums) of about six to nine members.

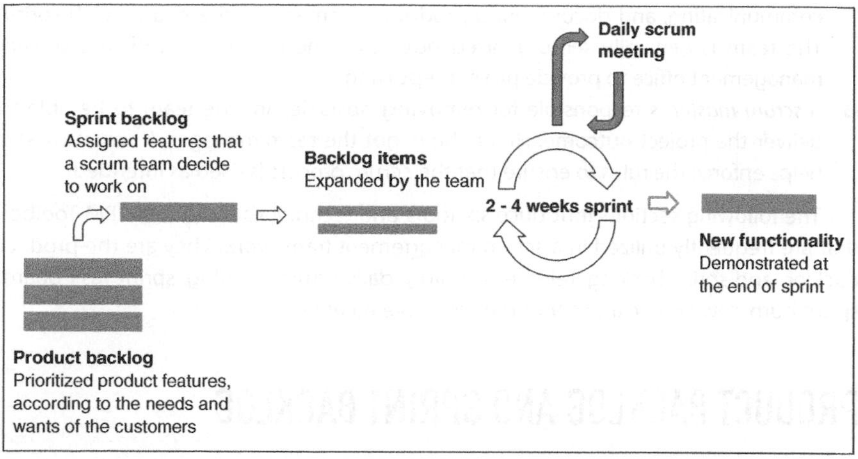

Figure 11.1: The Scrum Workflow
Source: Adapted from Eric J. Braude and Michael E. Bernstein. *Software Engineering*, 2nd ed. (Hoboken, NJ: John Wiley & Sons, 2011): 72.

Each team focuses on a self-contained area of work.[3] To develop software, a list of customer wants and needs is created, referred to as a *product backlog* (see Figure 11.1). Scrum uses fixed-length iterations, called *sprints*, which are typically two to four weeks long.[4] While in a sprint, scrum teams are responsible for taking on a set of features from the backlog and developing a deployable set of features that is properly tested.[5] The team is given full authority to successfully complete the sprint. Every day, an approximately 15-minute daily scrum meeting is organized to assess the status of the sprint, report on problems, and identify future tasks. At the end of a sprint, a customer demonstration is conducted. The leftover features and new tasks are gathered, a new backlog is created, and a new sprint begins.

Alternative to traditional project approaches, scrum is an iterative approach that allows the teams to develop a subset of high-value features as early as possible to incorporate early feedback from the customers. In essence, the scrum framework emphasizes empirical feedback, team self-management, and striving to build properly tested product increments within short iterations.[6] Within the scrum management framework, participants hold several key roles. They are, for example, product owner, scrum team, and scrum master.

- The *product owner* represents the customer of the project. Based on the product requirements, the product owner provides customer-centric items (user stories) to the team, prioritizes the items, and adds additional items to the product backlog. The product owner is accountable for ensuring the value of the project output to the business. New products will be promoted into production, only after the product owner accepts the output from the scrum teams after the sprint demonstration.
- The *scrum team* typically consists of six to nine members across disciplines. These members are responsible for analyzing, designing, developing, testing,

communicating, and documenting product increments at the end of each sprint. The team is generally self-organized but has some interaction with the project management office to provide project reporting.

- *A scrum master* is responsible for removing obstacles for the team to be able to deliver the project outcome. He or she is not the team manager. A scrum master helps enforce the rules to ensure that the scrum process is used as intended.

The following sections introduce six tools and techniques for use in a PM Toolbox that are frequently utilized in a scrum management framework. They are the product backlog and sprint backlog, release planning, daily scrum meeting, sprint task board, sprint burn down chart, and sprint retrospective meeting.

PRODUCT BACKLOG AND SPRINT BACKLOG

There are two common types of backlogs that are used in the scrum management framework: the product backlog and the sprint backlog. The product backlog is a prioritized listing of all items that may be developed at some point in the future. These items represent the customers' requirements. There is no commitment for when these items will be assigned to a team to be worked on. There is also no guarantee that these items will ever be assigned to a team.

A sprint backlog represents a list of items that a scrum team has taken from the product backlog and has committed to developing during the next sprint.

Information on the Backlogs

The product backlog is visible to any stakeholder and is composed of customer-centric features—descriptions of the desired product from the point of view of the customer or user. These could be in the form of user stories or use cases.[7] These features also have a level of priority assigned to them to create a forced-ranked list of desired functionalities. However, the priority of the items can be changed over time. There are no tasks listed on the product backlog. See Figure 11.2 for an example of a product backlog for a team that is developing a travel booking web site (the product).

A sprint backlog contains specific tasks that will be performed during a sprint in addition to the customer-centric features, from the product backlog, that are associated with those tasks. Items on the sprint backlog do not need to be prioritized because they have been committed to for a particular sprint release.

The sprint backlog is visible to the project team such that they can use it as a reference during daily scrum meetings. Figure 11.3 illustrates an example of a sprint backlog if the team decided to work on the first and third stories from the product backlog shown in Figure 11.2.

Populating Backlogs

Any team member, regardless of their role, can add items to the product backlog. However, the product backlog is owned by the product owner, who can assign priority

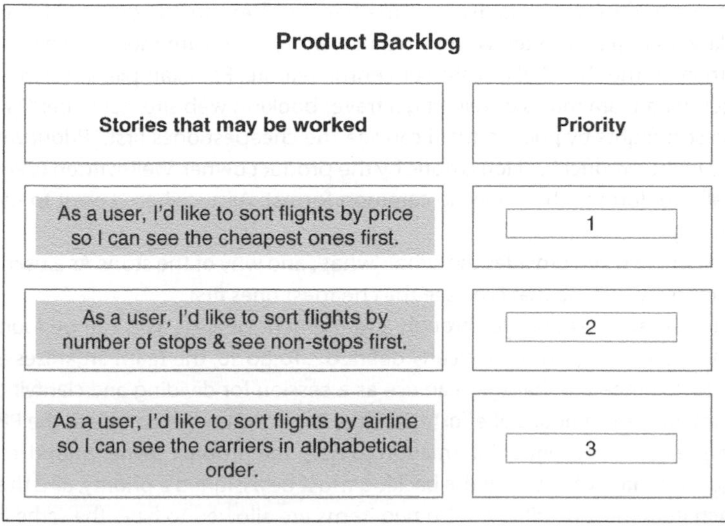

Figure 11.2: Example Product Backlog

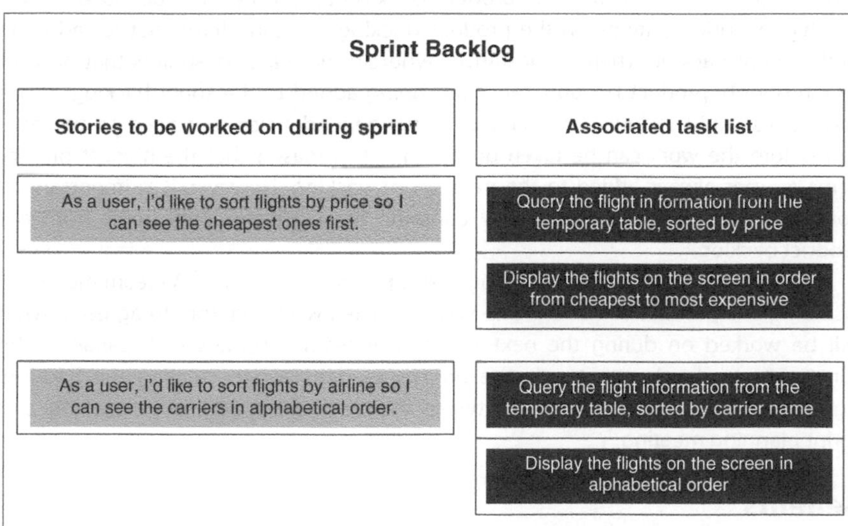

Figure 11.3: Example Spring Backlog

to the items or remove the items from the backlog.[8] As already discussed, the product backlog item (PBI) is often written in a user story form with more emphasis on the "what" than on the "how" of a customer-centric feature. For example, a story that may be written for a team that is developing a travel booking web site, could be: "As a user, I'd like to sort flights by price so that I can see the cheapest ones first." Prioritization of the items in the product backlog is done by the product owner. Well-written user stories will usually conform to the following common format: As a <who>, I want to <what>, so that <why>.

The example given provides the 'who', 'what', and why of the story: As a user, I want to sort flights by price so that I can see the cheapest ones first.

It is important that PBIs go through a refinement process as it can be found that some PBIs are too large and not clearly defined.[9] To do so, the team organizes a backlog refinement meeting that they can use as a session for dividing and clarifying large PBIs, estimating the amount of effort that they would spend to complete the PBIs and providing necessary technical information to help the product owner prioritize PBIs.[10] It is important that every item in the backlog must be assigned a priority, and the priority of each item must be different. No two items are allowed to have the same level of priority assigned. Estimated by the team, the level of effort to complete the item can be associated with each PBI.

Sprint backlogs, on the other hand, must be populated by each individual scrum team during the sprint planning meeting, discussed later. When populating a sprint backlog, teams look to the product backlog to find items to be worked on.[11] The highest-priority items on the product backlog are considered first for addition to the sprint backlog. There are instances wherein there are constraints that prevent a high priority product backlog item from being added to the sprint backlog. Those constraints could include, for example, some dependencies that need to be satisfied before the work can be taken on. It is for this reason that the highest priority items are not always added to the next sprint backlog. However, it is important to note that the sprint backlog usually contains the highest-priority items from the product backlog.

The sprint backlog is created during sprint planning sessions.[12] All team members, including the product owner and the scrum master, work together to agree to what will be worked on during the next sprint. The team considers their capacity, the constraints to development, and the priority of items when creating the backlog. Figure 11.4 shows the relationship between the product backlog, sprint backlog, and sprint planning meeting.

Benefits

The product backlog allows for a close alignment between what the project teams are working on and what the business sees as the most important items for delivering business value. The ability to list everything that is desired, along with the ability to prioritize and reprioritize the work as necessary, enables the business to have more strategic agility to meet the challenges that they face.

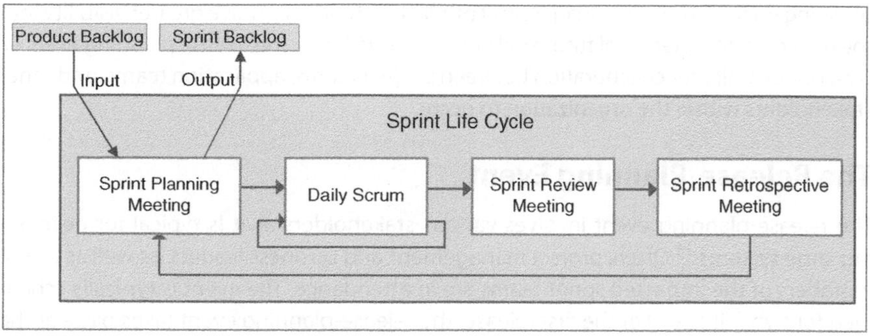

Figure 11.4: Product Backlog and Spring Backlog in Sprint Life Cycle

The creation of the sprint backlog necessitates engagement and collaboration between the business and project teams. It results in a common understanding between all parties on what the team has committed to delivering.

Both backlogs are simple to create and easy to read and understand by all stakeholders involved, but the use of the backlogs may come with some initial challenges (see "Overcoming Initial Challenges"). When used effectively, however, the backlogs provide a road map for how the team should execute their overall project objectives.

Overcoming Initial Challenges

Challenge 1: The knowledge gap. Product owners may struggle to fully define work and may require the expertise of team members to help address items like defining technical requirements.

Challenge 2: Initial difficulty with prioritization. Product owners may find it hard to prioritize the product backlog items at first; but once the product backlog is established, they will find it easier to add new stories and adjust priorities as appropriate.

RELEASE PLANNING

A release is a group of sprints. In developing complex and large (enterprise) systems, multiple scrum teams are needed. The teams often group sprints into releases to allow for planning. The number of sprints in a release is determined by the project organization, but is typically six to eight sprints in duration.

Release planning becomes a valuable technique when systems with interdependencies are being developed by multiple teams. In developing large systems, release planning enables project managers and other stakeholders to have greater visibility into the expected completion of future milestones. In addition, the release planning exercise is an opportunity for collaboration between project teams, application teams, and other stakeholders within the organization to occur.

The Release-Planning Event

The release-planning event involves various stakeholders as it is typical for developing large systems.[13] Often, project management and business leaders, as well as all the members of the impacted sprint teams are in attendance. The event is typically scheduled for two full days. For the first release, the release-planning event takes place at the beginning of the release. For subsequent releases, the event takes place at the end of each release. Figure 11.5 shows the life cycle of a four-week release.

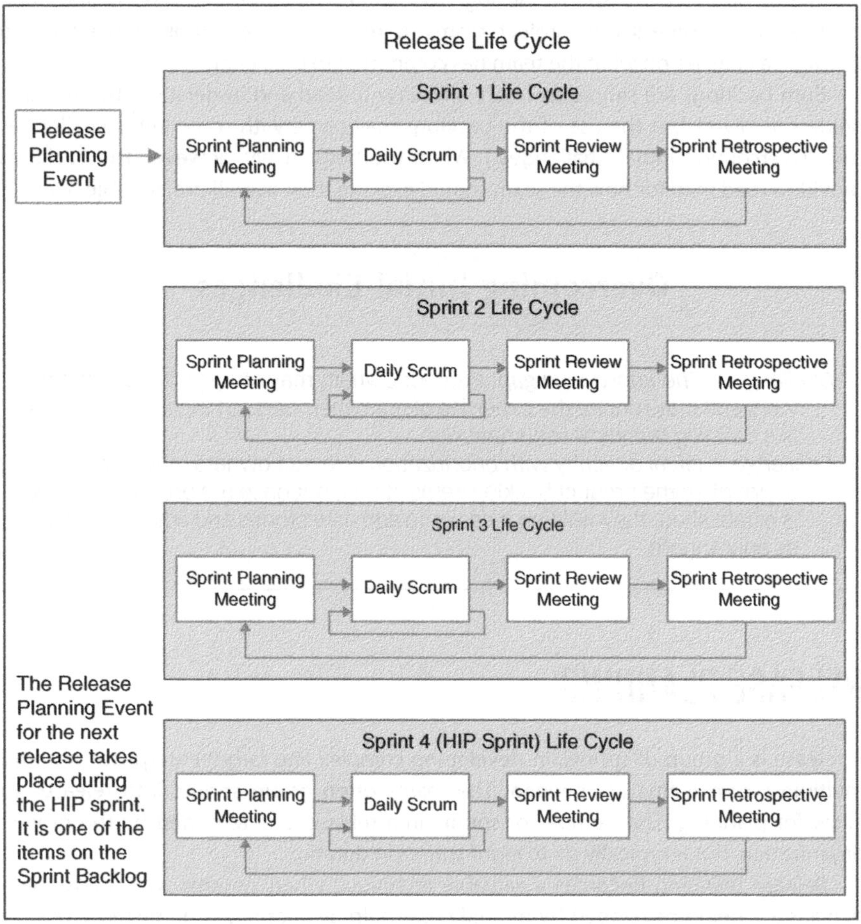

Figure 11.5: The Release-Planning Event

Initial Draft Release Plan

Before the interteam collaboration can begin, each scrum team must work together to create their own initial draft release plan. This involves assigning product backlog items to the sprints included in the release. For example, if a team is involved in a five-sprint release, they would consider dependencies and capacity when putting together a plan for which product backlog items they will develop in each of the next four sprints. The team would then display this plan for everyone to see (Figure 11.6). The last sprint is reserved as a HIP (hardening, innovation, and planning) sprint. No product backlog items are planned for this sprint during release planning. HIP sprints are discussed later in this section.

There are some risks involved when attempting to plan multiple sprints in advance. Priorities can change, requirements can change, and the effort needed to create the solutions could turn out to be greater than anticipated. To mitigate these risks, contingency time is built into the release plan. This is accomplished by progressively reducing the amount of scope planned for each sprint in the release. A common method for this is as follows. In a six-week sprint, the amount of product backlog items assigned to the first sprint represents 100 percent of the team capacity. The second sprint assigned product backlog items that should use 80 percent of the team capacity. The third sprint is filled to 60 percent capacity, and the fourth and fifth sprints are filled to 40 percent and 20 percent capacity, respectively. The result is that the team is able to commit to deliver on some longer-term business needs, while also maintaining the agility to handle changes in scope and direction.

Final Release Plans

After all the scrum teams have created their draft release plans, the plans are visibly displayed either physically or electronically in a way that allows for each scrum team

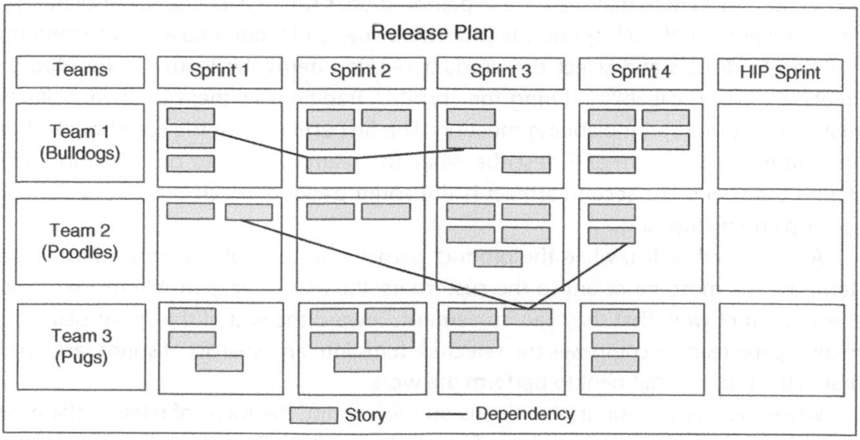

Figure 11.6: Release Plans Developed by Several Scrum Teams

the opportunity to review the release plans of all the other teams. Time is set aside for the teams to review each other's plans. During this time, the teams interact and are encouraged to ask questions to get a full understanding of what each team plans to work on during the next release. Through these discussions, each team member looks to find any dependencies between the planned activities of the teams. Once the dependencies between the teams have been identified, the teams discuss how to coordinate their activities. The draft release plans are then updated to account for any needed changes.

Time for another review of the release plans is given to the teams after they've updated their draft release plans. The process of review and update will continue until all the plans are believed to be coordinated properly. Before the release plans can be finalized, every member of every scrum team is asked to commit to the plan. If anyone, for any reason, is uncomfortable with any of the release plans, that person will explain their concerns and the entire group will discuss what changes should or should not be made. The release-planning event is not complete until everyone agrees to commit to the plan.

HIP Sprint

The HIP sprint is the last sprint of each release. This sprint is an opportunity for the teams to ensure that they are on solid footing for the next release. Hardening (H) refers to activities that reduce technical debt, such as system wide performance testing, infrastructure upgrades, bringing code up to architectural guidelines, and updating necessary documentation. Innovation (I) is encouraged during the HIP sprint. The agile teams can use this time to follow-up on any ideas they may have that would benefit the system. The HIP sprint is also the time to begin planning (P) the next release. The release planning event, for the next release, takes place during this sprint.

Release Planning versus Sprint Planning

It is important to note that the release planning does not replace the sprint planning. At the beginning of each sprint, the product owner and team hold a sprint planning meeting to discuss and select the product backlog items the team will develop to create a working capability during the sprint.[14] Traditionally, there are two artifacts that result from a sprint planning meeting. The first artifacts are the sprint goals that are written by the team and describe what the team expects to get accomplished during the sprint. The second artifact is the sprint backlog, which is the result of the sprint-planning meeting.

As discussed earlier, while the product owner is responsible for indicating which items are the most important to the customers, the team is responsible for selecting the amount of work that they can implement. Toward the end of the sprint-planning meeting, the team decomposes the selected items into an initial list of sprint tasks and makes the final commitment to perform the work.

While the focus of sprint planning is on each sprint, the focus of release planning is on the overall release consisting of multiple sprints.[15] During some releases, the team could decide to move product backlog items into earlier releases because they have the capacity to do so. They may also decide to rearrange the order of items or even reduce their priority and drop them. Having identified the dependencies during release planning, the team can communicate these changes to the other teams as appropriate.

Benefits

Release planning gives the scrum teams the ability to plan some longer-range activities. The release-planning event increases the collaboration among teams and allows for dependencies between teams to be identified early and accounted for in the plans.

A successful release-planning event can facilitate the alignment between the overall organizational goals as they progress and the individual team goals.

Release planning is also effective in identifying cross-team dependencies that could block or delay execution progress. By identifying dependencies early and accounting for them during release planning, it lessens the chances of unanticipated delays during development.

THE DAILY SCRUM MEETING

A daily scrum meeting is a fast paced meeting that begins once the sprint planning meeting has been finalized and agreed upon by the entire team (see Figure 11.7). The primary purpose for a daily scrum meeting is to answer the following three questions[16]:

1. What was completed yesterday?
2. What impediments were encountered that blocked me from being effective?
3. And, finally, what is planned for today or prior to the next scrum meeting?

By having each team member answer these questions, the team now has a complete picture of what's happening, the overall sprint progress toward its goals, and any modification that needs to be made to the upcoming day's work.

Within the daily scrum meeting, the scrum team will leverage a type of task board as a way of communicating sprint progress. A task board is a very effective method of illustrating the current state of the sprint backlog over time, discussed later in this chapter.

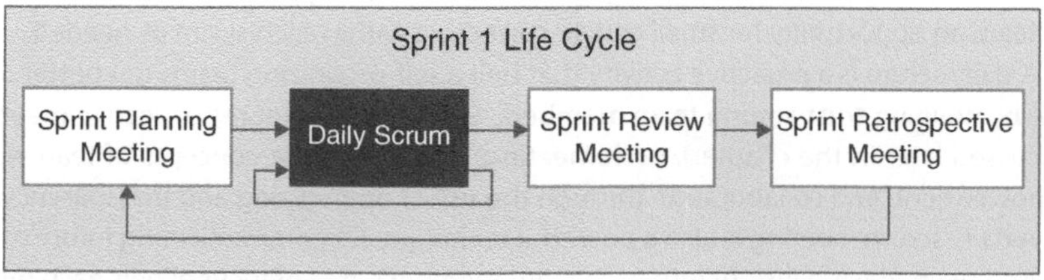

Figure 11.7: The Daily Scrum in the Sprint Life Cycle

At a high level, a task board contains four main pillars representing user stories or product backlog items, tasks to do, task in progress, and task completed; see the next section for more detail.

Organizing a Daily Scrum Meeting

A daily scrum meeting is typically held in the same location and time, each and every day. The meeting is most effective when conducted in the morning as it helps set the objectives for the upcoming day. The meeting usually lasts 15 minutes or less, in which each team member will update the team on what was completed yesterday, impediments encountered, and, finally, what they plan to do for today. Typically, the scrum master will begin the meeting by starting with the person to their left and working clockwise around the room until all stakeholders have reported their status updates. Team members should focus on answering the three questions above with brevity in mind and avoid discussions not pertaining to the agenda. It is common that all team members stand up during the meeting, as sitting down has a tendency to extend the duration of the meeting.

Participants

A daily scrum meeting involves the entire scrum team as well as the scrum master.[17] The product owner is considered optional and isn't required to attend every scrum meeting. In order to have an effective daily scrum meeting, it's important that all "participating" scrum team members come well prepared to quickly discuss the work they are responsible for completing.

It is important that the scrum master keeps the meeting on time as to avoid any distractions from the meeting agenda. Meetings that stray from the agenda will more than likely run over the allotted time, begin to distract team members from answering their three questions, and even steer the meeting into sidebar discussions around fixing problems. When these types of distractions occur, it's important that the scrum master brings the meeting back to focus as quickly as possible (see "Requirements for Effective Daily Scrum Meetings").

Benefits

The primary benefit of a daily scrum meeting is to provide each scrum team member an opportunity to communicate the status of their sprint backlog items, while keeping the team in sync on how things are going. In addition, by having a daily meeting it provides the team an opportunity for small course correction within each sprint as needed.

A daily scrum is a proactive activity that helps self-organizing teams get better over time by building trust among team members. Similar to a sprint retrospective meeting (discussed later in the chapter), each meeting is based on the concepts of teamwork, empowerment, and collaboration through the use of open dialog and transparency.

A daily scrum meeting is also a powerful technique for communicating information across a team. Having insight into each team member's area of focus allows each member to see the big picture while mitigating down stream risk.

Requirements for Effective Daily Scrum Meetings

1. *Resources*. A daily scrum meeting requires that the scrum team be dedicated to attending the meeting. Dedicating too many resources to a daily meeting can negatively impact the overall velocity of a project if not managed properly.
2. *Preparation and focus*. A lack of preparation for the daily scrum meeting can definitely impact the overall value of the meeting. Team members should be prepared to answer the primary three questions below while dedicating the necessary time after each meeting to address any gaps or concerns:
   - What work was completed the previous day?
   - What work will be proposed for today?
   - Did the team face any problems or impediments?
3. *Time*. A daily scrum meeting needs to adhere to the time allotted (15 minutes or less). If meetings are running late on a regular basis, the scrum master should step in to ensure that each team member is sticking to the defined agenda. This can be done by holding discussions after the scrum meeting to address any areas that may be pushing the meeting over the allotted time.

SPRINT TASK BOARD

For a scrum team, a sprint task board is used to organize tasks into categories based upon their stage of completion. Sticky notes, with tasks written on them, are typically placed on the board to give an easy visual representation of the work being done and the progress made towards completing those tasks. The task board is often referred to as the "kanban board."

A very simple example of a task board could have four columns labeled "User Story," "Tasks to Do," "Tasks in Progress," and "Tasks Completed" (see Figure 11.8). At the beginning of a sprint, all of the tasks would be in the "Tasks to Do" column. As soon as a team member begins to work on a task, the sticky note with that task would be moved to the "Tasks in Progress" column. Upon completion of the task, the sticky note would be moved to the "Tasks Completed" column. At the end of a sprint, all the tasks should be in the "Tasks Completed" column.

Using the Sprint Task Board

Sprint task boards are not limited to only four columns as illustrated in Figure 11.8. Adding more meaningful columns could give sprint teams a more precise visual status. A task board may also contain columns such as "Tasks to Do," "In Design," "In Coding," and "In Testing," or any other combination of columns that will show the tasks moving

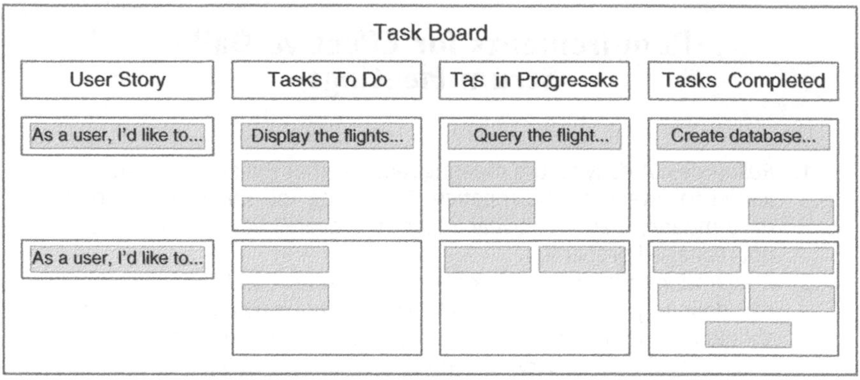

Figure 11.8: Example Sprint Task Board

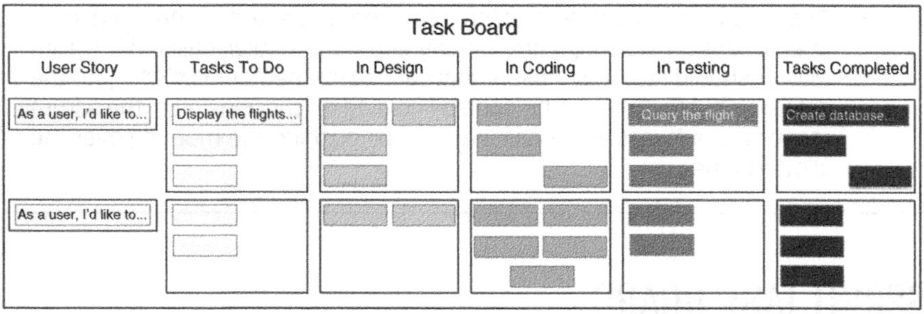

Figure 11.9: Variation of a Sprint Task Board

through the sprint life cycle. Figure 11.9 shows a sprint task board that was created to use the columns mentioned previously.

It is also recommended that dependencies between tasks be noted on the sticky notes and that the team member working on the task adds his or her name as the task is moved to new columns. This will allow for easier communication and collaboration when the need arises. As illustrated in Figure 11.10, the task "Display the flights on the screen in order from cheapest to most expensive" is a successor of "Query the flight information from the temporary table, sorted by price." Once the task began, Jeffrey Leach, who was responsible for design, added his name to the task. When the design was completed, James Henry, who is responsible for coding, added his name to the task and crossed out the name of the designer.

Benefits

The use of task boards makes the progress of the team visible to everyone. It will also facilitate team coordination that could occur when the team works on a series of tasks.

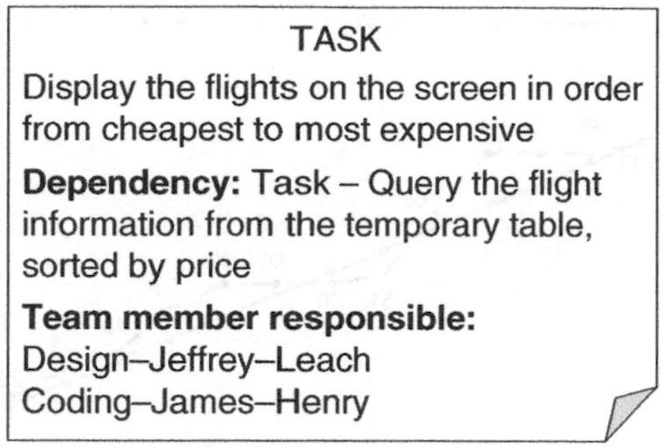
TASK
Display the flights on the screen in order from cheapest to most expensive
Dependency: Task – Query the flight information from the temporary table, sorted by price
Team member responsible:
Design–Jeffrey–Leach
Coding–James–Henry

Figure 11.10: Sample Sticky Note for a Task

If the sprint task board is used properly, everyone should know who is assigned to each task, dependencies between tasks are properly accounted for, and there should be no duplication of effort.

Additionally, a sprint task board is a simple and easy way to manage the flow of work and information throughout a sprint. It visually represents each team member's work and progress over the duration of a sprint. Having the task board available for team members to consume it now allows team members to know what other team members are working on.

THE SPRINT BURN DOWN CHART

A sprint burn down chart generally displays the total quantity of work or tasks needing to be completed versus the time allotted to complete a sprint (or an entire release consisting of several sprints). The chart is a simple and easy to use tool that provides relatively accurate estimates of the overall progress of a sprint or release progress.[18]

Having tasks broken down into sufficient detail is a necessary prerequisite for creating a burn down chart. This is normally completed during the sprint-planning phase where each task has been forecasted based on the hours it will take to complete the work. These estimates are normally determined by the entire team during the planning meeting.

Developing a Sprint Burn Down Chart

An example of the sprint burn down chart is shown in Figure 11.11. While the chart is rather simplistic, it is very beneficial to the team when utilizing it on a daily basis.

On the chart, the *x*-axis represents the amount of time that the team has defined for the entire sprint (20 days in this example). The *y*-axis represents the total amount of time (in hours or days) the team will take to complete all of the given tasks for that sprint (120 hours in this example). Since the amount of work to be complete will decrease

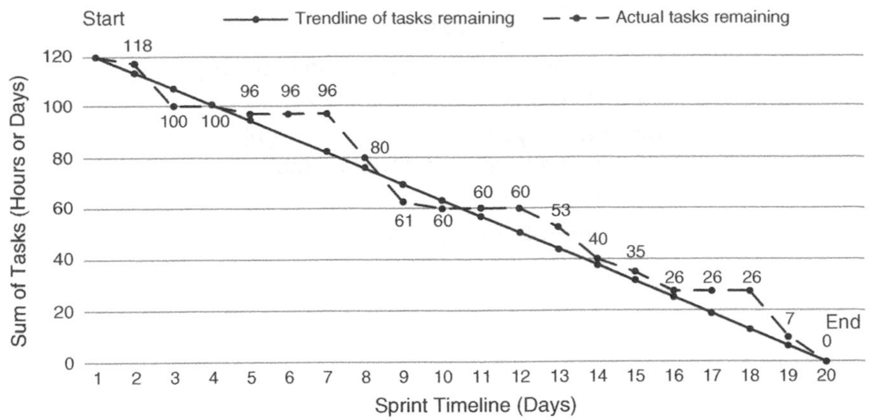

Figure 11.11: Example Sprint Burn Down Chart

over time, the general trend in the chart is to burn down to a point where zero work remains. A trend line can be calculated and drawn to illustrate projected outcomes, in other words, when work might be completed.

Using a Sprint Burn Down Chart

A sprint burn down chart is updated on a daily basis by each of the task owners. Collecting and communicating this information can be done by using many different techniques. Table 11.1 is a simple table that associates tasks to a user story along with a status, owner, projected hours, and remaining hours. In this scenario, after spending an entire day working on Task B, the business analyst (BA) has determined that another five hours of work will be required to complete the tasking. The BA then updates this information using the "Remaining Hours" field located in the table.

After each task has been updated, the data is then aggregated and plotted for that day to determine whether or not the sprint was progressing as planned (see Figure 11.12). If the "Actual Tasks Remaining" line was above the "Trend Line of Tasks Remaining," this means the sprint is progressing at a slower pace than expected and may not complete the defined scope of work. However, if the "Actual Tasks Remaining"

Table 11.1: Task Tracking Table

User Story	Task #	Task Description	Status	Owner	Projected Hrs.	Remaining Hrs.
Story 1	A	Develop requirements	Closed	SME	8	**0**
	B	Coordinate with customers	In Work	BA	6	**5**
	C	Code module 1	Open	Developer 2	10	**10**

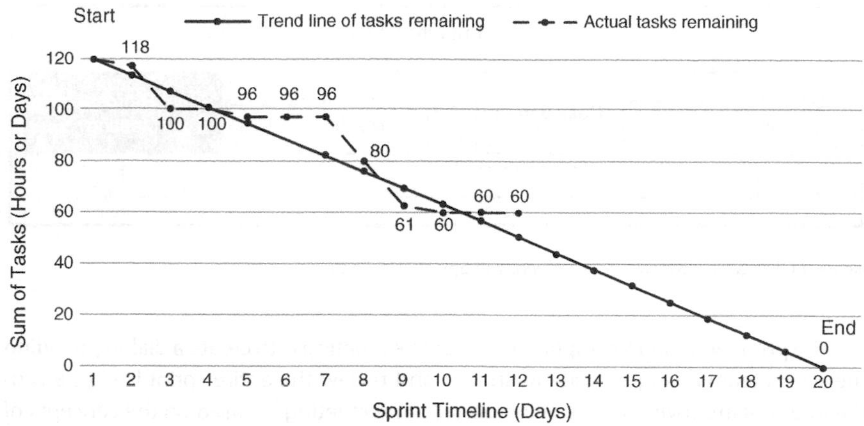

Figure 11.12: Plotting Work in Progress

line was below the "Trend Line of Tasks Remaining," this means the sprint is moving at a faster rate than projected and may be able to finish ahead of schedule.

Benefits

The sprint burn down chart provides each team member a visual representation of the total estimated time for all tasks to be completed (hours or days) versus the time permitted to complete a sprint or release. It allows team members to actively estimate the time it takes to complete each of their tasks. From there, data is aggregated so that team members can determine whether or not they are under- or over-executing across the sprint or release.

The chart also creates value by providing a powerful way to communicate the overall status of the sprint or release across the entire team. If the project starts to deviate in a positive or negative fashion from the trend line, the team will have this information to help mitigate potential risk in the future. Throughout this process, the chart facilitates effective communications across the team in a transparent fashion for all team members to consume.

Finally, a sprint burn down chart allows team members to make decisions based on accurate data resulting in informative decisions being made.

THE SPRINT RETROSPECTIVE MEETING

A sprint retrospective meeting follows the sprint review meeting and is typically the last step in the sprint life cycle. Upon completion of the sprint retrospective meeting, a new sprint is created beginning with a sprint-planning meeting, followed by daily scrums, a sprint review meeting, and another sprint retrospective meeting (see Figure 11.13).

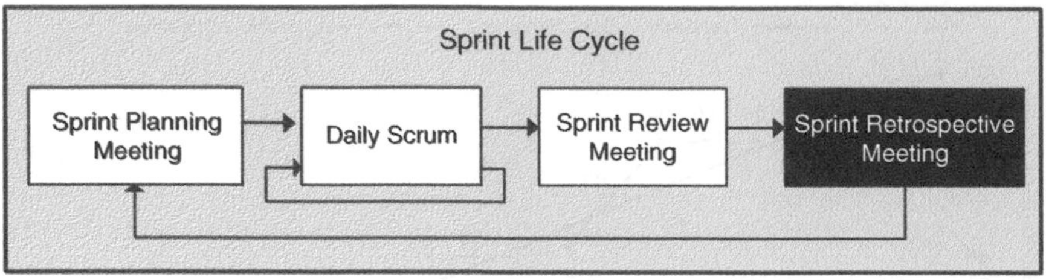

Figure 11.13: Sprint Retrospective Meeting in Sprint Life Cycle

The primary reason for a sprint retrospective meeting is to create a dialog providing the scrum team an opportunity to analyze and review the entire sprint life cycle with the intent of improving the overall process.[19] The meeting is based on the concepts of continuous improvement. At the end of each sprint, a retrospective meeting is held to analyze and review all the things that went well, as well as the things that didn't.

Within the meeting, the scrum team may choose to analyze existing processes, technologies, collaboration, communication techniques, and so forth. Once the sprint retrospective meeting concludes, the scrum team will then carefully select the process improvement areas of interest and plan for them in the upcoming sprint.

Organizing a Sprint Retrospective Meeting

An effective sprint retrospective meeting happens at the end of every sprint cycle and takes the necessary time needed to thoroughly discuss and document the sprint retrospective agenda. Typically, the duration of a sprint retrospective meeting is about an hour and a half but, in many cases, can require more time depending on the duration of the sprint, complexity of the project, the size of the team, or the team's overall experience with scrum methodologies.

Participants

A sprint retrospective meeting normally involves every scrum team member including the scrum master. In order to have an effective meeting, it is critical that all contributing team members feel comfortable and safe in their surroundings as they provide suggestions and recommendations on ways to improve the process. In most cases, the product owner is the only individual that may not be involved in the sprint retrospective meeting due to his or her potential decision-making authority (or power) over other team members, thus preventing the team from being honest and open in the process. The result of not having an open and honest dialogue is allowing process inefficiencies to continue to impact future sprints.

Using a Sprint Retrospective Meeting

As stated earlier, the goal of a sprint retrospective meeting is to determine what worked well and what could be improved upon. This process of information gathering can be done in many different ways.

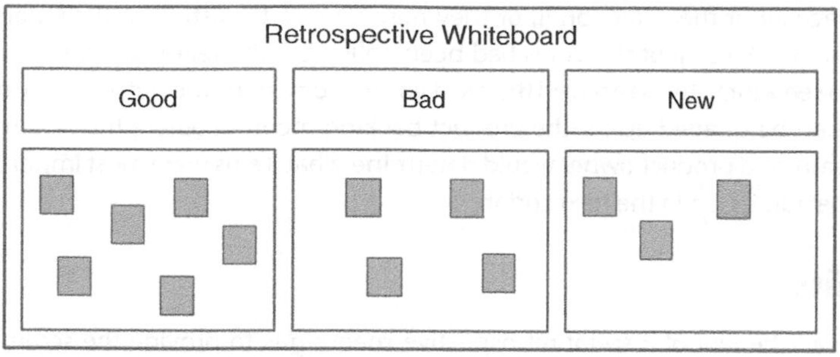

Figure 11.14: Retrospective Whiteboard

Visual learning is a great technique that can help the scrum master facilitate communications within the team. As an example, a simple whiteboard can be divided Into three areas: good, bad, and new (see Figure 11.14). Good items would identify areas that worked well over the sprint and didn't require any modifications. This would include things like continuing to limit daily sprint stand-up meetings to 15 minutes or to continue using custom client side queries. Bad items, on the other hand would include areas that did not work well resulting in process inefficiencies. Examples of this would include such items as defining the term *completed* or *done* more precisely for team members to better understand, segmenting product backlog items into smaller components during the backlog refinement phase, or finally shifting the daily scrum meeting from 9:00 A.M. to 8:30 A.M.

Finally, new items would focus on areas that hadn't been defined or were missing from the overall process. This would include such items as building a knowledge management system to inventory agenda items that fell outside the daily scrum meeting, or adding a video teleconferencing solution in the scrum room for external team members to enhance collaboration and communication.

There are many different techniques that a scrum team can utilize when gathering and categorizing information. As an example, sticky notes can be provided to each team member based on a color schema defined (green = good, red = bad, and yellow = new). Each team member would then fill out as many sticky notes as needed in order to communicate their recommendations. Once completed, each team member would place their sticky notes in a box for the scrum master to place on the whiteboard under the appropriate columns. Using a box ensures that all recommendations are anonymous and free from judgment, thus helping create a safe environment for team members to provide recommendations.

Selecting Improvement Areas

Using a whiteboard such as the one illustrated in Figure 11.14, the team would now need to identify and prioritize which improvement areas they'd like to include in the upcoming sprint. One popular technique is to provide each team member a few stars (3 to 5), which they would place on the items they feel would make the most significant impact to the project. If a team member felt very strongly about a particular area they could place all of their stars on it, or they may choose to distribute their stars across several areas. Once all of the votes had been collected, the team would then tally the results to see which items received the most votes. From there, the output of the meeting would then be queued up for the product backlog meeting where the scrum master, scrum team, and product owner would determine what items were most important and should be addressed in the next sprint.

Benefits

The primary benefit of a sprint retrospective meeting is to provide the scrum team a dedicated point in time after each sprint for team members to provide feedback on areas they feel improvements could be made. Continuous improvement is critical to the overall integrity and foundation of a project as it moves throughout its project cycle. Overlooking continuous process improvement results in projects repeating past mistakes and exponentially impacting the overall efficiency of the project.

Additionally, a sprint retrospective meeting is based on the concepts of teamwork, empowerment, and collaboration through the use of open dialogue and transparency. This is a powerful way to get each team member committed to the project as their feedback and recommendations are now being heard and applied across the project.

Things to Avoid

1. *Poor preparation*. A lack of preparation for a sprint retrospective will most definitely impact the overall value of the meeting. This would include things like not having a structured agenda, missing supplies or materials (Post-its, pens, whiteboard), and even technology hurdles not addressed within the sprint retrospective room.
2. *Complexity*. As mentioned earlier, a sprint retrospective can be very simple at times but if certain dynamics become too complex, the value of a sprint retrospective can quickly diminish. Examples of this would include the number of participants involved in the meeting or having to manage the priority of recommendations from past meetings (the backlog). Selecting backlog items with the greatest benefit to the project and the team can sometimes be difficult to determine due to the politics of a project.

CONCLUDING REMARKS

This chapter introduces six tools and techniques used within the scrum management framework used to execute agile projects. They are the product backlog and sprint backlog, release planning, daily scrum meeting, sprint task board, sprint burn down chart, and sprint retrospective meeting. As parts of agile project execution methodologies, these tools and techniques are introduced to facilitate the iterative workflow and incremental delivery of project work that can be completed in short iterations.

Practitioners are encouraged to adapt these tools and techniques contingently upon their needs. It is important to note also that there are other tools and techniques used with other agile methodologies beyond what have been presented in this chapter that practitioners may want to explore further.

References

1. Braude E. J., and M. E. Bernstein. *Software Engineering,* 2nd ed. (Hoboken, NJ: John Wiley & Sons, 2011).
2. James, M. "Scrum Reference Card," 2010. http://ScrumReferenceCard.com. Accessed February 2015.
3. Schwaber, K. *Agile Project Management with Scrum* (Redmond, WA: Microsoft Publishing, 2004).
4. James, 2010.
5. Cohn, M. *Succeeding with Agile: Software Development Using Scrum* (Upper Saddle River, NJ: Pearson Education, 2010).
6. Schwaber, 2004.
7. Cohn, 2010.
8. Schwaber, 2004.
9. Cohn, 2010.
10. James, 2010.
11. Schwaber, 2004.
12. Ibid.
13. Rubin, K. S. *Essential Scrum: A Practical Guide to the Most Popular Agile Process* (Ann Arbor, MI: Pearson Education, 2012).
14. Schwaber, 2004.
15. Rubin, 2012.
16. Schwaber, 2004.
17. Ibid.
18. Ibid.
19. Ibid.

PART V

Project Reporting and Closure Tools

12

PERFORMANCE REPORTING

Project performance reporting involves collecting key data concerning the performance status of the project at a particular point in time, synthesizing the data into meaningful information, and then communicating the performance information to project stakeholders to familiarize them with the project's progress toward its objectives. Effective project performance reporting supports two key components of successful project management: open and strong lines of communication and transparent communication of information.[1]

The intent of performance reporting is threefold: (1) to determine and communicate how project resources are being used to achieve the objectives of the project, (2) to provide information on current performance against the project plan and performance baseline, and (3) to use the information to enable informed project decisions.

The tools presented in this chapter are intended to assist project managers in effectively fulfilling this threefold intent. Just as performance reporting is a required activity of all project managers, the following performance reporting tools should be a part of every project manager's PM Toolbox. We begin by looking at the project reporting checklist.

PROJECT REPORTING CHECKLIST

There are many types of project status reports and great variation in the type of information contained in the status reports. This is due in large part to variations in reporting needs and desires of project stakeholders, variation of industry reporting standards and practices, and variations in the size and complexity of projects themselves. Project status report form and content is therefore situational, but needs to be standardized and consistent with the needs of the project stakeholders who are on the receiving end of the report. How does one establish and maintain standard messaging?

A simple and effective tool to consider is the project reporting checklist. The checklist assists the project manager in determining the correct status information to include in a report and to consistently provide the information over time.

Developing the Project Reporting Checklist

To be effective, a project status report must be current, concise, accurate, and contain only the information needed to keep stakeholders abreast of progress and the resources used to accomplish the project's objectives.

The project reporting checklist will be different for every organization because every organization has its own unique set of information required for project reporting. Developing a standard set of checklist items is good practice, as it drives consistency in project reporting format and content within an organization.

The items contained within the checklist are developed by first understanding the information required by the project sponsor and other key project stakeholders. Then, additional items can be included that are unique to a particular project or to the project's second tier stakeholders.

Table 12.1 illustrates an example set of project reporting items to consider as a reference and starting point for developing your own customized project reporting checklist. The checklist shown is somewhat extensive, so keep in mind that the best project status report is concise and to the point. Developing your custom checklist will involve using a subset of items shown in Table 12.1.

The "Status" column can be used to indicate that the information needed to develop a project status report has been collected. Some project managers add an additional column to the checklist labeled "Source" to indicate where the source of information resides, or to provide a hyperlink to the source data itself.

Using the Project Reporting Checklist

Most organizations have a defined point in time when reporting of project performance is expected to begin. This can be as early as once formal project initiation is determined. Whenever the point, the project manager should begin using the checklist to formulate the content that will be included in his or her project reports.

In practice, the content included in a project report is fairly repetitive over time. However, it is valuable to review the checklist periodically to serve as a memory jogger to provide additional information in a report, which may not be repetitive in nature. This is normally the type of information included in the "general" section of the checklist.

Additionally, different information may be required by the project stakeholders as a project progresses through the various stages of the project cycle. A review of the project reporting checklist during these stage transitions will help the project manager modify the reporting content accordingly.

Benefits

The greatest value that project managers gain from the use of the project reporting checklist is knowledge gained about the project reporting requirements of their project stakeholders. It helps to ensure they are reporting the most important project performance information, as determined by their stakeholders.

The checklist also helps to ensure that the information contained in a project report is both concise in nature and accurate at the time it is reported.

Table 12.1: Example Project Reporting Checklist

Status	Checklist Items
Project Scope	
☑	Have the project objectives been changed?
☑	Have the deliverables in the project plan changed?
☑	Have there been any changes to the project scope?
☑	Are there any scope changes awaiting approval?
Project Schedule	
☑	Has the schedule been updated?
☑	Is the project progressing on the critical path or critical chain?
☑	Does the time expended to date vary from the project plan?
☑	Do we have adequate resources to maintain the schedule?
☑	Are project subcontractors or partners on schedule?
☑	What is the estimated completion date?
Project Budget	
☑	Has the budget been updated?
☑	Is the available budget to date in alignment with the project plan?
☑	What is the average monthly budget burn rate?
Project Performance	
☑	Have all deliverables been met to date?
☑	Have all project milestones been met to date?
☑	What is the earned value (EV)?
☑	What is the schedule variance (SV)?
☑	What is the schedule index (SI)?
☑	What is the cost variance (CV)?
☑	What is the cost index (CI)?
Issues	
☑	Are there any current issues that need to be reported?
☑	What are the resolution plans for any open issues?
☑	Do our subcontractors or partners have any current issues?
☑	Do any issues require project sponsor or top management action?
Risks	
☑	What are the HIGH-level risks?
☑	What are the risk response plans for all HIGH-level risks?
☑	What is the overall risk profile of the project?
☑	Do any risks require project sponsor or top management action?

(continued)

Table 12.1: *(Continued)*	
Status	**Checklist Items**
General	
☑	Is the project being impacted by any external factors?
☑	Are there any quality issues associated with the project outcomes?
☑	Are we receiving payments as planned?
☑	Are there any actions or decisions needed on the part of the project sponsor or top management?

THE PROJECT STRIKE ZONE

The first principle of project performance reporting, as stated previously, involves understanding and communicating how well project performance is progressing toward achievement of the project objectives. Many times, project managers become overfocused on progress against their cost and schedule baselines, and forget that the real intent of a project is to achieve the business objectives driving the need for the project.

The project strike zone is an excellent tool for evaluating and communicating progress toward achievement of the *project objectives*. It is used to identify the critical objectives for a project, to help a project manager and his or her stakeholders track progress toward achievement of the key business results anticipated, and to set the boundaries within which a project manager and team can operate without direct top management involvement.

As shown in Figure 12.1, elements of the project strike zone include the project objectives, target and threshold values, an "actual" field that provides indication of where a project is operating with respect to the target and threshold limits, and a high-level status indicator.[2]

Bill Shaley, a senior project manager for a leading telecom company, described the culture within his company this way: "Managing a project in this company is like having a rocket strapped to your back with roller skates on your feet—there's no mechanism for stopping when you're in trouble." Sound familiar? The project strike zone is such a mechanism that is designed to stop a project, either temporarily or permanently, if the negotiated threshold limits are breached, at which point the project is evaluated for termination or replan and continuation.

Developing the Project Strike Zone

Developing an effective project strike zone is a critical activity for ensuring that the project manager, project team, top management, and other stakeholders all understand and agree on the objectives of the project. It is also critical for establishing the boundary conditions that will drive effective decision making on the project.

Defining a meaningful project strike zone requires quality information from a number of sources. The initial set of objectives is derived directly from the approved project

Project Strike Zone				
Project Objectives	**Strike Zone**		**Actual**	**Status**
Value Proposition	**Target**	**Threshold**		
• Increase market share in product segment				Green
• Order growth within 6 months of launch	10%	5%	7% (est)	
• Market share increase after 1 year	5%	0%	4% (est)	
Time-Benefits Target				Red
• Project Initiation approval	1/03/2018	1/15/2018	1/04/2018	
• Business case approval	6/01/2018	6/30/2018	6/01/2018	
• Integrated plan approval	8/06/2018	8/20/2018	8/17/2018	
• Validation release	4/15/2019	4/30/2019	6/29/2019	
• Release to customers	7/15/2019	8/01/2019	TBD	
Resources				Green
• Team staffing commitments complete	6/30/2018	7/15/2018	7/1/2018	
• Staffing gaps	All project teams Staffed as min level	No critical path resource gaps	Staffed	
Technology				Green
• Technology identification complete	4/30/2018	5/15/2018	4/28/2018	
• Core technology development complete	Priority 1 & 2 tech's Delivered @ Alpha	Priority 1 tech's Delivered @ Alpha	on track	
Financials				Yellow
• Program Budget	100% of Plan	105% of Plan	101% est	
• Product Cost	$8500	$8900	$9100 est	
• Profitability Index	2.0	1.8	1.9 est	

Figure 12.1: Example Project Strike Zone

business case (Chapter 3). To be able to establish and later negotiate the control limits for each objective with the project manager, the project sponsor also needs to know the project team's capabilities and experience and past track record, and balance thresholds against the new project's complexities and risks accordingly.

Identify Project Objectives

Identification of the project objectives begins during the initiation stage of a project. The factors represent a subset of the metrics normally tracked by a project team. The project strike zone should include only the measures that represent the high-level project objectives (often the business objectives). The project objectives will be unique to every organization, and are derived directly from strategic management and portfolio management processes (Chapter 2).

The strike zone is most effective when the objectives identified are kept to a critical few (usually five to six), as this focuses the project and top management's attention on the highest priority contributors to the success of the project. The factors deemed as "must haves" often include market, financial, and schedule targets, and value proposition of the project output.

Set the Recommended Target and Threshold Values

The target and threshold control limits shown in Figure 12.1 form the strike zone of success for each project objective. The target value for an objective is that which the project business case and baseline plan is based on. The target values should be pulled directly from the project business case.

The threshold values represent the upper or lower limit of success for the project objectives.[3] Some discussion and debate is normally required to get an understanding of how far off target an objective can range, and still constitute success for the project. For example, the target project budget may be set to $500,000. But if additional spending of 5 percent is allowable, then the budget threshold can be set at $525,000. This means that even though a project team misses the target budget of $500,000, they are still successful from a project budget perspective if they spend up to $525,000.

Negotiate the Final Target and Threshold Values

Once the project manager establishes the recommended target and threshold values for each project objective, the project manager presents the information to the senior executive sponsoring the project. Based on the complexity and risk level of the project, and on the capability and track record of the project manager and team, the project sponsor may adjust the values accordingly. For example, on a project that is low complexity, low risk, and is being managed by an experienced project manager, the range between target and threshold values may be opened up to allow for a higher degree of decision-making empowerment for the project manager. Conversely, on a project that is of higher complexity, risk, or is being managed by an inexperienced project manager, the range between target and threshold values will be tighten up to limit the decision-making empowerment of the project manager, at least initially.

Once the targets and boundaries are negotiated, the team should be empowered to move rapidly as long as they do not violate one of the strike zone threshold values.

Using the Project Strike Zone

The project strike zone provides many uses to project managers, the executive sponsor of a project, and to the project governance body. Project managers utilize it to formalize the critical project objectives for the project, to negotiate and establish the team's empowerment boundaries with executive management, to communicate overall project progress and success, and to facilitate various trade-off decisions throughout the project cycle (see "When Things Go Bad").

Executive managers utilize the project strike zone to ensure that a new project's definition supports the intended business objectives, and to establish control limits in order to ensure that the project team's capabilities are in balance with the complexity of the project. When used properly, it provides top managers a forward-looking view of project alignment to the business objectives. When problems are encountered, the tool's structure is intended to provide an early warning of trending problems, followed by a clear identification of "showstopper" conditions based on the level of achievement of the project objectives. If a project is halted, senior executives can either reset the project objective targets or thresholds, modify the scope of the project to bring it within the current targets, or, in the extreme case, cancel the project to prevent further investment of resources.

Executive managers and the project governance body set the boundary conditions (targets and thresholds) of the project strike zone between which the project manager can operate, thereby empowering the project manager to make decisions and manage the project without direct top management involvement. As long as the project progresses within the strike zone of each project objective, the project is considered on target and the project manager remains fully empowered to manage the project through its life cycle. However, if the project does not progress within the strike zone of each project objective, the project is not considered on target and the top managers must directly intervene.

When Things Go Bad

Santiam was the code name for a multimillion-dollar new product development project within a leading consumer electronics company. One of the primary strategic goals of the company was to move into a market outside of their traditional business. The product to be produced by the Santiam project team was the first introduction into the new market. To be successful, introducing the product into the market at the correct time, capturing a portion of the market share, and selling the product at a better price than their competitors were all critical project objectives needed to achieve the strategic business goal of entering the new market.

As most project managers know, however, the best developed plans are not immune to risks and alternative realities associated with doing business in a dynamic environment—the Santiam project was no exception. Six months into execution, word came from a supplier of a key component that technological difficulties had been encountered. One of the intermediate schedule dates identified in the project strike zone was in jeopardy.

During the next project review with her executive leadership team, Melissa Bingham, the Santiam project manager, presented the updated project strike zone with status on the product introduction date criteria presented as YELLOW (caution to management). Details of the current issue were discussed and risks associated with achieving the success criteria were reviewed. A mitigation recommendation to place a representative from the quality organization at the supplier's location to continuously monitor the situation and assist with solutions was approved by the executive team. The Santiam project team was given the go-ahead to continue development of the product but under a heightened state of risk awareness.

After an additional three weeks' time, it became clear that the problem had become a critical issue for the project when the supplier announced a six-week slip in delivery schedule. This six-week delay would cause a significant delay of the product into the market, turning the product introduction project objective in the project strike zone from "YELLOW" to "RED"—meaning the project needed immediate top management intervention. An analysis of the other project objectives showed that a delayed launch would jeopardize

(continued)

the desired market share capture and drive the profitability index below the acceptable threshold value of 1.8 (see Figure 12.1).

In effect, the business case for the project was in the red zone. Bingham had the information she needed to make a recommendation to her project sponsor. Her recommendation would be to cancel the Santiam project to prevent significant future losses to the business. Project discontinuance decisions are never easy, especially when thousands, or in this case, millions of dollars have already been spent. In the end, Bingham's executive leadership team utilized the information in the project strike zone as the basis for their decision to cancel the project.

Benefits

In practice, the value of the project strike zone is achieved through the direct communication and interaction between the project manager and the sponsor of the project in setting the vision and key success parameters. These parameters are then recorded in the strike zone and become the management and tracking focus for keeping the project aligned to its business and project objectives.

Use of the tool fosters a "no surprises to senior management" behavior by increasing the flow of relevant information between the project team and top management. This results in an efficient means of elevating critical issues and barriers to success for rapid decision making and resolution.

When used appropriately, it enables empowerment of the project manager and team by establishing the boundaries for authority, responsibility, and accountability. Too often, we hear project managers tell us that they have all the responsibility for driving project success but lack the authority. The project strike zone is the best tool we are familiar with to balance both sides of this equation.

THE PROJECT DASHBOARD

In today's frenzied pace of many projects, project managers need to understand how the project they are responsible for is performing with respect to the key performance indicators, but they rarely have sufficient time to read through a number of detailed status reports from their functional teams. From this time-versus-information dilemma grew the concept of the project dashboard.[4]

Much like the dashboard of an automobile provides the driver a quick snapshot of the current performance of the vehicle, the project dashboard provides the project manager an up-to-date view of the current status of his or her project. Unlike the project strike zone, which focuses on performance against the higher-level project objectives and business goals, the project dashboard focuses on the current state of the lower-level key performance indicators (KPIs).

The dashboard should be designed as an easy to read and concise (often a single page) representation of all KPIs as illustrated in Figure 12.2.

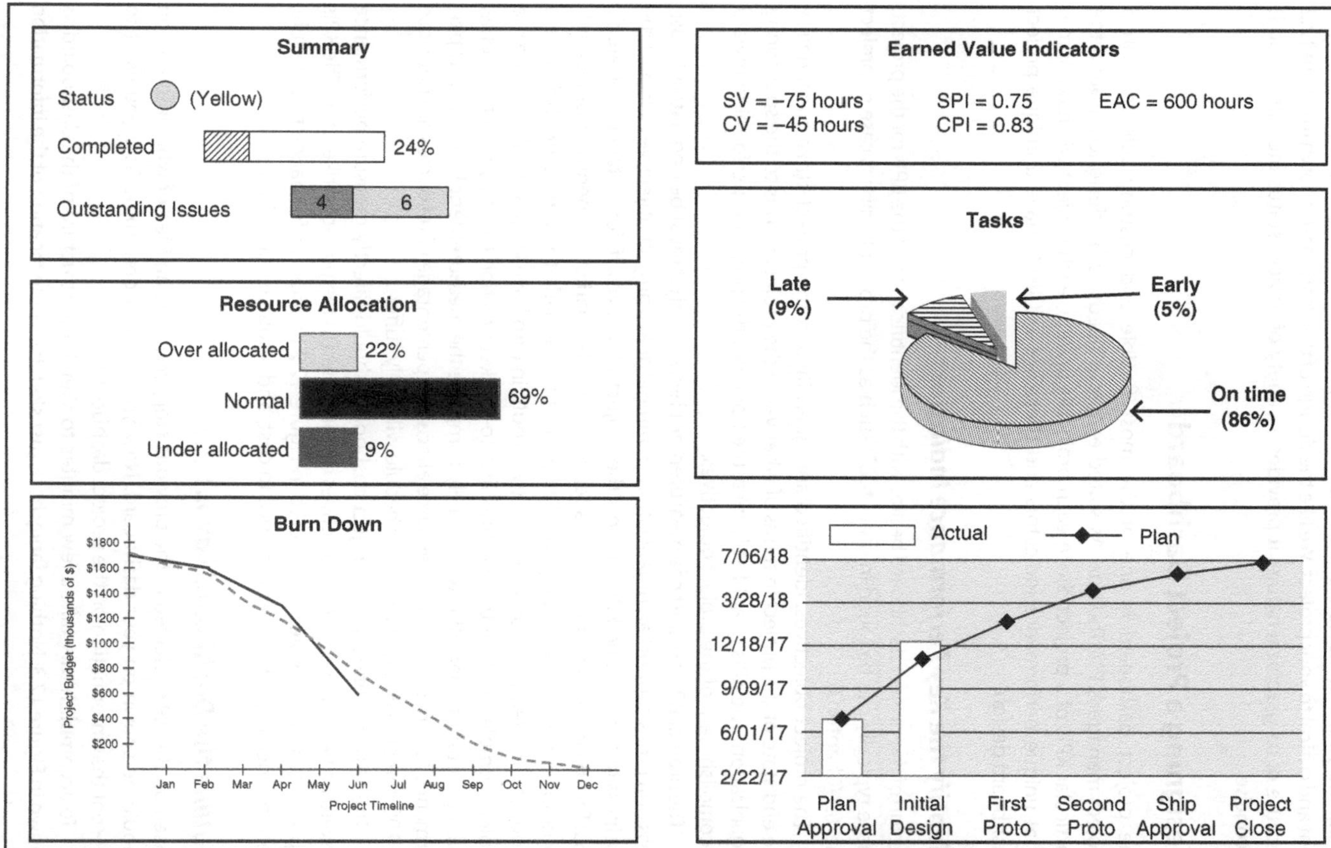

Figure 12.2: Example Project Dashboard

There are many types of project dashboards in use and available as reference for designing your own customized dashboard that represents the information most relevant and critical to your project. We like the design of the dashboard shown in Figure 12.2 because of its graphical nature, it provides a variety of project status measures, and it is concise.

Designing a Project Dashboard

The project dashboard is one of the most flexible and customizable tools in a project manager's PM Toolbox. As stated earlier, it needs to be designed around the particular KPIs of a project. Since each project is unique, each project will have somewhat unique performance indicators and therefore will likely have a unique project dashboard design.

Identify the Key Performance Indicators

Design of the project dashboard begins with the identification of the KPIs for the project. These typically can be found in other tools such as the project business case or project charter (Chapter 3).

The project objectives, identified and quantified in the project strike zone, define the end state of the project in terms of what value the project brings to the sponsoring organization. The KPIs quantifiably measure how well the project is performing toward accomplishment of the project objectives.

The project KPIs are part of a measurement hierarchy that must be understood. Business outcomes support an organization's strategic goals, project objectives support the business outcomes, and KPIs support the project objectives. If, for instance, a strategic goal for an enterprise is to be the leader in a particular market segment, a business outcome in support of that strategic goal would be first-to-market advantage with their new products or offerings. A project objective would in turn have to quantifiably define the project completion date that ensures first-to-market position for the project outcome. Two important project KPIs would likely complete the measurement hierarchy: (1) performance to schedule, and (2) resource allocation percentage (if resources are not close to 100 percent allocated to plan, schedule will likely suffer).

The KPIs identified in the project dashboard should directly measure performance toward achieving the project objectives documented in the project strike zone. The KPIs represented in the project dashboard in Figure 12.2 include performance to schedule, performance to budget, performance to cost, and resource utilization.

Outline the Dashboard Layout

Based on the KPIs selected in the previous step, you have an idea of what information should be shown on your dashboard. Now you need to determine *how* you want to present that information on the project dashboard.

To accomplish this, take a few minutes to sketch the structure of the dashboard as shown in Figure 12.3. Nothing fancy here, just sketch out the location of the information

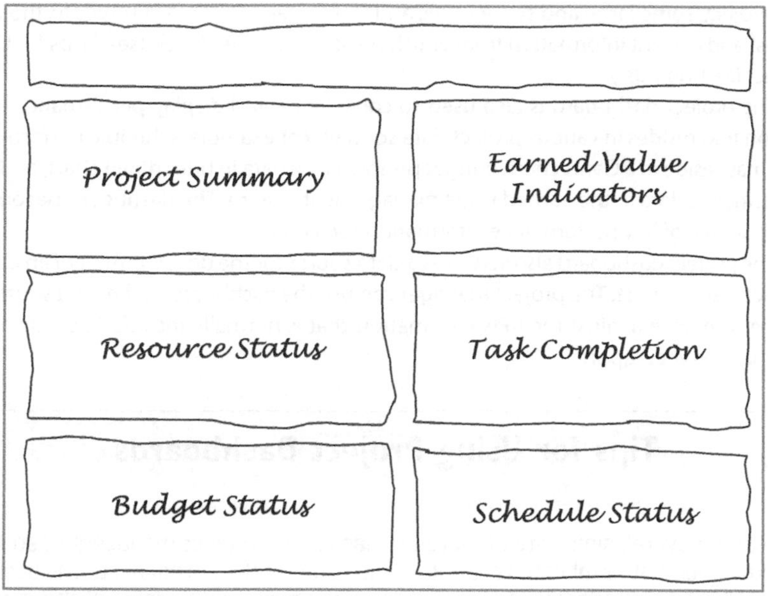

Figure 12.3: Example Dashboard Structure Layout

on the page. The goal is to design the dashboard so it is both comprehensive in content and is appealing to the eye of the recipient.

Populate the Dashboard

The final step in designing the project dashboard involves locating the pertinent performance data and representing it on the dashboard. Whenever possible, use graphical representations as they facilitate a speedier analysis of the current performance on the part of the recipient than a text-based representation.

Some project managers embed hyperlinks within the top-level performance graphics that link to detailed data about the KPI of interest. For instance, if additional detail is needed for the *performance to schedule* KPI, a link can be provided to a detailed Gantt chart, milestone analysis chart, or even the schedule section of the current detailed status report for the project.

Using the Project Dashboard

The project dashboard can be used as both a communication tool and decision support tool by project managers. By using the project dashboard to synthesize lower-level performance data into higher-level information, a project manager becomes armed with the right information he or she needs to communicate the current status of the project with respect to the KPIs. Additionally, many decisions have to be made during the course of a project, some large and some small, and the project dashboard serves as the basis of past and current information from which decisions can be driven (see "Tips for Using Project Dashboards").

The project dashboard is also used to consolidate and display performance information that resides in various project data sources. For example, schedule performance data may reside in a Gantt chart, budget performance data in burn down chart, and cost performance data in the earned value management system. The dashboard becomes a single source of key performance information for a project.

The project dashboard also serves as a data source for the development of an overall project status report. The project manager can use the dashboard as the data source for the performance against the KPIs information that is normally included as part of the summary project report.

Tips for Using Project Dashboards

With the overall simplicity of a project dashboard, project managers need to remember that dashboards are not, in and of themselves, a panacea. The dashboard is only as effective as the design of its structure, the value of the measures and metrics chosen, the accuracy of the data represented, and how effectively the dashboard is used to drive communication and decisions.

Project managers must avoid descending into a quantitative and analytical quagmire when using dashboards. There is a real return on investment that must be maintained in that the value gained from the use of the project dashboard must be greater than the cost of obtaining and analyzing the information contained within it.

Beware of false and conflicting information that may show up in a dashboard. Take the time to ensure that the information is current, accurate, and that it conveys an accurate message about the performance of the project against the KPIs. If not, the dashboard may do you more harm than good.

Benefits

The project dashboard provides many benefits for project managers; we discuss a few of the most significant. First, the dashboard helps the project manager focus on the key performance indicators and how the project is operating relative to the indicators. To gauge exactly how well a project team is performing, the dashboard allows the project manager to capture and report specific data points relative to the KPIs.

The visual nature of the project dashboard provides a concise snapshot in time of project performance that enables quick analysis of progress, as well as easy identification of data outliers and correlations (e.g., correlation between resource utilization and schedule performance).

If trend information is included in the dashboard, the project manager has the ability to make informed decisions based on both past performance and future prediction of performance. Used in this manner, the dashboard becomes a valuable decision support tool that is based on business intelligence.

THE SUMMARY STATUS REPORT

A project manager spends much of his or her time communicating the status of the project to various stakeholders. This reporting takes the form of both formal and informal communications, from hallway conversations to formal project reviews and decision checkpoint meetings with top management and other key stakeholders. Regardless of whether the status is formal or informal in nature, the message should remain consistent.

Consistent communication of project progress is a critical element of cross-project collaboration. Any significant deviations or changes to the project must be communicated by the project manager. In like fashion, any changes that have occurred on any of the functional teams must be communicated to the project manager. Late or ineffective communication can quickly result in rework, delays, and added cost to a project.

The summary status report is a document that highlights and briefly describes the status of the project, reporting on the scope, cost, and time variance, showing significant accomplishments, identifying issues, predicting trends, and stating actions required to overcome issues, risks, and reverse negative trends. Contrary to the beliefs of many that the report is about the history of the project, the report should be about the future of the project that is based on its past. An example of a summary status report is illustrated in Figure 12.4.

Developing a Summary Status Report

Producing a meaningful summary status report starts off with quality information inputs. A solid project baseline plan is the foundation for a good progress report. When the project baseline plan with various baselines—scope, cost, and schedule, for example—is available, then they are compared to the actual state of the project to assess its performance. The actual state is derived from work results and other project records. Through work results, for example, we report which tasks or deliverables are completed and resources expended, presenting them by means of schedule and cost control tools such as the jogging line and earned value analysis.[5] Other information describing the project execution may be included in project records such as correspondence, minutes of meetings, and progress statements.

Design the Reporting System

Defining the purpose, hierarchy, frequency, responsibilities, and distribution of the summary status report is what we call the *reporting system*. This provides consistency in style and format, enabling comparison with previous and future reports for the particular project and other projects. The purpose statement normally centers on whether the report is for internal or external stakeholders. In most cases, the detail and amount of information for internal reports will be different from external reports.

SUMMARY STATUS REPORT

Project Name: Silverbow　　Project Manager: Lance Martin　　Date: July 2016

Purpose: Design, develop, test and launch a new PC/Tablet convertible device

Baseline vs Actual

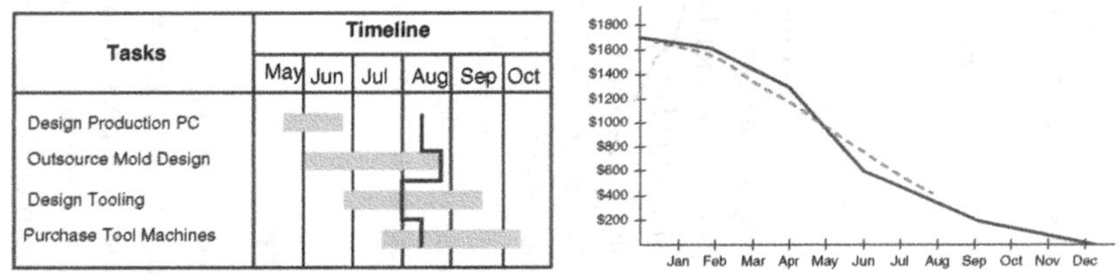

Accomplishments: Design complete, Tooling 40% complete, Purchasing 20% complete

Issues:

Additional project funding delayed two months

Test resources have not rolled off of previous project

Risks:

Power supply manufactureris our only supplier

Schedule is overly optimistic

Requirements not received from customers

Trend: Project is currently estimated to be completed 2 weeks later than the baseline plan

Budget spending is on par with the anticipated burn rate, expected to be within 3% of baseline

Actions Planned:

Explore the use of temporary personnel to perform system test activities until permanent staff is released from the Berkley project

Decisions Needed:

Approval to work at risk until additional funding is available

Approval to use temporary test staff

Figure 12.4: Example Summary Status Report

For example, external reports generally focus on helping the customer determine the status of work being funded and need to be structured accordingly. Each of the two types of reports may need a hierarchical structure, including the summary status report, the detailed progress report, and backup data.

The summary status report provides the crucial points of the overall report while enabling management to review performance progress and trends at a glance. The detailed report contains the general status of the project, major developments, significant variances, major problems, predictions of final schedule and cost, and specifications of corrective actions. Essentially, both the summary and detailed report address the same types of information but in a very different level of detail (see "The Case of Over Reporting"). Because of these similarities, we focus only on the methodology of the summary status report.

The purpose and hierarchy set the stage for the frequency of the report. The report cadence is often defined by the primary recipient, but we also see commonalities based on size and length of projects. For example:

- Weekly reporting for small projects that are 6 months in duration.
- Biweekly reporting for medium projects that are 12 months in duration.
- Monthly reporting for large projects that are 24 months in duration.

These experientially developed benchmarks are based on the belief that each of the projects should have a similar number of summary reports—26 for small, 26 for medium, and 24 for large projects. The point is not the sheer number of status reports, but rather a consistent number of control cycles. Added to these should be the so-called unscheduled reports, prepared in response to unexpected events of a critical nature that F. L. Harrison calls "Red Bandits" to aptly describe their potential to act as project showstoppers.[6] Finally, defining responsibilities in preparing the report and who needs to see it completes the design of the reporting system.

Determine the Variances

With the reporting system in place, it is time to turn the project data being collected into useful information about the performance of the project. Begin by determining the variance between the baseline and the actual project status. A comparison of the project baseline plan and the actual work results should easily yield the variance—that is, the difference between the two. Take, for example, the example progress report.

How the variance is collected is situational, and involves a number of other tools. If for example the project is a small departmental project, it might use the Jogging Line to indicate the schedule variance. Such a non-project-driven department might easily opt to show only this type of variance. In a different situation, where this would be a large project to design and deploy performance metrics throughout a project-driven firm, the milestone analysis might be used to identify the schedule and cost variance, or even the full-scale earned value system might be employed and supported by verbal descriptions about the quality and scope variances.

Ensure that you are reporting progress at the project level. A good strategy is to use the WBS as a framework, as described in the WBS section in Chapter 5. The process of reporting starts at the work package level, identifying the variances and aggregating them up the WBS hierarchy to establish the variances for the whole project. Subsequent steps of identifying issues, predicting trends, and specifying corrective actions should follow the same approach of using the WBS as the framework. One way to do that is to use the system of rehearsals and progress meetings described in Chapter 9.

The Case of Overreporting

This is a progress reporting story of a project manager in an enterprise information technology (IT) department that, unfortunately, is all too common. According to the project manager, "project managers in our group write a project status report every month for every project they manage. I am expected to show how much time I spend on each project, including all administrative work. Frankly, every project manager just reports 100 percent of their regular work hours, even though we may be working 120 percent or 130 percent of our regular work hours, just to minimize the amount of work we have to report on.

These are really very long reports, almost always 6 to 7 pages. I usually manage four to five projects at a time and do a report for each of them. It takes a lot of time to write them and frankly, it is time I should be using to manage my projects. I have a hard time believing that our managers really spend time reading 20 to 25 detailed reports each month."

This is a case of overreporting in which value is lost for both the project managers and their leaders. A better approach would be to provide a summary progress report that would add more value to everyone involved.

Identify Issues and Risks

If there is a variance, especially an unfavorable one, report the issues causing the variance. Also, identify the risks that may occur in the future, and report the impact they may have on the project. The first area (issues) probes to learn what present problems are at the root of the variance and what their impact on the project is. The second area (risks) looks into the future to predict possible troubles and assess their future impact (see the section on issues and risks in Figure 12.4 for an example). The point, of course, is to figure out how the project can deal with them now. For instance, consider a project where the project manager just learned that one of their major materials suppliers might be on the verge of bankruptcy. This possible event would be identified as a high-level risk, worthy of reporting on the summary status report. The team would immediately develop a contingency strategy instead of waiting to hear a few months later about the bankruptcy, at which point they might be helpless to correct the impact of the issue on the project.

Generally, issues impacting the project progress may come from any area of work. It may be useful to use an issues tracker to track all project issues (see "Keeping an Issues Database"). Issue logs are an effective way of tracking problems on the project. The issues that are impacting the project, or those that need top management involvement to resolve, are carried forward to the summary status report. Figure 12.5 shows an example issues log.

Project Issue Log					
Issue #	**Issue Description**	**Date Raised**	**Owner**	**Priority**	**Status**
1	Second round of funding not approved	10/8/2018	Williams	1	Closed
2	Research data not available until January	10/25/2018	Owens	1	Open
3	Quick set-up feature broken	10/26/2018	Powers	2	Open
4	Vendor missed 1st delivery	11/23/2018	Gupta	1	Closed
5	.....				
6	.....				

Figure 12.5: An Example Issues Tracker

Keeping an Issues Database

Learning from past experience is a way for many organizations to continuously improve. A technique that helps classify such learning and offers improvement strategies to future generations of projects is use of an issues database. Simply, it records three types of information: (1) issues of significant impact that occurred in past projects, (2) the nature of the impact that happened or was prevented from happening, and (3) what actions were or could be used to successfully resolve such issues.

How is the database developed? Searching through risk logs of past or current projects and post-project reviews helps identify the preceding types of information. Issues of a similar nature are then grouped. For example, groups may include *team* issues, *process* issues, *vendor* issues, *scheduling* issues, *risk management* issues, and so on. Computerized databases of this sort that are searchable are of special value.

What can the database do for a project? It can serve as a checklist for planning future projects. Also, it can serve as a predictive tool by identifying issues and risks in monitoring schedule, cost, and scope. Additionally, the database offers "premade" impact assessment and actions to mitigate impacts.

Predict Trends

This section of the report shows the predicted future performance based upon the current status of the project. Although forecasts of this type are not easy and are notoriously vulnerable, their essence is less in their accuracy and more in their creation of early-warning signals. For example, in the summary status report in Figure 12.4, "project is currently estimated to be completed 2 weeks later than the baseline plan" is a clear warning, one that mandates action to attempt to reverse the trend. The ability to forecast the trend, week after week, or whatever the report frequency, is paramount in building an anticipatory climate where project teams are alert about the project's past progress but even more about what the future bears.

Specify Actions

If trends are unfavorable, this section identifies actions that should be taken to prevent them and deliver to the baseline plan. With a look into the future, which is our trend, we need to specify corrective actions, assess their impact, and assign an owner in the report. Along with the trend, the specification of corrective actions is perhaps the most valuable part of the report because it enables project teams to be proactive. While the performance progress is important in telling where we are, it is no more than the project history—there is really nothing that we can do to correct or change it. Our only opportunity to change the project is in the future, and that is what the trend and corrective actions offer: an opportunity to anticipate and shape the future by acting—now.

Using the Summary Status Report

Whether small or large, projects need the summary status report. Pressed for resources, small projects—especially in a multiproject environment—will likely issue the summary report as their only report, doing away with a detailed report.

Although many will prepare the report in a formal, written format per predetermined frequency, it is not unusual for managers of small projects to report status verbally (see "The Case of Underreporting").

The Case of Underreporting

We heard this story during a ten-minute lunch with a project manager for a technology firm, Eric Biesot. According to Biesot, "we develop components for our internal customers who build them into their new products for external customers. With seven projects that I am managing right now, I don't really have time to write progress reports. This is really the case for all the project managers. All of us run multiple projects at a time, too many we believe, and no one has time for reporting since we typically work 70 hours a week.

My boss would like to have the reports, but knowing how busy we are, he doesn't require them. He was in our shoes before he was promoted to this position, so I guess he understands what kind of situation we are in. He does ask us in our weekly staff meetings if we have any problems he can help with. But he can't really help much because he has no resources to help out. I usually develop a Gantt chart for each project, but with this pace of work, I just don't have time to keep them updated."

This is a case of severe underreporting and a dangerous situation to be in. This project manager, his manager, and their organization will be in a continuously reactive or "fire-fighting" situation without the use of a streamlined summary status report for each project.

An hour may be sufficient to prepare a typical summary status report for a small or medium sized project. Even as time requirements go higher with the size and complexity of projects, it is clear that a few hours of a large project team's time should suffice for the summary report production. This assumes that extra time—perhaps running in tens of hours—was spent to generate the performance data that feeds into the report.

Benefits

If we view time spent on developing the report as an investment, then return on this investment can be very lucrative in multiple ways. First, the process that results in the report ensures a proactive cycle of project control, communicating information about project problems and status to all concerned, including top management, and taking actions to put the project back on track.

Second, the summary status report is a vehicle to secure stakeholders' involvement in the project. By feeding them with information about the past and future of the project, we help them see the big picture and understand the impact of their contribution. This, in turn, helps maintain their motivation and coordination with others, further strengthening the team cohesion.[7]

Third, the cycle of reporting instills discipline. Busy managers often get carried away by daily pressures, the regularity of reports is a forcing function that makes them sit down, collect data on project health, look into the project future, and form opinions dictating actions. Such work, unlike their daily firefighting of project problems, is the clear essence of project management—think, predict, act.

If a project manager functions in an *influence without authority* environment, the report offers a fourth benefit. The progress report gives the project manager a voice to top management. Having higher management's ear and demonstrating that he or she can clearly communicate the current progress of their project helps the project manager increase his or her influence.

THE PROJECT INDICATOR

In most organizations, the project manager is required to provide verbal status to top management on a regular basis, both in informal or more formal project reviews. The project indicator is used by the project manager to summarize the *overall* status of a project based upon the input and discussion with the project team. There is a direct correlation between the summary status report in the previous section and the project indicator. The project indicator is a presentation device to communicate the information contained in the summary report. The tool gives the project manager a high-level view of the total project and helps him or her to determine if the project remains successfully on track or if there are potential barriers and issues that must be addressed.

It is effective to have a common project indicator format in use on all projects for consistency and comparability of information. The reporting format should include all critical project elements that are important to top management so that they can quickly evaluate progress on projects and determine which need more of their focus and attention. An example project indicator is displayed in Figure 12.6.

The project indicator is brief and limited to one or two pages. It is meant to give a concise, but comprehensive description of current project status, key issues and changes

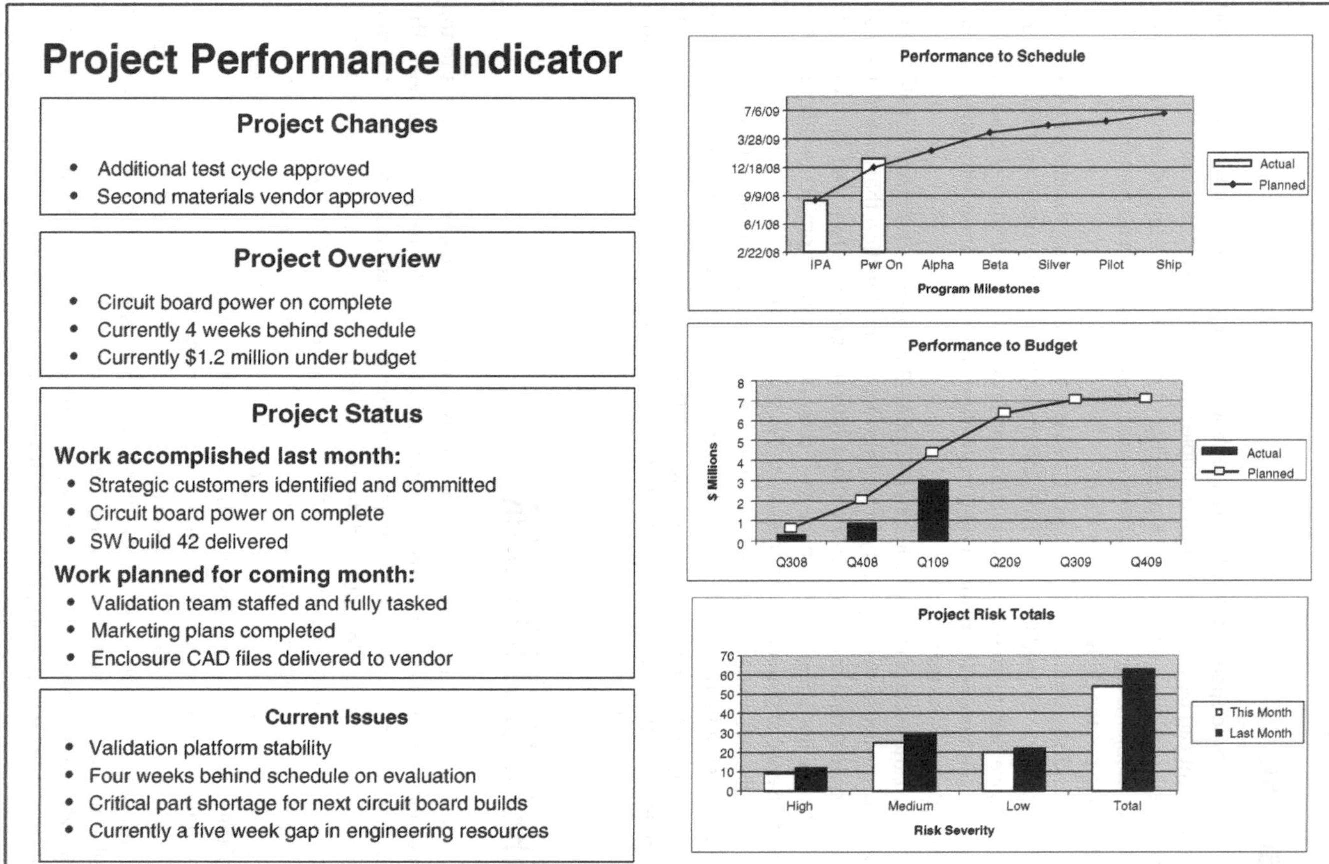
Project Performance Indicator
Project Changes
• Additional test cycle approved
• Second materials vendor approved
Project Overview
• Circuit board power on complete
• Currently 4 weeks behind schedule
• Currently $1.2 million under budget
Project Status
Work accomplished last month:
• Strategic customers identified and committed
• Circuit board power on complete
• SW build 42 delivered
Work planned for coming month:
• Validation team staffed and fully tasked
• Marketing plans completed
• Enclosure CAD files delivered to vendor
Current Issues
• Validation platform stability
• Four weeks behind schedule on evaluation
• Critical part shortage for next circuit board builds
• Currently a five week gap in engineering resources
Performance to Schedule
7/6/09
3/28/09
12/18/08
9/9/08
6/1/08
2/22/08
IPA
Pwr On
Alpha
Beta
Silver
Pilot
Ship
Program Milestones
Actual
Planned
Performance to Budget
$ Millions
8
7
6
5
4
3
2
1
0
Q308
Q408
Q109
Q209
Q309
Q409
Actual
Planned
Project Risk Totals
70
60
50
40
30
20
10
0
High
Medium
Low
Total
Risk Severity
This Month
Last Month

Figure 12.6: Example Project Indicator

that have been encountered, performance against project performance metrics, and the management of critical risk events.

Developing the Project Indicator

Creating a project indicator begins with understanding the information that should be included and communicated in the informal or formal project review with top management. As stated previously, it is useful to top managers if all project indicators are consistent in form and content. This is probably another application of the 80/20 rule; 80 percent of the information contained in the project indicator should be common to all projects being evaluated by the organization's top management, and 20 percent of the information should be unique to each project. In general, the project indicator should include the following information:

- Significant changes to the project.
- Work completed since the last review.
- Work planned during the next reporting period.
- Performance against plan.
- Issues encountered.
- Risks identified.

Changes to the Project

A brief description of the significant changes to the project that may have an impact on performance should always be included in a project indicator. Example changes include significant scope increases or decreases, changes in project budget or funding, changes in project resources, and changes to the project objectives.

Work Completed

This section of the project indicator provides an overview of the key elements of work completed since the last project report. This would include the achievement of project milestones, completion of project deliverables, the resolution of major blocking issues, removal of risk events, and the completion of key project events such as a customer review or signing of a partnership agreement.

Work Planned

A high-level description of the work that is planned between the current reporting cycle and the next should be included in the project indicator. This allows for a discussion on not just what has occurred in the past, but also what will occur in the near future, and what you as the project manager are anticipating in the next work cycle.

Performance Information

The project indicator should provide a concise description of how the project is performing against the key performance indicators. This can be accomplished either graphically (which is always preferred when communicating to senior managers), or in text. If the project is using earned value management (Chapter 10), the earned value analysis values should be included in this section.

Issues

This section of the indicator should concisely describe the major issues that the project team is working to resolve. Be short and concise on your descriptions. Remember that the project indicator is meant as a verbal communication tool, so issues are briefly described on the indicator and then explained to whatever detail is needed during the ensuing conversation with top managers. For every issue communicated on the project indicator, there must be an action plan employed.

Risks

Much like the project issues, a concise description of the critical risk events should be included in the project indicator. Try to limit the risks to the critical three to five events that the project team is working to eliminate or mitigate. Even better, a summarization of progress against risks may be a better representation of project risks, such as is shown in Figure 12.6.

Other Items

Since each project is unique, include a section in the indicator that addresses information that pertains strictly to the project at hand. For reference on what to include in this section, the "General" section in the project reporting checklist covered at the beginning of this chapter is a good source.

Using the Project Indicator

As stated earlier, the project indicator is a verbal communication tool. It is used to communicate overall project status to the top management of an organization. In the process it also facilitates the necessary discussion between the project manager and his or her senior leaders. For each item included in the indicator, ensure you have the backing details to engage in a conversation, or include the right member of your project team who can speak to the details.

The indicator can become the means to engage top management in the critical aspects of the project, and facilitate a request for assistance if and when needed.

The project indicator can also be used for effective communication of project status to the project team. Often, project team members are not privy to the overall status of the project they are a part of. To be most effective when used in this manner, the project team should be briefed *after* top management so pertinent aspects of the conversations with management can also be communicated to the team members.

To provide a complete overview of project status, many project managers present the project indicator along with the project strike zone. This provides a more holistic message that incudes operational status as well as a review of the project objectives and current performance against those objectives.

Variations

Some project managers use indicators within their project teams to facilitate intrateam progress reporting. In this use case, the functional project team leaders prepare and present a more detailed and focused functional indicator that reflects the work of each functional team (see Figure 12.7 for an example).

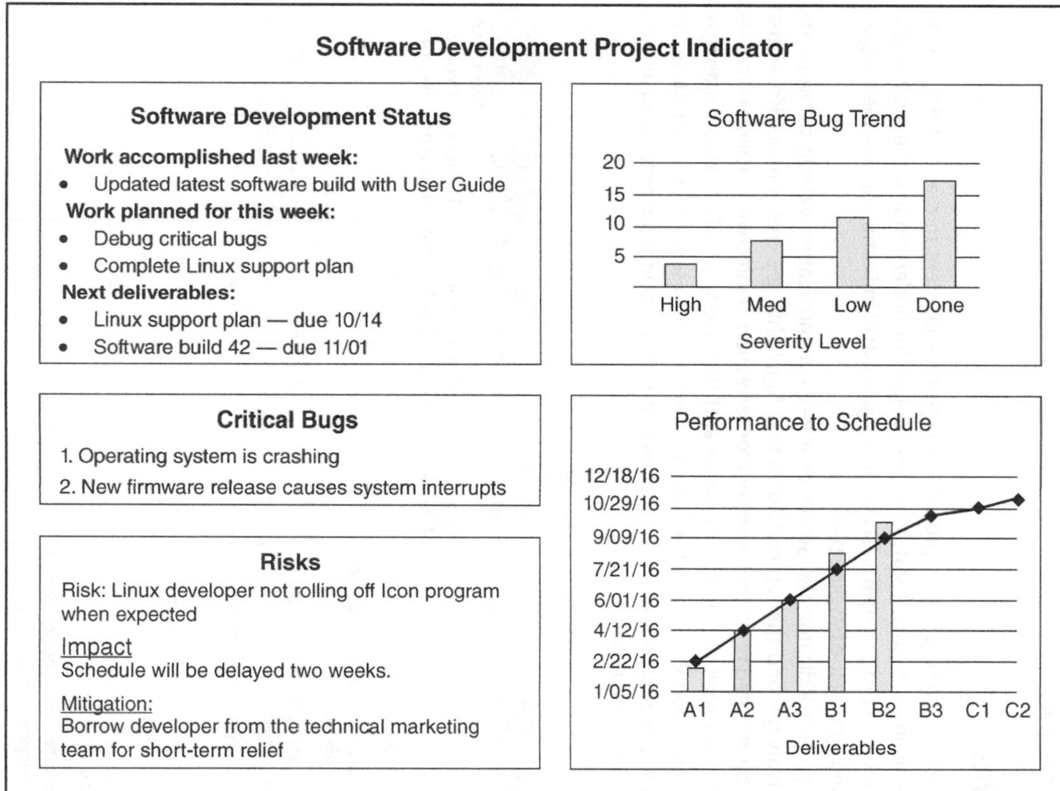

Software Development Project Indicator
Software Development Status
Work accomplished last week:
Updated latest software build with User Guide
Work planned for this week:
Debug critical bugs
Complete Linux support plan
Next deliverables:
Linux support plan — due 10/14
Software build 42 — due 11/01
Critical Bugs
1. Operating system is crashing
2. New firmware release causes system interrupts
Risks
Risk: Linux developer not rolling off Icon program when expected
Impact
Schedule will be delayed two weeks.
Mitigation:
Borrow developer from the technical marketing team for short-term relief
Software Bug Trend
20
15
10
5
High
Med
Low
Done
Severity Level
Performance to Schedule
12/18/16
10/29/16
9/09/16
7/21/16
6/01/16
4/12/16
2/22/16
1/05/16
A1
A2
A3
B1
B2
B3
C1
C2
Deliverables

Figure 12.7: Example Functional Indicator

The project manager must work with each functional team leader to determine the best format and content to present in the functional indicator. By requiring (or encouraging depending on a project manager's style) the functional leaders to keep their functional indicators current, the project manager will receive a comprehensive yet concise report on fuctional team status on a regular basis (recommended weekly). Additionally, the information in the functional indicators can be used as a data source for building the project indicator.

Benefits

The project indicator provides project managers a consistent format and data model for determining the current status of their project, for understanding the most critical challenges their project team is facing, and a gross-level indication of how their project is trending against the baseline plan.

The project indicator also serves as a key communication vehicle between the project manager and the top managers that highlights key cross-project issues that need to be elevated to senior leadership for resolution. It effectively facilitates focused discussion between the project manager and their managers. For project managers that are uncomfortable having these discussions with their senior leaders, or those that are new to the opportunity, the project indicator also provides a mental prop and cue card for discussion topics to cover.

Table 12.2: Project Reporting Tools

Situation	Reporting Checklist	Project Strike Zone	Project Dashboard	Project Indicator	Summary Project Report
Prepare information to report	✓		✓		
Tailor reporting information based on project life cycle	✓				
Communicate performance against project objectives		✓		✓	✓
Facilitates project-level decision making		✓	✓	✓	
Communicate performance against operational KPIs			✓		✓
Communicate functional status to project manager			✓	✓	
Communicate overall project status to top managers		✓		✓	✓
Describe current issues			✓	✓	✓
Communicate project trends and risks		✓	✓	✓	✓
Best for written status reporting					✓
Best for verbal status reporting		✓	✓	✓	

CHOOSING YOUR REPORTING TOOLS

The tools presented in this chapter are designed for various project performance reporting situations. Matching the tools to their most appropriate usage is sometimes a bit confusing. To help in this effort, Table 12.2 lists various performance reporting situations and identifies which tools are geared for each situation. Consider this table as a starting point, and create your own custom project situation analysis and tools of choice to fit your particular project management style.

References

1. Project Management Institute. *A Guide to the Project Management Body of Knowledge,* 5th ed. (Drexell Hill, PA: Project Management Institute, 2013).
2. Martinelli, Russ, James Waddell, and Tim Rahschulte. *Program Management for Improved Business Results,* 2nd ed. (Hoboken, NJ: John Wiley & Sons, 2014).
3. Martinelli, Russ, and Jim Waddell. "The Program Strike Zone: Beyond the Bounding Box." *Project Management World Today,* March–April 2010.
4. Kerzner, Harold. *Project Management Metrics, KPIs, and Dashboards: A Guide to Measuring and Monitoring Project Performance,* 2nd ed. (Hoboken, NJ: John Wiley & Sons, 2013).
5. Lientz, B. P., and P. R. Kathryn. *Breakthrough Project Management* (San Diego, CA: Academic Press, 1999).
6. Harrison, F. L. *Advanced Project Management: A Structured Approach,* 3rd ed. (New York, NY: Halsted Press, 1992).
7. Project Management Institute, 2013.

13

PROJECT CLOSURE

Contributed by

Tim Rahschulte

The value of continuous improvement that occurs from continuous learning is not ambiguous, nor is the concept for continuous learning new or novel. Rather, it has been known for years that team and organizational learning can create a competitive advantage. The concept, as well as tools and techniques, gained immense popularity in the 1990s, due in part to Peter Senge's book *The Fifth Discipline*. Since this seminal text, countless other authors offered publications and thought leadership on the topic of learning.

Learning is indeed an important variable that separates the best from the rest. Organizations that can obtain, codify, and make easily available data and information from individuals and teams in the form of knowledge and lessons learned have a competitive advantage. Often, such knowledge comes from project managers and their teams—those individuals working to create new products, offer new services, and help their organization innovate from where they are now to where they aim to be in the future. Those who do it well continually get better and those who struggle fall behind.

This chapter focuses on best practice project closure tools, techniques, and the functions that the project manager needs to facilitate to ensure project closure leads to continuous improvement from lessons and experiences learned. Three specific tools will be detailed in this chapter: project closure plan and checklist, project closure report, and project postmortem. Additionally, a section on project retrospectives is detailed. Prior to the tools, however, an overview of project closure is offered to ensure understanding with regard to closing a project.

UNDERSTANDING PROJECT CLOSURE

Projects vary in size, scope, and complexity. Regardless of such differences, however, all projects follow a rather similar, albeit generic, life cycle from start to finish. In general, a project cycle progresses from starting or initiating to planning and organizing to

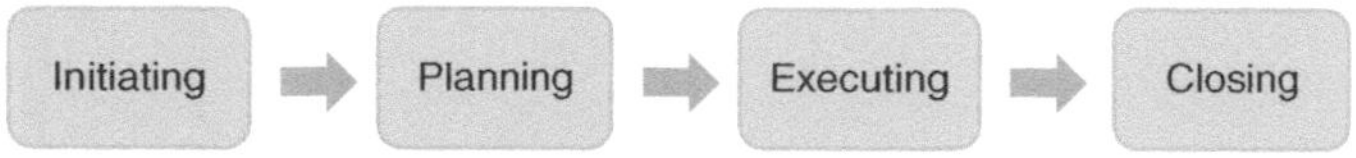

Figure 13.1: Generalization of a Project Life Cycle

executing and then to closing. Harold Kerzner and others have spent a significant amount of time writing books and articles detailing life cycles used by project teams in industries including engineering, manufacturing, software development and construction.[1] Because of Kerzner's fine research and contribution, there is no need to replicate it here in detail. The general process, which is outlined in Figure 13.1, is sufficient and enables the focus here to be on best practice tools and techniques that project managers can use to ensure effectiveness in their work.

It is important to note that a project life cycle is not synonymous with project process groups that many project managers and teams use to organize their work. This is expressly noted as such in the *Project Management Book of Knowledge* (PMBOK) Guide in which it states, a project life cycle "should not be confused with the Project Management Process Groups."[2] It is understandable that confusion occurs. The five process groups are initiating, planning, executing, monitoring and controlling, and closing. The nomenclature is nearly exact to that of Kerzner and others relative to project life cycles. Further, they are both referred to as a cycle or process, and the work effort (deliverables, decisions, and otherwise) varies in accordance to the stage of work under way.

Figure 13.2 illustrates each process group relative to the project's life cycle over time and the interaction of one group with the others. The differences between a project life cycle and project process groups becomes evident with the visualization of Figure 13.2. Whereas a life cycle is naturally explained and illustrated linearly, with definitive starts and finishes, the process groups overlap with one another. As can be visualized, there

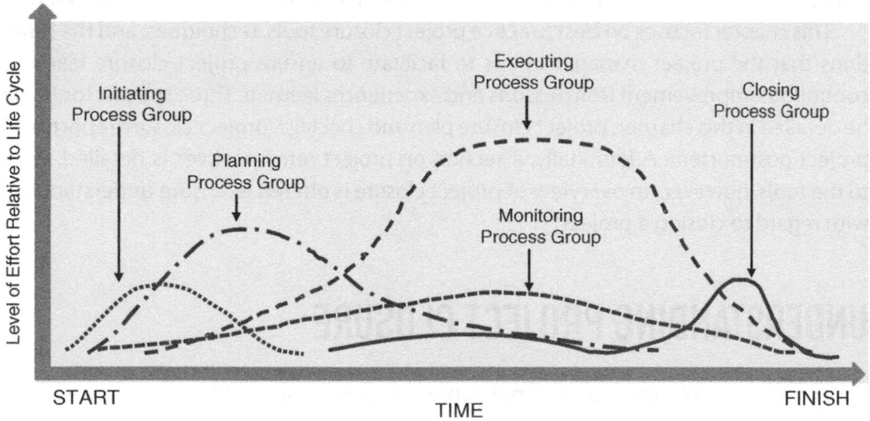

Figure 13.2: Process Groups Interact in a Phase or a Project

is simultaneous work associated with most process groups over the period of the entire project life cycle. Although this is illustrated, the simultaneous work is not always comprehended and applied. For example, although project closure activities are to be occurring in nearly every phase of the project's life cycle, often project managers and their teams conduct such activities only during the closing phase of the project. Focusing project closure activities only at the end of a project limits the learning that can occur and therefore increases project and enterprise risk. This will be explained in further detail throughout this chapter.

Focusing here on project closure, the closing process group consists of the work effort necessary to end, or close out, any pertinent project activities across the phases of work and process groups. Such work here is often associated with contractual obligations, procurement paperwork, handoff agreement(s) between the project and operations teams, and reallocation of resources, just to mention a few.

Project Closing Activities

When engaged in closing activities, the project manager's responsibility is to review the work of the prior phase (or phases if the project is beyond the first phase in the project's life cycle) to ensure all necessary and planned work is complete and objectives are met. The importance of the closing process group is twofold. First, is to ensure that the work planned was actually completed. During project planning work efforts, all pertinent expectations associated with resourcing, timing, scope, quality, costs, and other activities are detailed and sequenced along a timeline or schedule. Proper closing activities ensure that what was deemed necessary during the planning was carried in accordance to expectations. If there is any variance between what was planned and what was completed, the project manager must determine why and then determine if any adjustments need to be made for the remainder of the project based on why the variance occurred. Variances here could be either positive or negative to the successful completion of the project.

While the first point of the twofold importance to proper project closing is project-centric, the second part of the importance for proper closing goes beyond the value of the project team to being enterprise-centric. The most effective learning organizations have processes, systems, and a culture in place to foster knowledge sharing. It is incumbent on the project manager to broadcast lessons learned, best practices, and issue resolution tactics to peer project managers and project teams across the enterprise. Doing so shares knowledge, which in turn can enable project teams to more efficiently and effectively plan and execute their projects. The more efficiently and effectively projects are delivered, the more business value project teams generate for the enterprise.

The key takeaway here is that the work effort associated with project closing is to ensure that all pertinent activities are complete. However, importantly, closing is about learning. This requires the project manager to engage a number of team members and other stakeholders. This point also addresses a major oversight by most project managers and their teams. As noted earlier, far too many project managers focus closing on end of project work instead of throughout each process group.

Project management closing is not a work effort reserved for the end of the project, but rather closing is about reviewing and learning throughout the project's life cycle.

It is common knowledge that there is a lot of work involved with managing a project. It does not matter if the project is small or large, rather straightforward or complex. All projects require numerous resources, many deliverables, and even more decisions during the life cycle. Often project managers and their teams look to the deliverable as the finish line, the point of completion. More seasoned project managers and mature learning organizations know differently. A milestone, deliverable, or even a handoff is not an ending point. Rather, it is the process of closure that marks an ending point. Figure 13.3 isolates the closing work effort relative to a project life cycle.

As can be discerned from the graphic in Figure 13.3, project closing is not a work effort reserved for the end of the project, but rather closing is about reviewing progress, completing work, and learning throughout the project's life cycle. This is being emphasized here because when discussing project closing with many (dare we say, most) project managers, the conversation is often most concerned with the end of the project.

How Do You Learn from Your Projects?

A recent Project Management Office audit at a company in the Pacific Northwest uncovered a common theme when it came to project closeout work. Ashley, a seasoned project manager within this rather immature project organization, shared this comment:

> When it comes to learning from our project work, we do conduct post mortem exercises after the project ends with members of the project

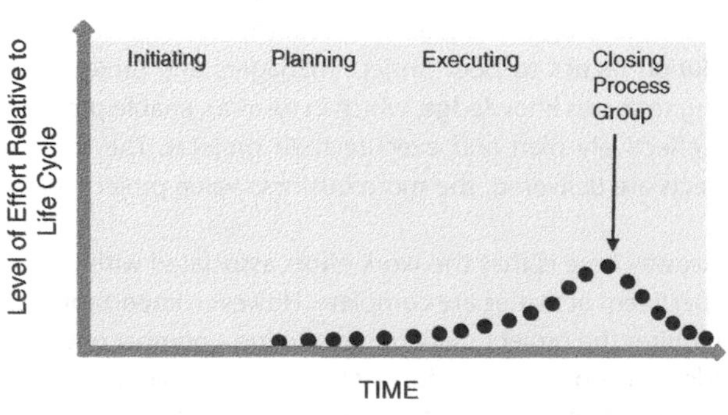

Figure 13.3: Project Closing Work Effort During the Life Cycle

team who are available [those not deployed on another project]. Unfortunately, we do not invest enough time, in my opinion, during the project to conduct learning initiatives necessary to capture new knowledge from lessons learned and adjust our work effort accordingly. We make a lot of the same mistakes, time and again, across projects. We are wasting resources by not learning.

Ashley's experience is a common one. With her company's approach to conduct lessons learned efforts at the end of the project, often a lot of time has passed since the start of the project and many lessons that may have been learned have since been forgotten due to the passing of time and not having codified, saved, and distributed the lessons experienced earlier and throughout the project. Additionally, lost was any opportunity to make the next phase in the project's life cycle more efficient and effective. When aggregated across multiple projects in an enterprise, this lack of learning becomes increasingly expensive and risky.

Waiting until the closing phase of the life cycle to reflect and learn becomes a problem not only for the project, but also for the entire company. Any project team that waits until the end of the project's life cycle loses out on opportunities to learn and improve. Highly experienced project managers and organizations with mature project management organizations know the value of intentionally engaging stakeholders in closing work throughout the life cycle of any project.

The remaining pages in this chapter detail how project managers can (and should) incorporate closing activities throughout their projects. The tools and techniques outlined hereto should help seasoned and aspiring project managers do just that—incorporate project closing activities and deliver greater levels of business value.

Summarizing Project Management Process Groups

As a quick reminder, here is an outline of the major project process groups.[3]

- *Initiating process group*. Those processes performed to define a new project or a new phase of an existing project by obtaining authorization to start the project or phase.
- *Planning process group*. Those processes required to establish the scope of the project, refine the objectives, and define the course of action required to attain the objectives that the project was undertaken to achieve.
- *Executing process group*. Those processes performed to complete the work defined in the project management plan to satisfy the project specifications.
- *Monitoring and controlling process group*. Those processes required to track, review, and regulate the progress and performance of the project;

(*continued*)

identify any areas in which changes to the plan are required; and initiate the corresponding changes.

- *Closing process group*. Those processes performed to finalize all activities across all process groups to formally close the project or phase.

PROJECT CLOSURE PLAN AND CHECKLIST

Proper project closure starts with proper planning. Planning is especially important when it comes to project closure because this phase in the project's life cycle and the work associated with it is often neglected. To ensure project closure is not neglected, best practices suggest starting the project's closure work during the planning phase of the project. Doing so ensures allocation of resources and clearly sets expectations of how project closure activities will be conducted throughout the project. This leads to the following two important questions.

1. What activities are associated with the project closure work effort?
2. When should the project closure work be conducted?

There are three major tasks for project managers to oversee during closeout work. These three tasks are the same (at a generalized level) regardless when in the life cycle they are administered. First, the project manager must evaluate if the outcomes, decisions, and deliverables of the project (at that point in the life cycle) met the expectations of all stakeholders. This work can be summarized by the following questions:

- Were all planned and scheduled deliverables and key milestones complete?
- Based on the work to date, is the overall health of the project and the team high and functional?
- Based on the work to date, are the stakeholders satisfied with progress and optimistic about the effectiveness of work relative to the next phase (or operationalization) of the project?
- Is all contract and procurement work finalized and up-to-date?
- Is the team adequately prepared, resourced, and optimistic about the effectiveness of work relative to the next phase of the project?

Essentially, these questions serve as a framework for project managers to begin project closure activities. Ideally, all questions would be answered "yes." If there is a "no" answer, the project manager needs to further investigate, determine what is needed to change the "no" to a "yes" in the next phase of work. In some cases, doing so will require support, guidance, and direction from senior leadership, the governing body, or sponsor of the project.

As project closure activities begin, a multitude of work lies ahead. The use of the project closure checklist is helpful for guiding the closure work and ensuring all areas are addressed. Table 13.1 offers a generic project closure checklist that can be modified to meet the needs of your particular organization and project.[4]

Table 13.1: Generic Project Closure Checklist

Status	Checklist Items
✓	Verify all project deliverables completed
✓	Conduct project closure meetings
✓	Develop a resource reassignment plan
✓	Close out all work orders, contracts, and subcontracts
✓	Prepare final reports (project, financial, quality, and so on)
✓	Submit final reports to customer and top management as required
✓	Finalize the project file
✓	Close out all financial documents
✓	Ensure all costs are charged to the project
✓	Submit final invoices to customer and pay final invoices from suppliers
✓	Document final change management log and final project scope
✓	Document actual delivery dates of all deliverables
✓	Conduct the final postproject review meeting
✓	Submit final customer or client acceptance documentation
✓	Officially notify the customer or client of project completion
✓	Officially notify vendors, suppliers, and partners of project completion
✓	Compile and store all required documentation for long-term data management
✓	Dispose of all equipment and materials
✓	Recognize the work of project team members
✓	Collect and document all project lessons learned
✓	Celebrate success

If the first part of the project manager's work in closing is to determine what worked and what didn't, the second critical aspect of closing is documenting your findings. Documentation here means updating project artifacts, planning documents related to upcoming phases of work, and templates and best practice documents for other project teams. This leads us to the third and final work effort for project managers in closing project work efforts—socialize lessons learned and key takeaways.

The socialization of lessons learned and knowledge goes beyond the updating of artifacts and templates. Here, socialization means broadcasting and distributing knowledge. This means that the project manager and their team serve as a subject matter expert by sharing knowledge: a researcher and a teacher, if you will, of key findings from experience. This requires project managers to communicate effectively and do so in a way that focuses on applying lessons from one project on to others. The more often this occurs, teams and organizations can become smarter, faster. The faster you can learn, the better off you will be. This work is where teams, and organizations at large, can build a competitive advantage. When knowledge sharing becomes a cultural expectation among project managers and their teams, intellectual capacity increases vicariously from one team to another and best practices, tools, processes, and know-how become expectations for continuous improvement.

Developing the Plan and Checklist

Proper planning is perhaps one of the most important aspects to ensuring project success. Project managers develop their closure plan in accordance with the overall project's scope and schedule. This enables the project manager to align the frequency of closure work with minor and major activities of the project. The project management plan typically addresses the following items in relation to a business problem or opportunity:

- Scope
- Time and schedule
- Costs
- Quality
- Communication
- Risk
- Procurement

During the planning phase of the project, closure activities can often be an oversight. This is unfortunate because closing a project has direct correlation and implication on each of the seven bulleted items just noted. Proper planning should include closing activities as part of the project's scope, time, and cost. Doing so certainly has implications relative to the project's quality, communication, risk management, and procurement initiatives. Therefore, the first step to ensure proper project closure is to develop the project closure plan during the project's planning phase.

Ensuring proper project closure is the process of identifying the requirements for closing out the most critical aspects of the project. These aspects could include decisions, deliverables, milestones, and (definitely) each major life cycle phase of the project. Figure 13.4 highlights the plan for project closure relative to inputs and outputs.

Essentially, the inputs necessary to clearly and completely plan for project closure include the core project management plans. These plans should be detailed during the planning phase of any project and from them, a project closure plan and its associated processes, practices, expectations, and checklists can be created.

Since every project is different, the detail (breadth and depth) of the project closure plan will likely be different for each project. The challenge faced by most project managers is to ensure project closure occurs (make sure the project is completing work as planned, document lessons learned, and share that knowledge to project teams) without being overprescribed in the process or overburdened by the tools used in the process. It is easy for project managers to simply say that project closure activities will occur at every milestone or deliverable in the project schedule. In reality, however, that is not needed. The project manager must discern at which critical points in the project should an investment of time, personnel, and perhaps other resources be made to reflect, learn, update documents and plans, and share findings. This is a balancing act for all project managers (see "Balancing the Art and Science of Project Management").

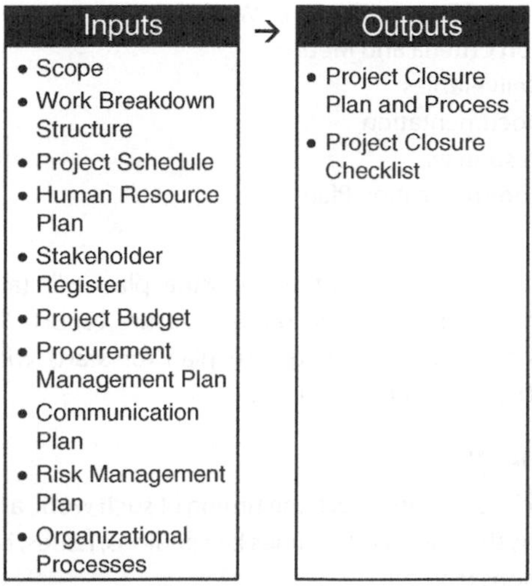

Inputs
• Scope
• Work Breakdown Structure
• Project Schedule
• Human Resource Plan
• Stakeholder Register
• Project Budget
• Procurement Management Plan
• Communication Plan
• Risk Management Plan
• Organizational Processes
→
Outputs
• Project Closure Plan and Process
• Project Closure Checklist

Figure 13.4: Project Closure Inputs and Outputs

Balancing the Art and Science of Project Management

There is a long-standing debate as to whether project management is a science. Indeed, over the years, there have been scientific protocols added to the work of project management. Consider the use of time estimating techniques, measuring project performance with earned value management calculations, and planning resource allocation needs relative to work effort. All of that is science. However, the sum of all of that does not make project management. There is also the importance for project managers to enable teams, navigate organizational politics, motivate and encourage people to thrive, and communicate effectively across diverse stakeholders and groups. That is all leadership and that is all art.

The reality is, project management is not an "either-or" equation, but rather a "both-and" need when it comes to art and science. The best project managers balance the science of management with the art of leadership. These managers know when to push for absolute use of project tools and when to lighten up to fit readiness of the team and organization's culture. That's balancing art and science.

Any project closing plan should include the following sections or categories of content:

- Executive Summary
- Project Scope and Business Objectives

- Start and End Dates of Project Phases of Work
- Project Completion Criteria and Metrics
- Project Closure Deliverables
- Project Closure Documentation
- Project Closure Resources
- Project Closure Communication Plan
- Final Approval

It is important to note here that the closure plan will (at the completion of the project) turn into the basis of the project closure report. So, in essence, using this outline enables the project manager to plan (or start) with the end in mind. The following describes each of the sections.

Executive Summary

It details the project closure work effort, the timing of such work, and the business value of its use. Additionally, this summary outlines key findings, issues, best practices, recommendations, and lessons learned.

Project Scope and Business Objectives

It summarizes the scope of the project and the key business objectives aiming to be met. This detail explains stakeholders and groups (inside and outside the organization) impacted by the project and aims to align deliverables and project work outcomes to business goals. It is important that any objective detailed in this section be plainly written, unambiguous, and specific to the point of being measurable with tools such as the project charter (Chapter 3), project business case (Chapter 3), and work breakdown structure (WBS) or program work breakdown structure (PWBS) (Chapter 5) contain the necessary information for this section.

Start and End Dates of Project Phases of Work

It summarizes the estimated start and end dates of the project. Some project managers may also choose to illustrate a milestone view of the project schedule in this section. Doing so helps to highlight the critical timing of the project work and at which points closure activities will be used. Being clear in this section about project closure events helps to ensure that project resources (especially personnel) are available for such work prior to being reallocated to other initiatives.

Project Completion Criteria and Metrics

This section details both project and business measures and metrics that will be used to evaluate project completion. A simple table (see Table 3.2) is most often used to denote the activity or deliverable being reviewed, the criteria used to determine completion, and any metrics used in the completion determination. Having this level of detail creates unambiguous means by which a project can be deemed successful.

Project Closure Deliverables

This section outlines two major closure needs. First, this section serves as a checklist for review of all project deliverables and whether or not each has been completed. Second,

Table 13.2: Project Completion Template

Activity/Deliverable	Completion Criteria	Completion Metrics

it identifies the handoff necessary to release project deliverables to operational owners. A simple table, such as the template shown in Table 13.3, is most commonly used to capture such detail. Highlighted will be any outstanding items such as those not complete or not yet ready for handoff. Such items will be accompanied with recommendations and actions for resolution.

Project Closure Documentation

Any project document should be reviewed for completeness, updated, and archived. To do so, a document or artifact register should be detailed at the planning phase of the project and updated throughout the project cycle. This section of the plan should include that register, clearly noting the owner(s) during the project and post project, the location of the artifact, and keep track of the updates or versions of the artifacts as the project matriculates its life cycle.

Project Closure Resources

This section outlines the resources needed for closure activities. Planning for the use of resources early in the project helps to ensure their availability at the time of closing activities. Again, a simple table (see Table 13.4) will suffice for planning purposes.

Project Closure Communication Plan

This section details the communication necessary to broadcast all lessons learned, key takeaways from the project work, new best practices, and any updated tools, templates, or other artifacts. A simple table (see Table 13.5) outlining key aspects of the communication plan can help serve as a summary view of all messaging (see Table 3.5).

Final Approval

This section is a placeholder for final sign-off of project closure. It is a formality that is often signed singularly by the project sponsor or jointly by the sponsor and operational

Table 13.3: Project Closure Deliverables Template

Deliverable	Operational Owner	Complete	Handoff Plan

Table 13.4: Project Closure Resource Template

Resource Name	Closure Activity Responsibility	Allocation Approved	Duration of Work Effort

Table 13.5: Project Closure Communication Template

Owner	Message	Audience	Distribution Method	Timing

leader assuming responsibility of the project deliverables. A name, signature, and date are the items most often captured and, once obtained, signal the formal closure of the project and release of all resources.

Using the Closure Plan and Checklist

It is incumbent on the project manager to use the closure plan and checklists as tools: tools to help negotiate the need for resource allocation for such work and tools to support the monitoring function of the project and expedite the review of work completion. As noted earlier, detailing the tool and using the tool is a balancing act between art (leadership) and science (management). The intent is to validate completion of work and readiness for handoff while simultaneously codifying lessons learned and sharing them for enterprise value. Use just enough rigidity necessary to capture this work, while maximizing latitude to project personnel necessary to complete the work.

Benefits

Handoffs, handovers, decisions, deliverables, and a myriad of other important activities occur throughout the project. Using the closure plan and checklist can mean the difference between project success and failure. Importantly, beyond the project itself, following a proper plan and checklist can create a distinctive advantage for a company by means of learning faster than competitors. Learning faster can translate into faster time to markets with products, higher quality of services offered to customers, and increased margins due to lower project expense ratios.

THE PROJECT CLOSURE REPORT

Most project sponsors expect a final project report. When completed properly, the project closure plan (discussed previously) manifests into the project closure report (see Figure 13.5). This tool is important in that it is the final report of the project that formalizes the closure of the project. It confirms not only the completion of the project, but also the acceptance of the project handoff from the project team to the operational owner or client.

Developing the Project Closure Report

The project closure report is a documented review of the entire project. It highlights the completion of project work. It should also highlight any variance between what was planned and what was actually accomplished. Such variances may be in the form of schedule, cost, resource utilization, and other pertinent measures and metrics. It should also highlight the likelihood of application of the project outcome(s) achieving the initially noted business case problem or opportunity. Noting all of this can follow the same outline and format as the project closure plan, which is outlined again here for convenience:

- Executive Summary
- Project Scope and Business Objectives
- Start and End Dates of Project Phases of Work

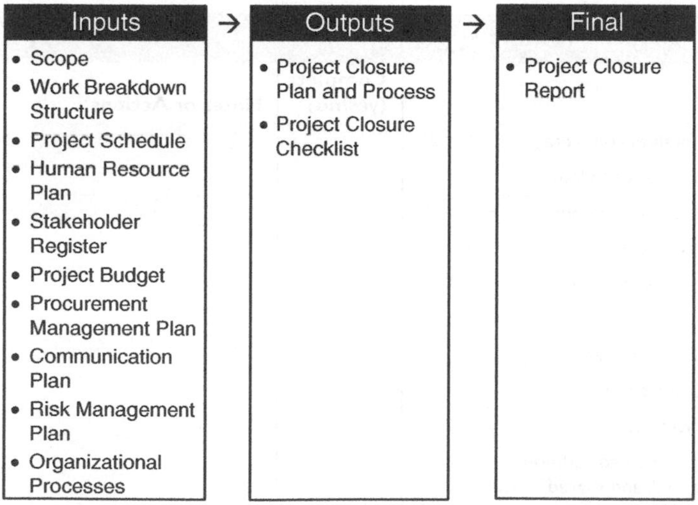

Figure 13.5: Manifestation of the Project Closure Report

- Project Completion Criteria and Metrics
- Project Closure Deliverables
- Project Closure Documentation
- Project Closure Resources
- Proejct Closure Communication Plan
- Final Approval

In addition to the tables and checklists outlined in the prior section, you may choose to summarize major closure activities into an executive level checklist. This executive level view of the project would summarize the milestones, deliverables, key decisions, handoffs and signoffs, and other pertinent aspects of the project. This level could aggregate such accomplishments as all deliverables or it may be necessary (depending on the project, the project manager's preference, and the sponsor's request) to detail each milestone and deliverable individually, rather than in aggregate.

Table 13.6 illustrates a sample executive level checklist used in a project closure report. As noted, the content and level of detail will likely vary from project to project, but this illustration provides a general view of the checklist.

In addition to the checklist and other aspects of the report already noted, project managers should also detail any business or stakeholder implications because of the project. There may be known positive or adverse effects to other projects, products, teams, or resources because of the outcome of your project. Conducting a thorough investigation and deriving a resolution to such implications may be outside of the scope of the project being closed out; however, detailing such implications to make them known is certainly within the scope of the project team and expectations of project work.

Table 13.6: Project Completion Template

Item	Complete (yes/no)	Notes or Actions
All deliverables complete		
All milestones complete		
All contracts closed out		
All issues resolved		
All handoffs signed off		
All payments made		
All invoices submitted		
All accounts closed		
All artifacts updated		
All lessons learned gathered, documented, and shared		
All personnel reassigned		
All excess materials disposed of or stored		
....		

Further, in developing this report, the sponsor or stakeholders may request a presentation to accompany the final report. Therefore, as the report is being planned and detailed, be mindful about how the report may be converted into a presentation format for use.

Using the Project Closure Report

A project closure report starts in the planning phase of the project's cycle. If nothing else, project managers use the planning phase to understand what the project sponsor and other stakeholders need at the closing of the project. This insight guides the project manager in knowing what to deliver in the project closing report. The primary use of the report is for sponsor signoff for project completion and for permanent documentation of project results.

Although the report can be planned at the beginning of the project, it is not until the end that it is fully documented to completion. In between planning and closing, the project manager must use discretion to periodically update the report. Doing so may expedite some time at the end. However, even if a decision is made to wait until the end, the project closing plan and checklists (outlined earlier) should be used throughout the project. At the end of the project, these checklists can be used to complete all necessary components of the closing report.

Seasoned project managers know that closing out a project can heighten emotions. The longer the project in duration, the more personnel have the opportunity to galvanize a culture, build relationships, and grow attached to "my" project. The closure of the project signifies a transition. What was once assumed as "my" project is now going to be closed. There may also be a sense of uncertainty in terms of resource allocation. The finish of one project usually means the start of something new and anything new can be emotionally challenging.

To address such concerns, project managers should work hard to meet with team members individually as well as in groups and the entire team. Spend time listening to each team member and help them celebrate the successes of the project, learn from the shortcomings of the project, and motivate them to embrace what is next. Remember that closing a project is more than just handing off ownership to an operational team or customer. As a project manager, it is also about helping your team move on to their next project assignment.

Benefits

One of the most frustrating aspects of managing projects is leaving work undone. The project closure report finalizes all work. There are benefits of the report for the project team and its stakeholders. From a sponsor perspective, the benefit of the report is having a summarized document of the entire project. For the end user or operational team, it is an official handoff of ownership. For the project manager and team, it outlines a clear point of transition to work on other projects. For other project teams, there is value in gaining lessons learned and best practices as well as identifying any implications from the closure of the project. Finally, as was noted in the project closure planning section, using the closure report can create a distinct advantage for a company by means of learning faster than competitors.

THE POSTMORTEM REVIEW

Sometimes called the post project review or post-implementation review, the project postmortem goes by many names. Regardless of what you call it, the project postmortem is a review of a project after the project closure acceptance, after all project closure activities are complete, and after the project has been in operational mode for a period of time. The purpose of such a review is to determine the following:

1. Was the project successful?
2. Were all closure activities handled properly, especially any final handoffs from the project team to the operations team or customer?
3. Were lessons learned captured and transferred to project teams across the enterprise properly?
4. Has the project achieved planned operational outcomes—the business goals and objectives outlined in the project's business case?

As can likely be imagined, a postmortem review can be rather stressful, especially if the project was not deemed successful. In such events, individuals could engage in finger-pointing, placing blame, dodging responsibility, skirting accountability, and miss all opportunity for learning, growth, and improvement. Because of the probability of such a negative event, it is the responsibility of the project manager to work with the project sponsor and operational team leadership to ensure that an honest postmortem review occurs without negativity.

To establish an honest and positive postmortem review, the project manager, sponsor, and team must work from the very beginning of the project to cultivate a positive culture: a culture of team, a culture of we are in this together as compared to "us" versus "them," a culture of how can we win together rather than how can I make sure I win and you lose. With this in mind, we can add one more element to our inputs-outputs illustration to capture how the postmortem work will leverage earlier project work. Figure 13.6 illustrates how early planning documents can be used for postmortem review work.

Conducting the Postmortem Review

The three steps most associated with the postmortem review are: (1) gather feedback from project and operational teams, (2) organize and facilitate a meeting among the teams and key stakeholders, and (3) capture the meeting outcomes in the form of a postmortem report.

Gathering Feedback

Feedback is important in general, but it is especially important for a postmortem review. Thoughts and opinions from team members are also important, but more so are facts about the project and operations. In order to get the facts, the project sponsor and operations leadership should identify a well-respected and nonbiased person to gather information among the project team members and stakeholders. This same person may

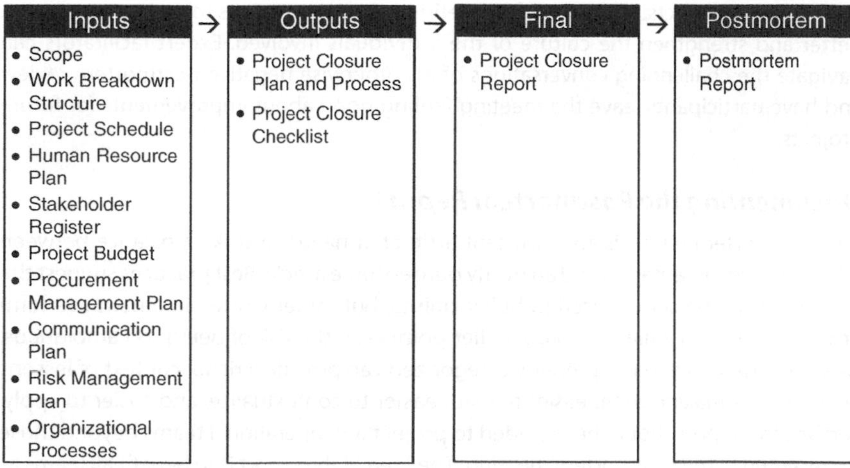

Figure 13.6: The Cycle of Closure Work

be used to facilitate the postmortem meeting. We do not recommend that the project manager be the postmortem review facilitator.

Gathering feedback ahead of the meeting will help to expedite the meeting and provide the facilitator with a sense as to how the meeting will go—will it be emotionally charged and negative or will it be productive, future focused, and positive? The three standard questions used when gathering feedback in preparation for the meeting are:

1. What went right that should be considered best practice and used during every project?
2. What went wrong and should be used as a leading indicator metric of project problems?
3. What should be done differently on the next project?

There are a number of different ways in which these questions can be worded and facilitated, but in general, these are the three primary questions used in postmortem reviews. This information can be gathered from face-to-face interviews, e-mail or web-based surveys, and from the review of prior lessons learned documentation if gathered throughout the project's life cycle.

Conducting the Meeting

The postmortem meeting can either feel like a celebration or a poor performance review. It will be a celebration if things have gone well. Due to poor execution however, missing business goals while in operations, personality conflicts, organizational politics, or a number of other variables, the postmortem meeting could be negative. It is for this reason a neutral and expert facilitator is recommended to oversee such meetings. This individual can summarize feedback gathered ahead of the meeting and dive into the details necessary to obtain real meaning and insight on what to do better during future projects. The aim of the meeting is to uncover ways to make project work better and strengthen the culture of the individuals involved. Expert facilitators can navigate the challenging conversations of an otherwise negative postmortem review and have participants leave the meeting feeling good about improvements for future projects.

Documenting the Postmortem Report

The postmortem report is an important artifact. It needs to strike a balance between pithy-level documentation and an overly burdensome article. Best practices suggest the document not be summarized in bullet points, but rather written in a narrative form of what to do and what not to do. Bullet points run the risk of being too ambiguous, whereas a narrative that is properly categorized can provide a richer context of lessons learned, thus making them easier to read, easier to contextualize, and easier to apply. Further, this report should be provided to project and operational teams beyond those represented in the postmortem meeting. The idea of the report is to benefit all teams as an enterprise asset, not just those involved in the postmortem.

The Top-Ten Best Practices for Postmortem Success

1. *Start early*. Make sure the postmortem work effort is part of the project's resource plan and schedule.
2. *Establish your team culture*. Establish ground rules and expectations of one another as you onboard project team members and stakeholders.
3. *Know your measures of success*. Work with your team and customer to identifiy the questions that will be raised during the postmortem so that there are no surprises.
4. *Use experts*. Since it is often difficult to facilitate a meeting and document the meeting and do so without bias, use an expert facilitator, scribe, and others during the postmortem review.
5. *Ensure representation*. Make sure the key members from the project team and operational team are present and comfortable in sharing their thoughts, ideas, and opinions openly and productively.
6. *Everyone contributes*. Conduct the postmortem in a way that everyone participates without peer pressure or senior leader persuasion.
7. *Work from facts*. Make sure facts are known and have any comments facilitated to the point of it being fact-based rather than subjective opinion.
8. *Focus on the future*. While the postmortem is a reflection of past events, the primary focus is on how to make future projects better and therefore the majority of time should be spent on what should be done differently next time.
9. *Detail the conversation in narrative form*. Rather than high-level bullet-point summaries of the conversation, have the meeting detailed

in a narrative report because stories make for easier learning than bullet points.

10. *Broadcast your results*. Be sure to share the postmortem report with other project teams and archive in an easily accessible database for other project managers to use.

Many facilitators of postmortem reviews find that using a checklist can be helpful throughout the postmortem process—from gathering feedback to conducting the meeting to documenting the final report. The sample checklist in Table 13.7 offers questions that go beyond the three (high-level) standard questions noted earlier in this section.

Depending on the response to each question in the postmortem checklist questionnaire, follow up or probing questions can (and should) be asked. For example, if any question is answered "no" the facilitator should ask, what should be done differently on the next project to ensure this occurs? The facilitator could also ask if there were any leading indicators of a problem.

Using the Postmortem Review

Like the project closure report, the best postmortem review meetings and final reports start in the planning phase of the project. Experienced project managers know to plan properly at the beginning phases of the project's cycle to help ensure that resources (time and personnel) are allocated for work after the project becomes part of operations (or customer) use. The actual postmortem work effort is conducted three to six months after the final handoff from the project team to the operations team or customer. The timing between the project handoff and postmortem work is necessary in order to allow the operations team enough time to realize the value and benefit from the project.

It is important to note that while the tools and templates outlined for postmortem work facilitate a reflection of the past, the focus during this work is on the future. The reflection of the past can mostly be gleaned from the project closure report and a short conversation with key project team members and stakeholders. Therefore, the effort and workload associated with postmortem activities should be minimal. While the workload is often minimal, the results from the work can be very dramatic. The tools, and especially the postmortem review meeting, should focus on leveraging experiential knowledge to make future projects better (more efficient and effective), create higher-performing project teams, and increase business value.

Most postmortem meetings have a duration of anywhere from one to four hours. The larger and more complex the project, the longer the postmortem meeting may take. Although it may be assumed at these meetings, it is important for the facilitator (or the project sponsor) to emphasize the need for an honest, candid, and objective discussion and to focus on process and not people. These behavioral ground rules serve as guidelines for participants and the facilitator. Such ground rules establish a

Table 13.7: Project Postmortem Checklist Questionnaire

Category	Question	Notes
Project planning	**1.** Did the business case clearly detail the problem or opportunity? **2.** Were business goals and project objectives clear and measureable? **3.** Was the project scope, schedule, budget, and quality clearly detailed? **4.** Were project plans detailed, accurate, and usable?	
Customer focus	**5.** Was the voice of the customer evident in all phases of the project? **6.** Did deliverables and milestones meet customer (and stakeholder) target expectations? **7.** Was communication with customers effective?	
Deliverables	**8.** Did the outcome (all outputs) of the project meet your expectations? **9.** What gap or variance (if any) exists between your expectations and the final deliverable? **10.** Were project monitoring and controlling efforts effective?	
Scheduling	**11.** Was planning and scheduling resources for the project effective? **12.** Was there a visible schedule available to "see" milestones and critical project work?	
Resourcing	**13.** Was the project resourced properly? **14.** Was the team effective? Would you describe the team as "high performing"? **15.** Was there a skills assessment conducted to determine resource need? **16.** Was there clarity in roles, functions, responsibilities, and proper awareness of interdependencies? **17.** Was there active (proper) senior management support?	
Managing risk	**18.** Was there a deliberate risk management plan and process in use? **19.** How effective was the planning and management of risks? **20.** Were contingency plans in place and effective for all risks?	
Communication	**21.** Was there an effective communication plan for the project? **22.** Was there proper stakeholder management (analysis and monitoring)? **23.** Was communication (to the team, to sponsors, to all stakeholders) effective?	
Decisions	**24.** Was there a clear decision making process (including an escalation process) in place and used? **25.** Were decisions made fast enough?	
Other	**26.** What are your most critical lessons learned from this project? **27.** … **28.** …	

constructive atmosphere for discussion and learning. These ground rules also help to prevent personal attacks that disengage participants and disable learning.

In addition to the information gathering ahead of the meeting and ground rules for use during the meeting, a prerequisite for an effective postmortem review is a well-crafted agenda. The agenda will certainly vary based on findings from the information gathering work effort. The following bullet points, however, offer a sample agenda for a postmortem review:

- Welcome and introduction of everyone.
- Review of the ground rules by the facilitator.
- Summary of information gathering findings from the facilitator (this information should be sent to participants in advance of the meeting if at all possible).
- Review and rank issues and critical success factors.
- Create a "what went wrong" list.
- Create a "what went right" list.
- Detail opportunities for improvement with specific (actionable) recommendations.
- Outline specific points of communications and next steps.

The facilitator can manage this agenda in a number of ways. For example, all items can be addressed in an open, round-table conversation. As an alternative, the facilitator could conduct this meeting as a workshop in which case participants are much more active. However the postmortem review is conducted, it should fit the organizational culture and conversational tone necessary to achieve the end results, which is individual, team, and organizational learning.

Variations

It has been noted that project managers often address project closure work at the end of projects rather than throughout the life cycle of projects. One of the best practices for periodically evaluating a project's progress toward meeting business goals is the use of a retrospective process.

Retrospectives are a series of events where team members who have a perspective to share meet at critical points during the project cycle to discuss what is working and what needs to be improved. The intent is to capture key lessons while a project is in flight, and apply improvements during the remainder of the life cycle.[5]

Retrospectives differ from the more traditional postmortem activities. In many companies, a project postmortem is held following the conclusion or cancellation of a project. Because this meeting occurs at the end of the project, it is too late to implement corrections on that project. At best, learnings can be applied to the next project. In our experience, the postmortem is many times held as an afterthought without a defined, objective process and does not lead to actionable change to project team practices.

There are significant advantages of using the retrospective process. First, consider that many times teams focus only on the negative aspects of their performance and forget the positive. Retrospectives also explore what is working well on a project, and ensures the practices are reinforced and repeated. Second, retrospectives allow an organization to make positive changes to projects that are currently in flight; therefore, they don't have to wait several months or more to realize the benefits of their learnings. Third, teams learn best when they solve immediate problems. As part of the retrospective method, teams focus on the few but critical opportunities to improve, and develop specific action plans to drive changes in their practices. These action plans are owned by various members of the project team who drive them to implementation.

One of the reasons retrospectives are quite effective is the use of a trained facilitator who is an expert in extracting the key issues and learnings from a diverse and distributed workforce. When applied most effectively, a trained, objective facilitator—someone other than the team leader—guides the team through an analysis of what is working well and what is not, and then helps the project team generate ideas for improvement and what they want to do differently moving forward.

The project retrospective is an organizational learning activity. Whether an organization uses the retrospective method or another method to capture its organizational learnings is not the most critically important factor in creating a learning organization. What is important is that the organization consistently invests the time and effort to stop, reflect, learn, and improve continuously over time.

Establishing a consistent practice of holding retrospective reviews will lead to the identification of many improvements needed for project success. Execution of the improvements should be approved by the project governance body, and then implemented within the organization and on targeted projects in a methodical manner. Improvements should be implemented as quickly as possible to reap benefits, but care must be taken to prevent more change than the organization is able to absorb at any one time.

Benefits

Projects are difficult to initiate, plan, execute, and close without issue. Mistakes, missteps, and errors will happen. When such issues arise, it is important to stop, reflect, and learn how to prevent it from happening again. Many great things happen on a project as well and should not be lost, but instead repeated on future projects. Without learning, we are doomed to repeat mistakes and forget success. Learning is a core function of project closing. Transferring learning from one project to the next will continuously increase the probability of success.

With the lessons learned from past projects helping to increase the probability of future project success, we can see the postmortem work serving as a linchpin in an organizational knowledge management process. With this understanding, the project life cycle can be reconfigured from a linear view to a process view, as illustrated in Figure 13.7.

The opportunity for learning from the postmortem review process is significant. It is because of the learning aspect of postmortems that most organizations associate these processes as being part of knowledge management.[6]

As noted, learning is a core function—and benefit—of postmortem work. This can be viewed as a long-term benefit. The short-term benefit is making sure the final project handoff to the operations team (or customer) was done so effectively. The relatively low work effort associated with postmortem activities is shadowed by the value added from the effort. Postmortem work can establish greater levels of team performance,

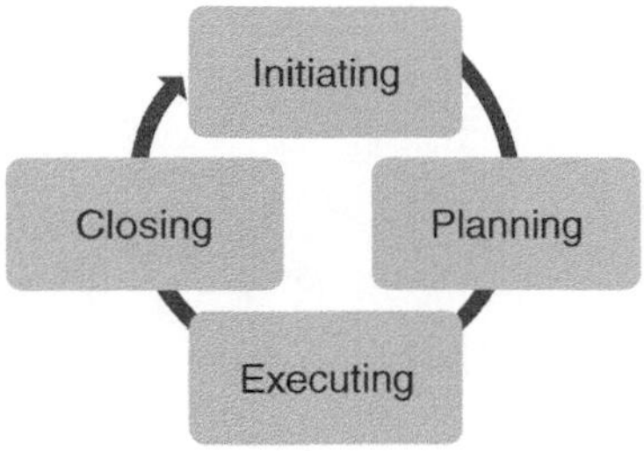

Figure 13.7: The Life Cycle of Project Learning

strengthen core principles and values of an organization, and foster a continuous learning and improvement mindset from project work.

CONCLUDING REMARKS

In this chapter, we presented three tools—the project closure plan and checklist, project closure report, and project postmortem. Additionally, a section on project retrospectives was detailed.

Each of the tools is designed with a distinct purpose. Collectively, as part of project closing work, they all serve to ensure that proper completion of work and handoffs occurred and importantly that lessons learned were captured and shared as an enterprise asset—a competitive distinction. As a set of complementary tools, they enable monitoring, reporting, and controlling of project risk. They also help us study the implementation and learn lessons for continuous improvements.

References

1. Kerzner, H. R. Jr., *Project Management: A Systems Approach to Planning, Scheduling, and Controlling,* 11th ed. (Hoboken, NJ: John Wiley & Sons, 2013).
2. Project Management Institute. *A Guide to the Project Management Body of Knowledge,* 5th ed. (Newtown Square, PA: Project Management Institute, 2013).
3. Ibid.
4. Archibald, R. D. *Managing High Technology Programs and Projects* (Hoboken, NJ: John Wiley & Sons, 1976).
5. Lavell, D., and Russ Martinelli. *Program and Project Retrospectives: An Introduction. PM World Today,* 10 (1), 2008.
6. Lientz, B. P., and P. R. Kathryn. *Breakthrough Project Management* (San Diego, CA: Academic Press, 1999).

PART VI

Risk and Stakeholder Management Tools

14

MANAGING PROJECT RISK

For many companies, establishing industry leadership means assuming a higher level of risk due to the uncertainties associated with navigating unchartered territories. Developing new capabilities in today's environment is risky business by nature, especially if a company wants to establish or maintain a leadership position. However, risk taking does not mean taking chances. It involves understanding the risk-reward ratio, then managing the risks associated with a project.[1]

Project risk is the potential failure to deliver the benefits promised when a project is initiated.[2] By understanding and containing the risk on a project, the project manager is able to manage in a proactive manner. Without good risk management practices and tools, the project manager will be forced into crisis management activities as problem after problem presents itself, forcing a team to constantly react to the problem of the day (or hour). As one well-known author stated, "If you don't actively attack the risks, the risks will actively attack you."[3] Risk management is a preventive practice that allows the project manager to identify potential problems *before* they occur and put corrective action in place to avoid or lessen the impact of the risk. Ultimately, this behavior allows the project team to accelerate through the project cycle at a much faster pace.

Understanding the level of risk associated with a project is crucial to the project manager for several reasons. First, by knowing the level of risk associated with a project, a project manager will have an understanding of the amount of schedule and budget reserve (risk reserve) needed to protect the project from uncertainty. Second, risk management is a focusing mechanism that provides guidance as to where critical project resources are needed—the highest risk events require adequate resources to avoid or mitigate them. Finally, good risk management practices enable informed risk-based decision making. Having knowledge of the potential downside or risk of a particular decision, as well as the facts driving the decision, improves the decision process by allowing the project manager and team to weigh potential alternatives, or trade-offs, to optimize the reward-risk ratio.[4]

The tools presented in this chapter are instrumental in identifying risks to the project, assessing their potential impact, developing actions to mitigate them, and monitoring risk dynamics. Although there is certainly no shortage of project risk management tools to include in a PM Toolbox, we have included the tools that we see most widely used and that provide the broadest application to project types and sizes. We begin with the risk management plan.

RISK MANAGEMENT PLAN

The risk management plan serves to establish the framework and methodology in which the project team will identify, monitor, and manage the risk associated with a project. Developing a risk management plan at the onset of a project can help to eliminate potential issues from emerging, or at least minimize the impact that they have on the project if they do occur. Having a well-thought-out plan will help project managers deal with project uncertainties when they arise and in the best case head them off *before* they impact their project.

Developing a Risk Management Plan

Making decisions is perhaps one of the toughest jobs that project managers have to take on. Decision making wouldn't be very challenging in a situation of total certainty, where all information that is needed for decision making is already available and the outcomes of decisions are predictable. Project managers' lives, however, are much more complex, and most of their decisions are made with incomplete information and uncertain outcomes. This is the realm of project risk management. Beyond it lies the region of total uncertainty, with complete absence of information, where nothing is known about outcomes. This total certainty (knowns)—risk (known unknowns)—total uncertainty (unknown unknowns) continuum is illustrated in Figure 14.1.

Risk Management Approach

The risk management plan is a document developed in the beginning of the project that provides a framework for dealing with risk throughout the project's life. Included in the plan is a general description of the approach used to identify, assess, manage, and monitor project risk events.[5] It should include information such as the:

- *Risk management methodology*. Identify and describe approaches, tools, and data sources that may be used to handle risks.

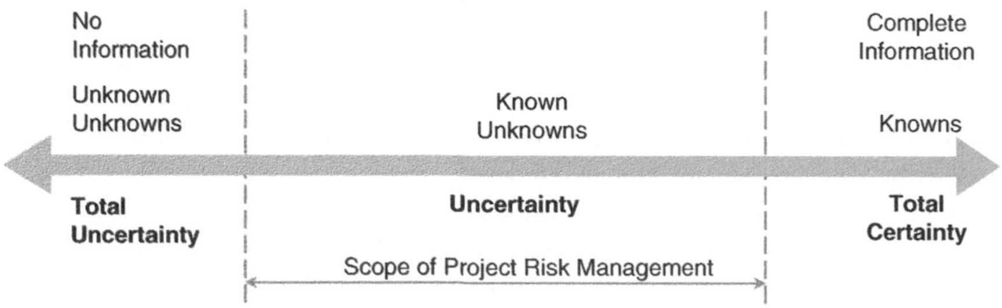

Figure 14.1: The Project Risk Management Continuum

- *Roles and responsibilities*. Define who does what in risk management on the project, from project team members to members of the company's risk management teams.
- *Budgeting and timing*. Specify the budget for risk management for the project, as well as the frequency of the risk management processes.
- *Tools*. Describe which specific tools for qualitative and quantitative risk analysis to use and when to use them.
- *Reporting and monitoring*. Define how risk will be reported and communicated to the project stakeholders, how risk events and triggers will be monitored throughout the project cycle, and how the information will be preserved for purposes of lessons learned.

The risk management approach is fairly well documented and used across most industries, and focuses on the simple cycle of risk identification, assessment, response planning, and monitoring. Basic risk management terminology is defined in "Basic Risk-Related Definitions."

Basic Risk-Related Definitions

Project risk. The cumulative effect of the chances of uncertain occurrences adversely affecting project objectives.[6]

Risk event. The description of what might happen to the detriment of the project.

Risk probability. The likelihood that a risk will occur.

Risk impact. Severity of its effect on the project objective. Also called risk *consequences* or *amount at stake*.

Risk event status. A measure of importance of a risk event. Also called its *ranking*.

Contingency reserve. Also called *risk reserve*, this is the amount of money or time normally included into the project cost or schedule baseline to reduce the risk of overruns of project objectives to a level acceptable to the organization.[7]

Risk Identification

This section of the plan describes the process for identifying all the potential risks that may influence the success of the project. It should describe the methods for how the risks are identified and the format in which the risks are recorded.

Multiple ways for accomplishing this step are available, ranging from engaging the project team in a brainstorming session, to consulting experienced team members, and to requesting opinions of experts not associated with the project. Typical methods of identifying risk are expert interviews, reviewing historical information from similar projects, conducting a risk brainstorming meeting, and using more formal techniques such as the Delphi method (see "The Basics of Delphi").

The Basics of Delphi

The Delphi technique is useful in situations where outcomes or trends are uncertain. This is particularly true in situations that have not been encountered previously.

The technique involves soliciting the opinions of a group of experts related to the future situation. It is best if forethought and preparation are exercised to identify specific knowledge gaps relating to the future situation under scrutiny. The knowledge gaps will guide the selection of the right panel of experts, and will focus the discussion topics on the right areas.

The technique can be applied informally, where the panel of experts is gathered together to discuss and debate the topics associated with the future scenario, focusing on the knowledge gap areas. Group opinions and ideas are then gathered and summarized. A second or third round of discussion ensues until convergence of opinion begins to occur.

Applied formally, questionnaires are given to each panel expert in two or more rounds. After each round, expert opinions are collected and summarized, and then sent out to each panelist to consider in his or her next response. The process can continue until convergence or a solution arises.[8]

Be acutely aware that complete consensus is highly unlikely and not very useful. Judgment must be applied on the part of the facilitator to gauge when convergence (not consensus) has occurred. Used to identify potential risk events, the value of Delphi comes from the *emergence* of ideas, not in the *convergence* of opinion.

When identifying risks, several things have to be taken into account. First, risks vary across the project life cycle. Typically, risk levels tend to be relatively high early in the project because so much is unknown. Similarly, later in the project many of the unknowns are turned into knowns and risk levels are relatively lower. Also, some risks occur only in certain project stages; for example, risks related to project acceptance tests are typically encountered near the end of the project. Sometimes, even assumptions may become a source of risk (see the example titled "Is an Assumption a Risk?").

The dynamic nature of risk makes the identification process iterative, requiring that once risks are identified early in the project, they need to be continuously reviewed, with appropriate adjustments.[9]

Is an Assumption a Risk?

"Is an assumption a risk?" This was a question that a project manager asked in a risk identification meeting. Assumptions are factors that are not entirely known or are uncertain but for planning purposes are considered to be true or certain. For example, consider a firm that launched a project to develop and market a product in a Pacific Rim country. A major assumption was that the country's annual market growth rate would continue to be around 10 percent. Per its assumption management practice, the firm first documented the assumption by defining it, nominating its owner, and identifying a monitoring metric.[10] Next, the project manager instructed the owner to periodically test the metric in order to ensure that no change of assumption occurred. Seeking to be proactive, the owner defined at which time the assumption becomes a risk (trigger point) and potential risk response actions may be needed.

A few months later, the country was hit by a recession and the growth rate declined significantly. The project team revisited the assumption and, since the recession was expected to last for some time, decided that the assumption changed into a risk, immediately invoking the risk response plan. So "is an assumption a risk?" It is not; rather, it is a source of potential risk.

Second, risk events rarely strike independently. Rather, they tend to interact with other risk events, combining into larger risks. Looking for such interactive possibilities is important in risk identification. Finally, since risks come in all types of packages, planners should conduct risk identification in a systematic way so that no stone is left unturned internally in the project and externally in the environment, including management of stakeholders. A huge help in this respect may be received from risk categorization.

Risk Categorization

Project planning outputs—scope, cost, time, and quality baselines—are what is at risk. Having full knowledge of them is crucial in developing response plans to counter risks to which the outputs will be exposed. These risks can be organized into different categories. For example, risks can be classified according to their effect on the project—scope, quality, schedule, and cost risks (in other words, failure to complete the project tasks within planned scope, quality, schedule, and cost performance). Another way is to categorize risks per their primary source into external (but unpredictable), external predictable (but uncertain), internal nontechnical, technical, and legal.[11] This perspective looks for the balance between the internal and environmental impacts. The point is that the firm and its projects need a consistent risk categorization schema, suiting its business and culture, which can serve as a framework for a systematic identification and treatment of risks.

Assess Risk Qualitatively and Prioritize

Once risks have been identified, it is important to determine both the probability that each of the risks will occur, and the impact to the project if they occur. In order to determine the severity of the risks identified, a probability and impact factor has to be assigned to each risk. This process allows the project manager to prioritize risks based on the effect they may have on a project.

A usual problem that arises is that a large number of risks might be identified when the initial risk identification is performed. However, which risks deserve the most attention? To answer that question a project team needs to analyze impact, probability, and severity (criticality) of each risk.

In the qualitative assessment, we tend to use a nonnumeric probability scale—for example, a five-level scale, where 1 = very unlikely, 2 = low likelihood, 3 = likely, 4 = highly likely, and 5 = near certain.[12] If you don't have much experience or data to reliably assess quantitative probabilities—addressed later in the quantitative assessment—qualitative scales are sufficiently good. Consequently, you will qualitatively assess each risk's probability on this nonnumeric scale.

The next step is to assess the impact of each risk, again on a discrete scale. One example is a scale such as 1 = very low impact, 2 = low impact, 3 = medium impact, 4 = high impact, and 5 = very high impact. To illustrate its use, let's assume that a risk to be assessed has three impacts: Project costs can increase, schedule can slip, and performance of the capability can be reduced. For each one of them, the scale can define the levels of impact (see Table 14.3 for an example). After each of the three impacts are rated, the overall impact rating of the risk is the largest of the three impacts.[13]

When all risks are assessed in this manner, it is time to use a formula to combine their risk probability and impact to establish a measure of severity. Although nonlinear formulas can be employed, linear formulas such as Severity = Probability + $N \times$ Impact are easier to apply.

For example, N can be equal to 2, meaning that impact is twice as important as probability in establishing risk severity. In this case, the assessed probability and impact for each risk would be entered in the formula Severity = Probability + $2 \times$ Impact. The outcome of the calculation will determine the priority of the risks.

The question still remains: How many of the highest-ranking risks should we deal with? Some larger projects commonly focus on the top ten highest-ranked risks. In contrast, some smaller projects decide to manage the top three risks, arguing the lack of resources to take on a larger number of risks. Both may be dangerous. If these projects have more than ten and more than three risks, respectively, they are bound to disregard some critical risks. However, if only one risk is truly a high-probability, high-impact risk, they are wasting resources looking at the top ten and three risks.

So, what is a reasonable way out? Tools such as the risk register and risk assessment matrix discussed later in this chapter provide a visual representation of risk prioritization and are helpful in determining which risks need action and resources applied.

If you don't have much data to reliably assess quantitative probabilities, qualitative assessment of risks based on nonnumeric scales is sufficient to develop your risk responses for the most critical risks. For projects where sufficiently reliable data is present, the next action is quantitative risk assessment.

Assess Risk Quantitatively

Quantitative risk assessment numerically analyzes the probability of each risk, its consequences on project objectives, and the extent of overall project risk.[14] It can be used separately or together with qualitative assessment.

The process begins from the results of the earlier risk identification step. For each of the identified risks, you need to quantify the probability of occurrence by asking, "What is the probability that this risk will happen?" "Ninety percent," the team decides.

This means that there is a 10 percent probability that the risk will not occur. Clearly, the probability that the risk will occur plus the probability that it will not occur equals 1. Assessing the probability is no more than an estimate based on solid historical information from similar experiences in past projects or considerate opinion of experts.

The next step is to determine the risk impact. "What will happen if this risk occurs?" is the question that should be asked. While the impact may be expressed in almost any units, from percentage of lost market share to loss of revenue, the real emphasis here is to estimate schedule or cost severity of the risk. For example, if the highest priority project objective is schedule, then the risk event status would be calculated in terms of time.[15]

$$\begin{aligned} \textit{Risk Event Status} &= \textit{Risk Probability} \times \textit{Impact} \\ &= 90\% \times 60\,\textit{days} \\ &= 54\,\textit{days} \end{aligned}$$

In this example, the potential schedule impact for a particular risk event has been determined to be 60 days, with a 90% probability of occurring at this point in time. The current risk event status is therefore 54 days, which is the current risk exposure for this event.

When the status is calculated for all risk events, the natural question is: Which risks are really vital and deserve attention and which are trivial? To answer this question, we will use principles similar to those on the issue of severity in qualitative assessment. First, establish numerical intervals of severity that determine whether a risk event status is critical (potential showstoppers), near-critical (soon to be potential showstoppers), or noncritical (minor risks). For example, in a smaller project, the risk event status exceeding 15 days was critical, between 7 and 14 days near-critical, and below 7 days noncritical. Second, respond to the highest-ranked risks, down to an agreed level.

Determine Risk Response

Once risks have been qualitatively or quantitatively assessed, the project ream must determine how to address the risks that have the greatest potential for impacting the project. This section of the risk management plan explains the response options and actions that are available to the project team in managing the risks.

Any suitable risk response action essentially falls into one of the four broad categories of response strategies: avoidance, transference, mitigation, and acceptance of risk.[16] Changing the project plan or condition to eliminate the selected risk event is risk avoidance. For instance, if faced with the risk of not having an available expert to perform a quality business process analysis, the risk can *avoided* by hiring such an expert.

Risk transfer simply involves shifting consequences of a risk event to a third party, along with the ownership of the response. If, for example, projects within a firm have historically been exposed to a risk of slow quality testing from their internal capabilities, the risk can be *transferred* to a third party by hiring a professional firm to do the testing.

The intent of risk mitigation is to lower the probability or impact (or both) of an unfavorable risk event to an acceptable threshold. A fairly common risk for many projects is the potential decision delays caused by the busy schedules of the executive sponsor. This risk can be *mitigated* a number of ways, such as reducing the number of major milestone decision points or the delegation of decision authority to one of the executive's direct reports.

The three response strategies—avoidance, transference, and mitigation—are deployed when risks they are responding to are among the highest-ranked risks. Obviously, these responses will be incorporated in the project plan.

For those risks that are not among the highest-ranked risks, or for risks that have no other viable response strategy, a risk acceptance strategy is used. This implies that project managers have decided to not change the project plan or are unable to articulate a feasible response action to deal with a risk.[17] A typical example of the risk acceptance is the establishment of contingency allowances. For an explanation of how the allowances are formed, see "How Much Reserves and Allowances to Plan For?"

How Much Reserves and Allowances to Plan For?

Let's think back about total certainty (knowns)—risk (known unknowns)—total uncertainty (unknown unknowns) continuum. What kind of reserves do we need in order to respond when any of these categories hit?

First, because of their totally certain nature, the knowns do not require any reserves. How do you allow for risk consequences of the known unknowns? Many firms add them to the baseline estimate as a separate fund for schedule and cost *contingency allowances*. Others incorporate them into the individual activities. While we favor the former, the latter approach—which, by the way, is too risky to use because of activity owners' tendency to use up allowances liberally—appears to have wider presence.

How is the fund formed? Popular methods include applying standard allowances and percentages based on past experience.[18] We argue that the use of the risk response plan may be a very appropriate way to compute the fund. Take a risk from the plan that is *not* among the highest-ranked risks—let's call it a lower-ranked risk. Multiplying its risk probability by risk impact provides the risk event status, which may be expressed in cost or schedule terms. These numbers are essentially cost and schedule reserves or allowances for the risk event. Adding up allowances for all of the lower-ranked risk events in the plan creates a project contingency allowances fund.

Finally, what about reserves for the unknown unknowns? Although they are absolutely not possible to foresee, such things may happen. Therefore, some firms develop *management reserves* involving cost, schedule, or both to allow for such future situations when cost or schedule objectives may be missed. Once the reserves are used, the cost baseline gets changed. Managing management reserves is in the domain of top management, typically the project sponsor.

An integral part of the response development is the identification and assignment of risk owners—individuals or parties responsible for each preventive action, trigger point, and contingent action. In so doing, one should recognize that while some risks are independent, leaving their owners fully responsible for their management, some risks might be interdependent. If so, their preventive actions, trigger points, and contingent actions should be developed and owned interdependently.

Risk Monitoring

Most of a project managers attention with respect to risk management tends to focus on the activities associated with risk identification, risk assessment, and risk response planning. Where project managers historically spend less time and focus are the activities associated with risk monitoring. Not uncommon, project managers therefore continue to be surprise when a risk event they had identified earlier, but were not monitoring, suddenly turns into an issue. To protect against this, diligent risk monitoring must be a part of every project manager's activities and he or she must have tools in their PM Toolbox to effectively perform this function.

There are four primary elements involved with risk monitoring activities: (1) systematically track the status of risks previously identified; (2) identify, document, and assess any new risks that emerge; (3) effectively manage the risk reserve; and (4) capture lessons learned for future risk identification and assessment efforts.

This section of the risk management plan should discuss how the project risks will be monitored on an ongoing basis. The key to risk monitoring is to ensure that it is used throughout the project cycle and includes the identification and use of trigger conditions that will accurately indicate if the probability of a risk occurring is increasing or has passed.

As stated previously, it is advantageous to the project manager to assign risk owners to the highest-level risks. A primary role of the risk owner is to continuously monitor the status of the risks he or she are responsible for, and periodically report that status to the project manager and team.

Since there is a time element to when risk events may affect a project, not all risks should be reported upon in each status meeting. Rather, as risk event triggers approach on the project schedule, the project manager should ensure that the appropriate risk owner provides the status updates at the appropriate time.

Using the Risk Management Plan

There is no project that cannot benefit from developing and using a risk management plan because all projects contain an element of uncertainty about the future. Small projects typically rely more on the qualitative assessment of risk, often deciding to handle only a few highest-ranked risk events. Not surprisingly, the dominant mode of risk management planning is informal, as is the periodic reevaluation of the plan throughout the project.

Although at times it may be overly simplistic for large and complex projects, the risk management plan is nevertheless widely used, with more formality and stronger orientation on quantitative risk assessment and prioritization. Focused on the larger number of highest-ranked risk events, larger projects also tend to do more formal, periodic reassessments of the plan.

Although the use of the risk management plan should become institutionalized on a project, development of the plan can vary greatly. For smaller projects, only a few hours may be required to conduct a planning session and develop a plan. This time proportionately rises as projects get bigger and more complex. Tens of hours may be necessary to devise a quality risk management plan for a team in charge of a large and complex project.

Benefits

The benefits gained from the development and use of a risk management plan are many. The plan helps sift through the myriad of uncertainties, pinpoint and highlight the project areas of highest risk, both before work has begun and throughout the project. This offers a project manager an opportunity to identify effective ways of reducing those risks in a proactive manner, rather than being confronted by them when they turn into issues (see "Issues and Risks").

The risk management plan provides a systematic response. By developing a risk management framework and methodology, a project manager is able to respond to risks calmly and systematically. This decreases the necessity for making unplanned decisions and actions on the fly.

Projects normally involve a number of stakeholders that span across an organization or between multiple organizations. When a project manager has developed a risk management plan and is performing to that plan, they build confidence among all stakeholders that their interests are properly guarded against risk. This will serve to protect the project against unwanted and unneeded interference by stakeholders or persons who are not part of the project team. Proper risk management allows the project manager to maintain better control of the project and the project decisions.

Finally, a well-documented risk management plan provides a great opportunity to capture key learnings for future projects. Many risks are associated with the particular environment an organization does business within as well as the common policies and practices associated with the firm itself. These risks tend to affect every project within an organization and can be addressed and dealt with directly in a project's risk management plan.

Issues and Risks

What is the difference between issues and risks? Without a pretense to deal with the semantics of the difference, let's take a look at their use in industry. First of all, the terms are commonly used interchangeably. For example, according to the Project Management Body of Knowledge (PMBOK) Guide, "Reports commonly used to monitor and control *risks* include *Issues*

Logs...."[19] Some project managers believe that risks and issues define different concerns and should, therefore, be defined into different categories that need different managerial responses. We agree with this belief.

An issue is an event that has already happened. Its time horizon includes the past and the present. For example, a loss of a team member is an issue that led to a month delay. In contrast, risk can be characterized as what could happen to the detriment of the project. For example, "a possibility of losing the project manager could cause a late completion of the project." Risks are in the future. Consequently, while we strive to *resolve* an issue, our managerial response is to *prevent* a risk or its impact.

For questions such as "What are the issues causing the variance?," the aim is to identify what has happened that caused the schedule variance. On the other hand, answers to a question such as "What new risks may pop up in the future and how could they change the preliminary predicted completion date?" seek to find future candidate events that need to be acted on in order to mitigate their impact on the project.

THE RISK IDENTIFICATION CHECKLIST

The first step in the risk management process is to identify all of the events that could possibly affect the success of the project. Although risk identification is the first step in the risk management process, it is not a one-time event. Risk identification is an iterative process that occurs throughout the project cycle. Some project teams begin by identifying the categories of risk, such as technology risk, market risk, business risk, and human risk, and then use brainstorming and other problem identifying techniques to identify all potential risk events within each category.

The key element of this step is to attempt to identify *all* potential risks. Do not make judgment at this step on whether a risk is of real concern or not, that is the next step in the process. When risk identification is done well, it can be overwhelming, especially early in a project when the number of uncertainties is at the highest. Remember that the goal of risk identification is to flush out as many potential risks as possible in order to get them on the table for discussion.[20]

The risk identification checklist is a good tool to use as a guide and framework for identifying different categories of risk as well as a number of common risks that plague many projects.

Developing a Risk Identification Checklist

The risk identification checklist will be unique for every organization because every organization has a unique set of uncertainties associated with the business environment it operates within, the policies and practices which guide it operation, its constraints that affect its project teams, and its ability to access and use information needed to inform the team about future events. Table 14.1 contains a sample risk identification checklist that can be used as a starting point and reference for developing your own checklist.

Table 14.1: Sample Risk Identification Checklist

Project Management Risks	
✓ Schedule activities are overly optimistic	✓ Timeline assumes the use of specific resources who may not be available
✓ Effort is greater than estimated	✓ Target end date has moved up with no adjustment to scope, time, or cost
✓ Requirements have not been baselined and continue to change	✓ Budget is not based on structured estimates
✓ Functional requirements lack user involvement and input	✓ Risk response plans have not been developed
✓ Lack of performance measures and/or performance reporting process	✓ Project scope, vision, and objectives are not clearly defined
✓ Project does not have senior management or customer buy-in	✓ Other similar projects have been delayed or canceled
✓ Person-hours (hours per month) are not reasonable for the work estimated	✓ All dependencies between functional groups have not been identified
Resource Risks	
✓ Hiring is taking longer than expected	✓ The personnel most qualified to work the project are not available
✓ There is tension between the project team and the client	✓ Unexpected training is needed to build required skill
✓ Estimated staffing profile does not seem reasonable given project scope and/or complexity	✓ No resource ramp time was included in the project schedule
Stakeholder Risks	
✓ End user rejects project outcome, resulting in rework	✓ End user input is not solicited
✓ End user or client will not participate in review cycles	✓ Communication time between project team and end users or client is slower than expected
Technical Risks	
✓ Necessary functionality cannot be implemented using selected technologies	✓ Components developed separately cannot be easily integrated
✓ Quality assurance activities are being ignored or minimized	✓ Inaccurate quality checking may result in quality problems
✓ The programming languages or other technologies are unfamiliar	✓ Development tools are not in place or not working as expected
✓ There is a dependency on a technology that is still under development	✓ The technology being used or developed is new to the organization
Environmental Risks	
✓ The project depends on government regulations	✓ Project depends on industry standards which may change
✓ Project deliverables are developed by third parties (subcontractors)	✓ Project interdependencies with external parties exist

Developing a standard set of risk identification categories and items is a good practice as it drives consistency in identifying various risks which can affect project outcomes. The questions contained in the checklist can be developed by first understanding the various work activities contained in the project work breakdown structure (WBS), the constraints within which the project will have to operate, and information contained in other guiding project artifacts such as the project business case and project charter (Chapter 3). Additional questions can then be developed by tapping into historical learnings through the review of risk events and issues that affected previous projects.

Using a Risk Identification Checklist

Effective risk management begins with thorough risk identification. Many argue that risk identification is in fact the keystone to good project planning and execution.[21] They base their argument on the idea that all projects face any number of uncertainties that will turn into issues and roadblocks if ignored or left unmanaged in a predictive and proactive manner. The ability to deal with uncertainties ahead of time is contingent on a project team's ability to predict their potential occurrence. That ability is rooted in risk identification.

The risk identification checklist should be developed during the initiation stage of the project or, better yet, as a standard job aid created by members of the project office. Regardless of how it is developed, it needs to be implemented in the earliest stages of a project.

The checklist is used as a guide for all project participants who will be involved in trying to predict possible risk events. It is meant to assist the project manager in ensuring that various perspectives of risk are considered such as risks associated with managing the project, environmental risks, resource and collaboration risks, risks associated with stakeholders, and certain technological risks when applicable.

The checklist should be distributed to all members of the project team, as well as key stakeholders, who are charged with identifying potential risks for the project. Best practice usage involves continued use of the checklist as a project progresses through its natural project cycle as new uncertainties arise.

To become institutionalized within an organization, the risk identification checklist should be periodically updated to reflect common risk events encountered across an organization's projects. An opportunistic time to update the checklist is during the postmortem review process for each project (Chapter 13). As the issues and risks encountered for a project are reviewed, those issues and risks that have a systemic propensity to affect future projects can be added to the risk identification checklist for future reference and use.

Benefits

As stated earlier, good risk management practice is rooted upon good risk identification: Risks have to be identified before they can be actively managed. The risk identification checklist provides value by assisting a project team in their risk identification activities by ensuring risk is viewed from multiple perspectives. Left unprompted, many project teams focus almost exclusively on project risk. The checklist helps to broaden their view by prompting a team to also consider areas of risk coming from the business environment, resource and collaboration activities, and stakeholders to name a few.

If periodically updated, the risk identification checklist can also add value by serving as a risk knowledge database for the organization. Many risks and issues are systemic in nature for every organization. By documenting the sources of these systemic risks and issues in the checklist, each project will benefit from knowledge gained on past projects.

THE RISK REGISTER

The risk register provides a record of identified risks relating to a project and serves as the central repository for all open and closed risk events.[22] The risk register typically includes a description of each risk event, a risk event identifier, risk assessment outcome, a description of the planned response, and summary of actions taken and current status. Many times, the risk events are prioritized in the risk register based on the risk assessment score or qualitative analysis. Table 14.2 demonstrates an example risk register.

Creating a Risk Register

The risk register is arguably the most crucial tool for managing project risk. A good register contains all the necessary information about the project risks, provides a comprehensive catalog of the risks, provides a severity determination, and describes the possible responses to the risk events.

The information in the risk register can be represented in a number of ways, such as a database, a paragraph-style document, or a spreadsheet. The spreadsheet style is far and away the most commonly used format because it presents all the information pertaining to project risks without the user having to scroll through several pages.

There are no standard information components to include in a register. We suggest you search a number of risk register examples, then adopt and adapt the content for your particular need. We encourage you to keep your register simple, however. The more complicated a register becomes, the more time you will have to spend managing your document, therefore leaving less time to manage your project. The following elements should be included in a risk register.

Risk Identifier

Each risk event must have a unique identifier for cataloging and monitoring purposes. The most common approach is to assign each risk that is identified a chronological number. Another approach is to map risk events to the WBS element for which they are associated. For instance, risks associated with a Level 3 WBS element might have identification numbers of 3.0.1.1, 3.0.1.2, 3.0.1.3, and so on.

Table 14.2: Example Project Risk Register

Risk Ref	Risk Description		Dates			Analysis			Response & Action	Owner	Status
	If	Then	Opened	Trigger	Closed	Likelihood	Impact	Severity			
1	User experience designers are not released from their current project in two weeks…	The project kickoff will be delayed two weeks	3/12/17	3/28/17	3/22/17	5	2	10 (HIGH)	Avoid: Request release of resources at next portfolio approval meeting	Ranger	Closed: Portfolio decision body approved the hiring of three additional people
2	Insufficient digital data storage capacity is available for weekend customer transac-tions…	The system will experience unscheduled down time of up to 60 hours	2/28/17	4/22/17		3	4	12 (HIGH)	Mitigate: Enable system transaction limits over each weekend until additional storage is available	Jordan	Active: Transaction limit feature in development. Request for quotes for additional storage have been released
3	Primary stakeholders do not agree on proposed product price…	Features will need to be removed from the design	1/29/17	5/1/17		2	2	4 (LOW)	Accept:	Harkin	Inactive: Risk deemed as low risk. Will continue to monitor on a monthly basis

Risk Description

The main component of the risk register is the risk description, at least as it related to the identification of risk events. We recommend using an "IF/THEN" format for your risk descriptions (see Figure 14.2). The format not only describes the risk, but also describes the potential consequences: "IF" *this* occurs (risk event), "THEN" that will be the outcome (consequences).

Dates

For risk timing, aging, and tracking purposes, the risk register must have a date component. The most common and useful dates are the date that the risk was identified, the risk trigger date (when the risk is likely to occur), and the closure date.

Severity

In order to prioritize the risk events (remember, you can't address every risk event identified), a severity component needs to be included in the risk register. Either quantitative (1, 2, 3) or qualitative (high, medium, low) representation of risk severity is an acceptable approach. Specific definitions for the numerical values or qualitative values must be documented.

Remember to evaluate the severity of risk from two perspectives: (1) the probability a risk event will occur, and (2) the severity of the impact if it does indeed occur. Total risk severity must factor in both probability and impact perspectives.

Response

For each risk event a project team decides to manage, a response approach must be decided upon and documented in the register for reference and tracking purposes. For low-priority risks and others that the team decides not to manage, the default response is *acceptance*. The risk register must contain a field to identify the chosen response for each risk event.

Owner

Every risk event, regardless of priority, must have an owner assigned. The risk register therefore must provide an owner component. The risk owner is the person who is responsible for monitoring the risk event and initiating the risk response action if and when it is necessary.

Status

Risk events are dynamic by nature, meaning they can change state over time. To facilitate communication, a risk register should include a risk status field. The most common risk statuses include *open*, *monitoring trigger event*, *response initiated*, and *closed*.

The risk register is a very flexible tool in that it can be constructed with any number of components as stated previously. Upon initiation of a project, take time to design the risk register format and components that support the risk management methodology described in the risk management plan.

Using the Risk Register

As the central tool for managing risk on a project, the risk register has many valuable usages. First, the register serves as the central repository for all risk events. Since it catalogs all project risks, its use must be initiated in the earliest stage of a project and used throughout the project cycle. Identifying new risks and updating the risk register should be part of an ongoing risk management process.

Since all risk events for a project are contained within the register, the opportunity exists to use the tool to prioritize the risk events. Since most projects contain more risk than a project has resources to manage, trade-off decisions have to be made concerning which risk events to manage and which to either accept or simply monitor. The risk register provides the necessary data and structure to represent risk event priority. Normal practice is to put focus on the top three to ten risks at any one time.

The risk register also fosters risk-related communication with project stakeholders. This can either be accomplished by using the register itself as a communication device, or by using select information within the register to feed other project communication tools. Used in its entirety, the risk register can be used to communicate the overall risk profile for a project. This is not our recommended approach however, as too much risk information tends to scare the hell out of project stakeholders. Rather, we recommend selecting information pertaining to the highest priority risk events that are active during a particular reporting period, and record that information in the risk register and current summary status report (see Chapter 12).

Since most risk events have consequences, the risk register is also an effective tool for assisting project managers in developing budget and schedule risk reserve, and driving the reserve into the project plan. By effectively using the IF/THEN approach discussed earlier to describe risk events, evaluation of the potential exposure of the highest priority risks provides a minimum, most likely, and maximum range for exposure for the project.

Lastly, the risk register is used to periodically monitor the status of risk events that have been identified. Project team members are balancing many tasks at any given time, so they need an ongoing process and tool to remind them of their risk management duties. With effective use of dates within the register, the project manager is able to track the overall risk trend of a project (Figure 14.2).

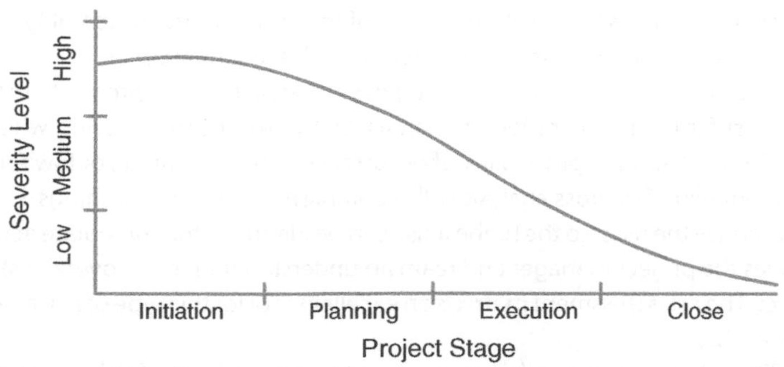

Figure 14.2: Risk Trend Chart

Risk trend charts track the overall project risk profile over time. If risk is being managed effectively, the project risk severity should decrease over time as illustrated in the figure. If not, additional resources may be needed to specifically resolve risk events, or an evaluation concerning project termination should be considered.

Benefits

The benefits gained from the use of the risk register are significant. The register provides a project manager and the organization as a whole a central risk knowledge repository. Not only can the knowledge about project uncertainty contained in the risk register be used to manage risks on a particular project, it can also be used to mitigate systemic risks which plague many projects within an organization.

The register also provides value to project managers by assisting them in resource allocation activities. Since it is not generally possible to manage all risk events, the prioritization information contained in the register provides valuable insight on where to apply resources to prevent the greatest risks to the projects.

By using the risk register to estimate risk reserve, the tool provides value by helping to create more realistic project plans.

The greatest benefit however, is the risk register's contribution to helping project managers protect their project objectives and business goals from being negatively impacted by the myriad of uncertainties that surround a project.

THE RISK ASSESSMENT MATRIX

When you have identified risks that could affect your project outcomes, you need to determine which ones you will spend project budget, time, and resource effort managing as not all risk events require action. The risk assessment step is needed to sift through all of the risk events identified and determine those that pose the most serious threat to the success of the project. The result is a prioritized "short" list of project risks that the team can then manage.

Scenario analysis is one of the most common methods for assessing risk events.[23] Scenario analysis involves analyzing each risk event in terms of the outcome of a risk event's occurrence, the severity of the impact of the outcome, the probability that a risk event will happen, and understanding when the risk event may occur.

We recommend that a project team begin with a qualitative approach to risk assessment, at least for the first iteration of analysis. By this we mean assessing whether the severity of impact and the probability of occurrence is high, medium, or low for each of the risks identified. This gross analysis will accomplish two important things. First, it will quickly prioritize the risks, so the highest risk can be identified for immediate action. Second, it gives the project manager and team an understanding of the overall risk level of the project. The risk assessment matrix is an excellent tool for this type of risk assessment (see Figure 14.3).

If additional analysis is needed, more quantitative methods of risk assessment can be utilized on the next iteration. For the more sophisticated project, we recommend

PROBABILITY (P) ↓	Severity = P + (2 x I)				
NC = 5	7	9	11	13	15
HL = 4	6	8	10	12	14
L = 3	5	7	9	11	13
LL = 2	4	6	8	10	12
VU = 1	3	5	7	9	11
	VL = 1	L = 2	M = 3	H = 4	VH = 5
	IMPACT (I) ⟶				

High Severity
Medium Severity
Low Severity

Figure 14.3: Example Risk Assessment Matrix

the Monte Carlo analysis technique for quantitative analysis. The Monte Carlo analysis technique is the next tool described in this chapter.

Developing a Risk Assessment Matrix

Good project management decision-making practices involve full understanding of the data representing the state of a project at the given time a decision is made. Better project management decision making involves also understanding the critical uncertainties or risks that can affect the decision outcome. Trying to factor all project risks into a decision is an impossible undertaking, however. Rather, a project manager must have a way to filter the critical risks from the noncritical risks. By focusing on risk severity, the risk assessment matrix provides such a filtering service.

The first step in this process is designing a risk assessment matrix that fits the needs of the project manager.

Design the Matrix Format

Risk assessment matrices come in several different formats that project managers can search and adopt for their use. The format shown in Figure 14.3 is by and large the most common.

The matrix is simply a 5 × 5 (sometimes 4 × 4) matrix, with risk probability represented along one axis (in this case the vertical axis) and risk impact along the other axis.

For each of the cells representing the probability that a risk event will occur, the probability levels (or scale) must be represented. In like manner, the impact that a risk event may have on one or more of the project objectives must be represented in each of the risk impact cells.

When first constructing the matrix, the cells that intersect the various levels of probability and impact will remain blank; population of the remaining cells occurs in a following step.

Define the Rating Scales

The next step is to define the scales for which risk probability and risk impact will be assessed. Remember that risk assessment at this stage is qualitative, not quantitative, even though numerical values are used to represent qualitative scales. For that reason, the rating scales must be simple and explicit to enable consistency in qualitative assessment for all risk events. In the example in Figure 14.3, we use a five-level scale for probability of occurrence: Nearly Certain (NC), Highly Likely (HL), Likely (L), Low Likelihood (LL), and Very Unlikely (VU). These qualitative values are not sufficient, however. A description of each value has to be defined by the project team as shown in Table 14.3.

In the same fashion, a discrete scale has to be determined for the impact of project risk. This scale has to be based on specific details of the project. We suggest referring to a tool described in Chapter 12 called the project strike zone, which defines the project goals from various perspectives. Using the highest-priority objective, for example, project completion date, develop the risk impact scale in relation to that objective. Table 14.4 illustrates how a risk impact scale might be constructed based on the project timeline.

The final scale to be defined is the risk severity scale. This is commonly accomplished through the use of a formula and calculation that incorporates the values of risk probability and risk impact. Although nonlinear formulas can be employed, linear formulas such as *severity* = [*probability* + (*N* × *impact*)] are easier to apply. For example, *N* can be equal to two, meaning that impact is twice as important as probability in establishing risk severity. In this case, the assessed probability and impact for each risk would be entered in the formula, *severity* = [*probability* + (2 × *impact*)], and the obtained value would be entered into the probability-impact (P-I) matrix. This is the formula utilized for the risk severity calculations in Figure 14.3.

Populate the Risk Assessment Matrix

Using the risk severity formula developed in the previous step, calculate the risk severity values for each cell in the matrix. Next, segment the matrix into three severity levels—high, medium, low—based on the organization's threshold for risk severity. The higher the value in a matrix cell, the higher its rank and severity of its potential impact to the project. For instance, a risk that has a risk severity score of 15 is more critical than a risk that has a score of 8, and should therefore be prioritized higher.

The severity levels will be unique to how the severity score is calculated and how the project manager decides to segment the numerical values in the matrix. For the example shown in Figure 14.3, the project manager determined that high-priority risks are those

Table 14.3: Example Five-Level Scale of Risk Probability

Scale	1 Nearly Certain	2 Highly Likely	3 Likely	4 Low Likelihood	5 Very Unlikely
Probability description	81–100% Probable occurrence	61–80% Probable occurrence	41–60% Probable occurrence	21–40% Probable occurrence	1–20% Probable occurrence

Table 14.4: Example Five-Level Scale of Risk Impact on Schedule

Scale	1 Very low	2 Low	3 Medium	4 High	5 Very high
Risk impact on schedule	Slight schedule delay	Overall project delay <5%	Overall project delay 5–14%	Overall project delay 15–25%	Overall project delay >25%

with a severity score of 12 or above. Medium-priority risks score between 9 and 11, while low-priority risks are those scoring 8 or below.

With the risk assessment matrix developed and populated with data, it is now ready to be put to use on a project.

Using the Risk Assessment Matrix

The most challenging aspect of performing a qualitative risk assessment is adequately defining the rating scales. Once that has been accomplished, however, the rating can be used for the duration of the project to effectively manage project risk.

If risk identification is adequately performed, it is common for a large number of risks to be identified, depending upon the type of project. The challenge in front of the project manager is to identify those that have both the highest impact on the project and those that are most likely to occur. This is where the risk assessment matrix provides its utility.

With limited resources, the project manager has to use the matrix to determine which risk events warrant the application of project resources the most. Some larger projects commonly focus on the top 10 highest-ranked risks. In contrast, some smaller projects decide to manage the top three risks, arguing the lack of resources to take on a larger number of risks. Both approaches may be ad hoc in nature. So, what is a reasonable approach? The answer is in the matrix. Respond to the highest-ranked risks in the matrix, down to an agreed level.[24] For example, focus on handling risks down to a risk score of 11 (Figure 14.3), and treat other risks as noncritical. With this approach, one neither squanders resources nor disregards significant risks. It should be noted that noncritical does not mean not important. Rather, it means that scarce project resources are not immediately needed to address the risk event, but may be needed in the future.

With the project risks adequately prioritized and the risk events targeted for action chosen, the project team can now develop risk response strategies and actions. Refer to the risk management plan for guidance on the preferred risk response approach to take given the particular stage the project is in, and the current project situation. Keep in mind that the project manager's ability to put actions in place to mitigate or avoid a risk event may be constrained by where a project is in relation to the project cycle. Specifically, if a project is nearing the end of project execution, it may be too late to affectively mitigate or avoid a risk that emerges. In this case the response option may only be to accept the risk or terminate the project.

Variations

The outcome of the risk analysis process is a shortened list of critical risk events that can be actively monitored, managed, and communicated to stakeholders. A variation of the risk assessment matrix, called the risk map, is slightly more effective for monitoring and communicating risk.

Shown in Figure 14.4, the map displays the most critical risks from the two dimensions of likelihood of occurrence and risk impact. The scales displayed on the *x*- and *y*-axes are those used to evaluate each risk event. We show numbered scales, but others may prefer to choose more qualitative values such as high, medium, and low.

Normally, a threshold line is drawn to visually indicate the separation of the critical risks events from those at a lower severity level. The line is determined from the description of the scale levels discussed earlier. It is important to include risk events that are below but near the threshold, as these are the risks that are most likely to move to a critical state.

The final piece of project intelligence that can be included on a risk map is change in likelihood of occurrence or impact.[25] This is indicated by arrows on Figure 14.4. We encourage project managers to include this piece of information because it reinforces the behavior of ongoing risk monitoring and analysis.

Benefits

The risk assessment matrix helps sift through the myriad of uncertainties to pinpoint and highlight the project areas of highest risk—both before work has begun and throughout the life of a project.[26] This offers an opportunity to focus project resources and to

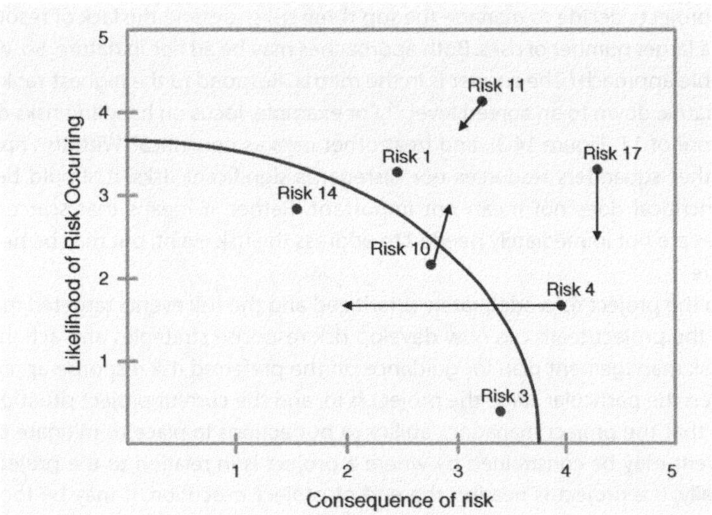

Figure 14.4: Example Risk Map

identify effective ways of reducing risk events in a proactive manner, rather than being confronted by them if they turn into issues later in the project.

In addition, the matrix generates information for more reasonable contingency planning and effective project decision making. It is impossible to predict all possible risk events. Therefore, best practice organizations include the use of schedule and budget contingency based on a risk analysis. The amount of contingency can be derived from risk impact information contained in the risk assessment matrix.

Much of the tool's utility comes from its visual representation of risk severity and simple design, making it more than adequate for situations where detailed quantitative risk assessment is not required. This is particularly the case in the early stages of the project cycle.

Finally, the risk assessment matrix is useful for increasing visibility and awareness on the part of top managers within an organization of the critical risks associated with a project. This enables sound decisions pertaining to the risk to be made in the appropriate context.

With the risk register and risk assessment matrix in a project manager's toolbox, he or she is equipped with the means to identify, document, assess, prioritize, and monitor any future event which may prevent a project team from fully realizing the project objectives. However, none of these activities meet the real intent of risk management. The real intent of project risk management is to use the information obtained to proactively make adjustments to the project plan to account for the uncertainties that surround every project, and to make better risk-informed decisions. The next two tools are effective in assisting project managers in these two endeavors. We begin with the Monte Carlo analysis.

MONTE CARLO ANALYSIS

The presence of uncertainties converted to risks decreases the probability of successfully completing a project as defined by the documented success criteria. If uncertainty did not exist, it would be easy to repeatedly meet the schedule, cost, and performance goals of the projects within our portfolios. However, with uncertainty being a part of every project, project managers have to find ways to handle the uncertainties that pose the greatest risk to their projects.

While managing a project, a project manager will face a situation where they have a long list of risk events, and little clue of the impact they may have on the project goals. Once the risk events have been prioritized, a Monte Carlo analysis can be performed to quantifiably evaluate the potential impact of the critical risks.

Monte Carlo analysis is effective in determining the impact of identified risks by running mathematical simulations to identify a range of outcomes associated with various confidence levels relating to probability of success. The simulation furnishes the project manager a range of possible outcomes and the probability they will occur for any choice of action. This process provides a valuable tool to compensate for the impact of critical risk events by determining the amount of risk reserve needed to increase the probability of success given the known uncertainties facing a project.

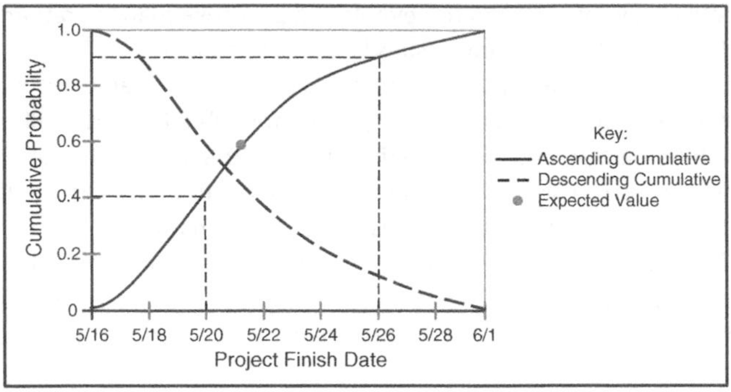

Figure 14.5: Cumulative Distribution of Project Duration Produced in Monte Carlo Analysis
Source: David Vose, *Risk Analysis: A Quantitative Guide*. Copyright © 2000 by John Wiley & Sons Limited. Reprinted with permission of John Wiley & Sons.

Monte Carlo analysis randomly samples a probability distribution of the critical project risks to simulate project scenarios in response to the risk impacts hundreds or even thousands of times.[27] This provides statistical distribution of the calculated project durations and approximates the expected value of the duration, as illustrated in Figure 14.5. With these distributions, you can quantify the risk of various schedule scenarios, alternative implementation strategies, activity paths, or even individual activities. For example, as Figure 14.5 indicates, there is a 40 percent probability that the project will be finished before or on May 20.

Performing a Monte Carlo Analysis

Typically, Monte Carlo analysis deals with schedule, cost, and cash flow risks, although other facets such as the quality of the final project output can at times be analyzed. Taken overall, performing a schedule risk analysis is more complex than a cost analysis, simply because dependencies between project activities need to be established in order to identify the critical path. For this reason, our focus is on looking at the Monte Carlo analysis process in schedule risk analysis (see Figure 14.6).

A number of important inputs are needed to perform a successful Monte Carlo analysis. First, the risk management plan should provide guidance on when and how to apply Monte Carlo on a project. The critical project risks and their estimated impact to the project provide the risk impact from which the simulations are built upon. Finally, the project schedule or budget for the project is needed to establish the baseline from which probability impact scenarios will be added. When analyzing impact to project timeline, the time-scaled arrow schedule format is the most useful (Figure 14.7).

This information will be fed into the Monte Carlo analysis to generate a range of possible project durations (see "Basic Terminology of Monte Carlo Analysis").

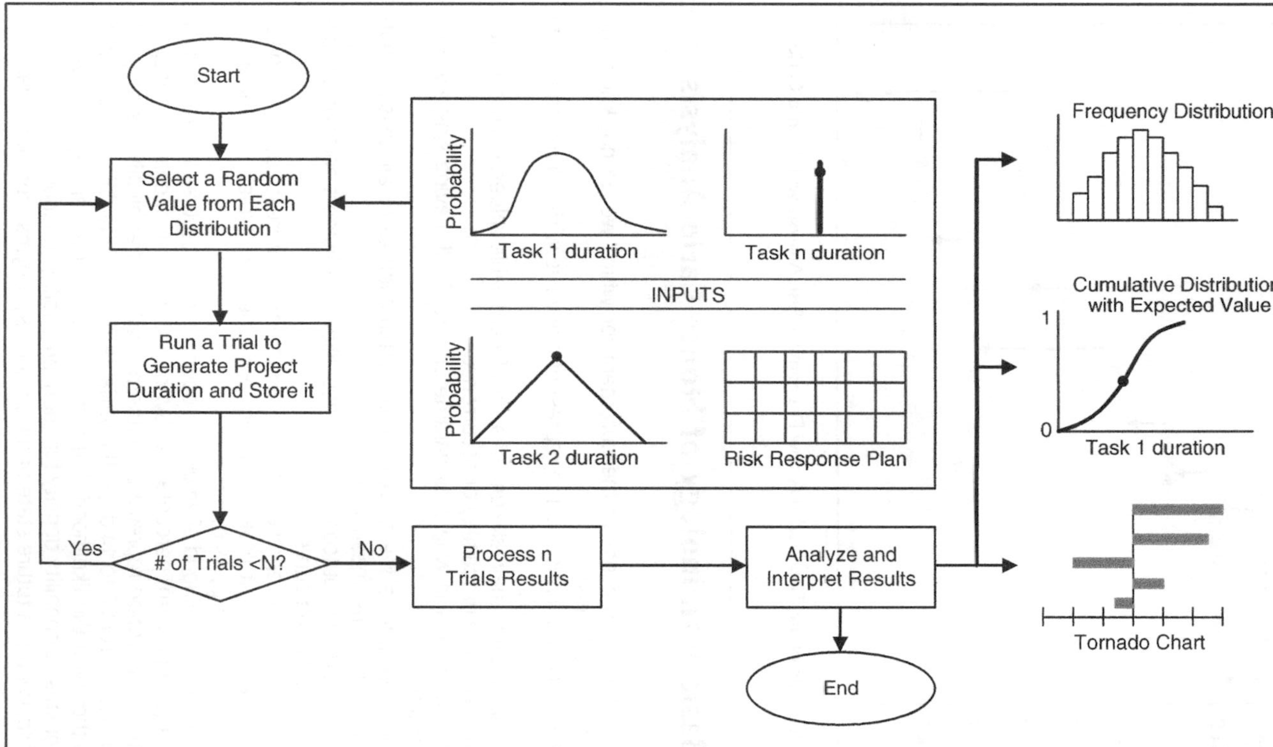

Figure 14.6: Monte Carlo Analysis Process for Schedule Risk

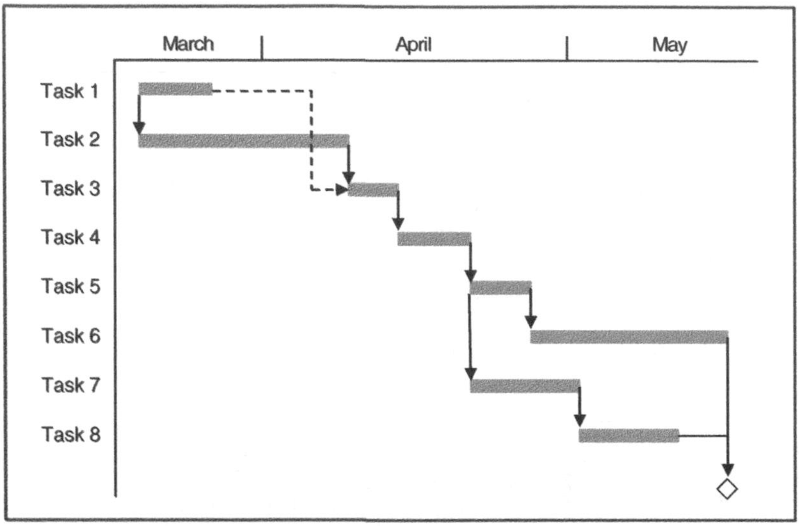

Figure 14.7: Example of the Time-Scaled Arrow Diagram for Risk Analysis with Monte Carlo

Basic Terminology of Monte Carlo Analysis

Chance event is a process or measurement for which we do not know the outcome in advance.

Continuous distribution is used to represent any value within a defined range of values (domain).

Discrete distribution may take one of a set of identifiable values, each of which has a calculable probability of occurrence.

Deterministic model is where all parameters are fixed, having single-valued estimates.

Expected value (EV) is the probability weighted average of all possible outcomes. Synonyms: mean, average.

Mode is the particular outcome that is most likely; the highest point on a probability distribution curve.

Model is a simplified representation of a system of interest such as project critical path chart. It projects project outcome (e.g., project duration) and outcome value (e.g., 18 months).

Probability is the likelihood of an event occurring, expressed as a number from 0 to 1 (or equivalent percentages). Synonyms: likelihood, chance, odds.

Probability distribution represents mathematically or graphically the range of values (e.g., from 2 to 14 days) the variable (e.g., activity duration) can take, together with the likelihood that the variable will take any specific value. Synonyms: probability density function, probability function.

Project scenario is a future state of the project. Synonyms: iteration, trial.

Random sampling is a process generating a random number between 0 and 1, which determines the value of the input variable from the probability distribution.
Random variable is a measure of a chance event. Synonyms: chance variable, stochastic variable.
Single-valued estimate has one value only. Synonym: point estimate.
Standard deviation is the square root of the variance.
Stochastic model is a model that includes random variables. Synonym: probabilistic model.
Variance is the expected value of the sum of squared deviations from the mean.

Generating a range of possible project durations and their probabilities is not possible without preparing probability distributions for project activity durations. This preparation process may begin with the question "How long does it take to complete a project activity?" Let's assume that you performed an activity many times, and each time it took ten days to complete. If asked to estimate the duration of that same activity in a future project, you would likely put it at ten days. If each project activity would have such a single point estimate (also called single value estimate) as an input to calculate project schedule duration, the duration would also have only one value. There is not much uncertainty in project activity durations in this single-valued deterministic model—they are all fixed. In the majority of today's projects, such a scenario is not realistic. More realistic is the following probabilistic (stochastic) model.

Imagine that you repeated an activity (call it Activity 1) an extremely large number of times (trials, iterations, scenarios), and its duration ran from 5 to 39 days (range of outcomes) due to various issues that arose. You recorded the fraction of times that each duration value (outcome) occurred. The fraction for a particular outcome is approximately equal to its probability (p) of occurrence for Activity 1. When we have these approximate probabilities (the more trials you do, the closer the fraction becomes to the true probability) for all possible outcomes, we can chart them as probability distributions (see the Task 1 duration curve in Figure 14.6). Assume that experience-based probability distributions are also available for some other activities in the project as well (see Task 2 duration in Figure 14.6). If we really had such probability distributions, they would be close to objective probabilities, which are defined as being determined from complete knowledge of the system and are not affected by personal beliefs (see "Frequently Used Probability Distributions").

Frequently Used Probability Distributions

Three values are used to describe a very simple and popular triangular distribution (see Figure 14.8a): Triangular (5, 10, 20); minimum (L = 5); most likely (M = 10); and maximum (H = 20). Numbers in parentheses are project task durations in days. The mean is calculated by (L + M + H) / 3.

(*continued*)

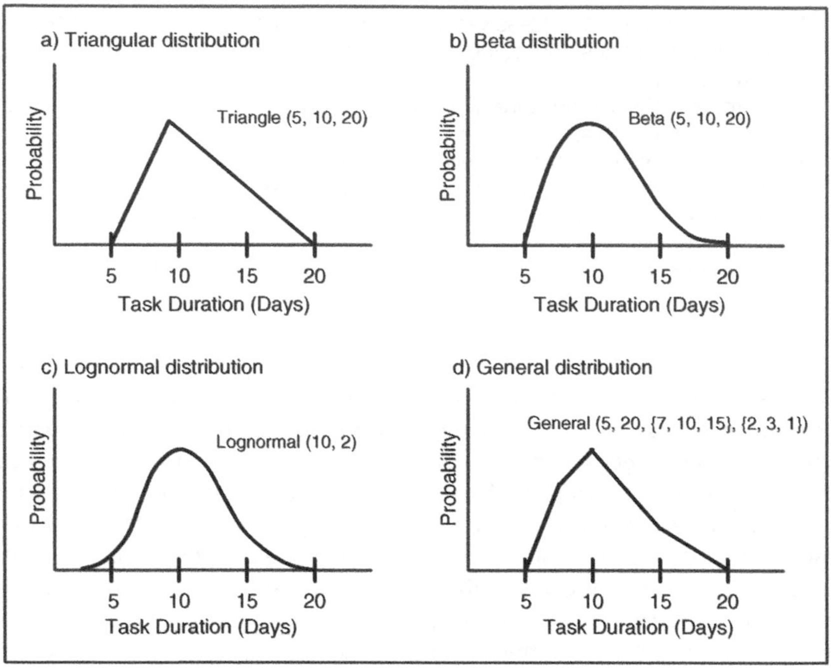

Figure 14.8: Frequently Used Distribution

The Beta Distribution (see Figure 14.8b), which has been used for a long time to estimate task durations in the Program Evaluation and Review Technique (PERT), requires the same three parameters as triangular distribution: minimum (5), most likely (10), and maximum (20). The mean is calculated by (L + 4M + H) / 6.

Two parameters describe a lognormal distribution (see Figure 14.8c)—mean (10) and standard deviation (2).

Known for its flexibility, the general distribution (see Figure 14.8d) allows shaping the distribution to reflect the opinion of experts. It is described by an array of values (7, 10, 15) with probabilities (2, 3, 1) that fall between the minimum (5) and maximum (20).

Although some companies do have experience-based databases with the approximate distributions of their project activities, that is the exception rather than the rule. What, then, do we do to prepare probability distributions for activity durations? We will do what is a dominant practice in real-world projects—prepare and rely on subjective probabilities derived from someone's belief whether an outcome (activity duration) will occur.

The most adequate way for this is to enlist the help of experts or experienced project participants. Brainstorming with activity owners, studying durations of similar activities in past projects, and consulting other specialists in the company who are not involved in the project all help determine the probability distributions or single values for the activity durations.[28] The single-value estimates are incorporated into the baseline project schedule.

Now it is time to apply risk impact. The baseline schedule that was generated from single-value estimates represents the best-case scenario. Monte Carlo Analysis provides a best-case to worst-case distribution of potential outcomes based on a probability distribution; therefore, we must also define the worst case. This is where the high-level risk events come into play. Referring to the risk assessment matrix, locate all the high-severity risks. The *combined* schedule impact of the high-severity risks represents the worst-case scenario.

Randomly Select a Value from Each Distribution

When the probability distributions are available for the project tasks (variables), the stage is set for the next step. Within the specific range of duration values bound by the best- and worst-case values, select one duration value randomly. The key word here is *randomly*. Using a random sampling technique, Monte Carlo analysis generates a random number between 0 and 1, which is fed into a mathematical equation that determines the task duration value to be generated for the distribution.[29] All selected values constitute a random sample of values that will be used to generate project durations. Sampling can also be done with other efficient methods such as Latin hypercube sampling. Whatever the method used, random sampling from probability distribution is performed in a manner that reproduces the distribution's shape.

Run a Trial to Generate Project Duration

Having a random sample of task duration values means that for each activity in the project schedule there is one value only. Plugging this combination of activity duration values in the project network diagram will produce a scenario for project duration. In essence, this is a deterministic schedule with a single value for project duration, built on single-value durations for each activity. At this time, we will store this project duration until the time comes to use it again.

Repeating this sequence of random sampling many times and running a trial will produce as many scenarios for project duration, each one plausible. This prompts the question, "How many trials do we need?" Typically, trials (iterations) go until the predetermined number is reached (number *N* in the decision box in Figure 14.6). That number depends on the number of variables (activities) and the degree of confidence required but typically lies between 100 and 1,000.[30] The idea here is that a sufficient number of trials preserves the characteristics of the original probability distributions for activities and approximates the solution distributions for project duration.[31]

Process Results

When the trials are complete, our "storage" will contain *N* project durations. Each one is a possible case for the behavior of the project schedule. Processing them by means of a software program can produce many forms of results, whereas our focus is on the following (see the right-hand side of Figure 14.6):

- *Expected value of the project duration*. Averaging trial values for project durations approximates the expected value, the probability weighted average of all possible outcomes. However, the higher the number of trials, the higher the precision of the expected value and the approximations of probability distribution shape for project durations.
- *Frequency distribution*. This is a histogram plot showing relative frequency obtained by grouping the data generated for project durations into a number of bars or classes. Frequency is the number of values in any class. Dividing the frequency by the total number of values will produce an approximate probability that the project duration (output variable) will lie in that class's range (see Figure 14.9).
- *Cumulative frequency*. Cumulative frequency can be expressed in either an ascending or descending format. It has two formats (refer again to Figure 14.6). The former indicates the probability of the project duration being less or equal to the value on the *x*-axis. Conversely, the latter shows the probability of the project duration being greater than or equal to the value on the *x*-axis. Expected value is marked on the plot with a black dot.
- *Tornado chart*. This chart shows the extent to which the uncertainty of the individual activities' duration impacts the uncertainty of project schedule duration (see Figure 14.10). Specifically, the bar represents the degree of impact the activity (input variable) has on the project schedule (model's output). Therefore, the longer the bar, the greater the impact that a project activity has on the project duration. Per standard practice, bars are plotted from top down in decreasing degree of impact. When

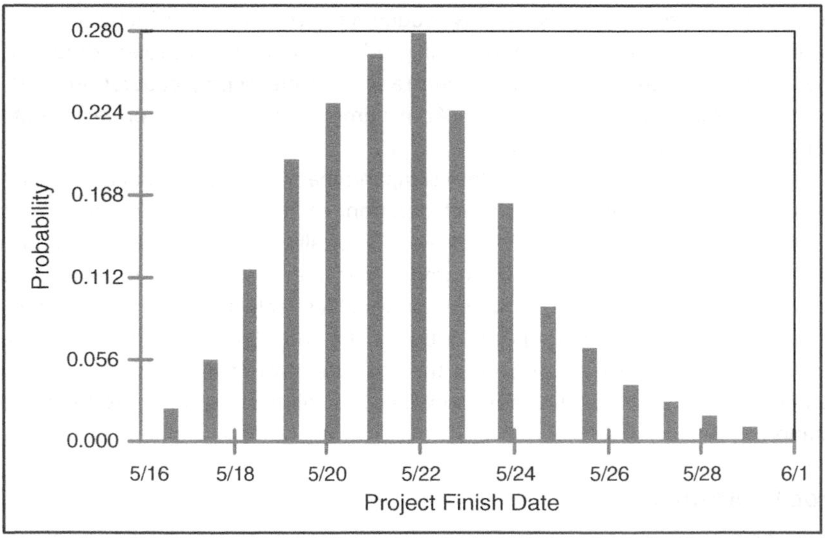

Figure 14.9: Frequency Distribution Histogram of Project Duration

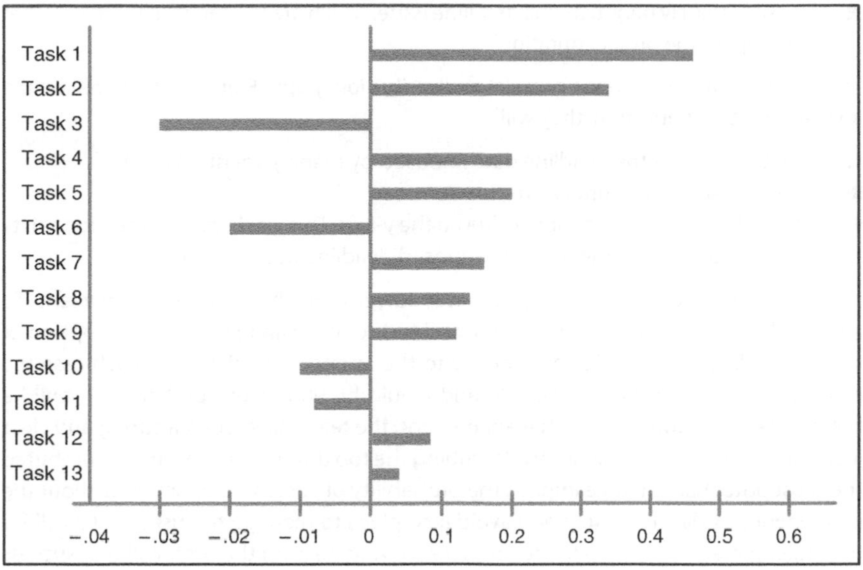

Figure 14.10: An Example Tornado Chart
Source: David Vose. *Risk Analysis: A Quantitative Guide*. Copyright © 2000 by John Wiley & Sons Limited. Reprinted with permission of John Wiley & Sons.

there are both the positive and negative impact, the chart is a bit reminiscent of a tornado, hence the name. To avoid the chart looking overly busy, Vose suggests limiting the plot to those activities (variables) that have an impact of at least a quarter of the maximum observed impact.

Analyze and Interpret Results

The results of the schedule risk analysis must be interpreted in a way that clearly provides answers to the questions the analysis was initiated to answer. For that reason, it is beneficial to follow four principles for schedule risk analysis:

1. Focus on the problem.
2. Keep statistics to a minimum.
3. Use graphs whenever appropriate.
4. Understand the model's (e.g., time-scaled arrow diagram chart) assumptions.

To demonstrate these principles, let's assume that the project team sets out to perform a schedule risk analysis to answer these questions:

- How likely is it that the team will achieve the project deadline (May 20) imposed by management?
- If the probability is lower than 90 percent (the team's preferred probability), what do we do to negotiate the deadline with management?

- If we successfully negotiate the deadline issue, which are the top three activities that most impact the project duration?

First, the team goes to the cumulative distribution graph (Figure 14.5). To obtain the answer to the first question, they will:

- Enter the *x*-axis at the deadline date imposed by management (May 20).
- Move upward to the cumulative curve.
- Move left to the corresponding value on the *y*-axis. This *y*-axis value is the probability of completing their project on the imposed deadline date (40 percent).

Clearly, the probability is very low, way lower than the preferred 90 percent. To answer the second question, the team decided to ask management for an option of adding six days of schedule contingency to the imposed deadline, in which case the project would be finished by May 26 and would be 90 percent probable. To build a better case for negotiations with management, the team developed a strong and clear justification for the contingency by describing the top three risk events that contributed the most potential schedule impact, the probability of success assessment without the contingency (40%), mitigation and avoidance plans to manage the risks, and a call for management action to help the team succeed by increasing the probability of success to 90%.

Finally, the team created a tornado chart (see Figure 14.10) that reveals the three key tasks with the highest impact on the deadline—these are the top priorities to monitor during implementation, an answer to their question three.

Using the Monte Carlo Analysis

Traditionally, it has been the large and complex projects that have most often enjoyed the benefits of Monte Carlo analysis. The belief was that, unlike smaller projects, larger projects had more important goals and could also afford necessary resources for performing Monte Carlo analysis. It took some significant events for this view to start changing. First, the trend toward "management by projects" led to a proliferation of smaller but important projects. Also, very powerful desktop computer programs for Monte Carlo analysis have become affordable. Finally, project offices capable of supporting many projects with Monte Carlo analysis have become a frequent organizational unit. All of these events helped put Monte Carlo within the reach of smaller projects, changing the application pattern of Monte Carlo analysis in corporations.

Today, Monte Carlo analysis is used in both larger and smaller projects to respond to certain situations. For example, if a project is sensitive to a completion deadline, Monte Carlo analysis is a preferred option. Similarly, if there are many project scenarios and what-if analyses to explore, Monte Carlo is favored over other analysis techniques.[32]

Roles in Monte Carlo analysis crucially influence how much time it takes to perform an analysis. A typical approach is to task the project office specialists to perform Monte Carlo analysis using the data provided by the project team. For a smaller project of 50 activities, for example, data entry and running Monte Carlo may take 10 to 30 minutes. Assuming that a project logic diagram exists, preparing activity probability distributions through team brainstorming and formatting them into a table to be fed to the project office may take an hour or two. A growing size and complexity of the project is bound to increase the time it takes to perform a Monte Carlo analysis.

Benefits

Original project schedules and budgets are often unrealistic or, more precisely, inadequate. A major reason for this inadequacy is the uncertainty surrounding the project activities. In response to this uncertainty, many project teams assign an arbitrary duration or cost to the activities and hope for the best.[33] Contrast this approach with Monte Carlo analysis, which allows richer, more detailed representation of the risk problem by using risk-factored scenarios based upon probability of success distributions. Therefore, arbitrary risk contingency is replaced by a quantitative analysis for determining the amount of contingency needed based on probability projections.

The Monte Carlo approach also generates graphical representation of multiple outcomes, especially if a commercially available software application is used. Graphical analysis enhances evaluation and communication of decision choices.

By taking into account variable factors (such as schedule tasks or cost elements), Monte Carlo makes it possible to spot which variables have the greatest impact on project results. In like manner, the technique makes it possible to analyze the effect of combining multiple variables. For instance, performing project tasks concurrently if additional staff is added versus performing them in sequence.

In summary, Monte Carlo analysis's value is in its ability to examine each project scenario, including the extreme scenarios, to see what conditions give rise to their results. That helps not only to validate the project realism but also to differentiate between what is possible and what is not possible, and most importantly, how to change what is not possible into what is possible.

THE DECISION TREE

A decision tree is a graphical tool for analyzing project situations that involve uncertainty or risk. Reflecting the decision process, the tree displays sequential decisions in the form of branches of a tree, from left to right, originating from an initial decision point and extending to the end outcomes (see Figure 14.11).[34] The path through the branches represents the sequence of separate decisions and chance events that occur. Decisions are evaluated by calculating the expected value and probabilities of each path.[35] For a description of a typical decision tree's components, see "Five Components of a Typical Decision Tree."

Five Components of a Typical Decision Tree

Decision nodes. Also called decision points, these are points in time when decisions are made or alternatives chosen. Shown as square boxes, they are controlled by the decision maker. The starting node is called the *root*.

Chance nodes. These represent times when the result of a probabilistic event occurs. Decision makers have no control over them. Chance nodes are also called probabilistic nodes or points, and are represented by a circle.

(*continued*)

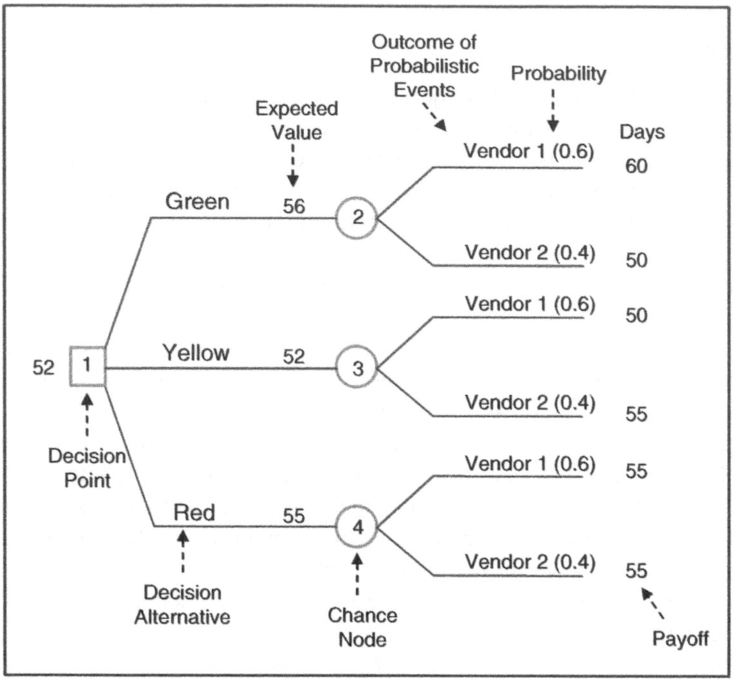

Figure 14.11: Decision Tree for a Project Situation under Risk

Branches. Lines connecting the decision and chance nodes in a sequential manner. The branches leading out of a decision node represent possible decisions, while those stemming from chance nodes represent possible outcomes of probabilistic events.

Probabilities. The probabilities of the probabilistic events shown on the branches representing those events. Mostly they are conditional and for any particular chance node must sum to 1.

Outcome values. Also called the payoff, outcome values of each alternative are placed at the end of the branch. They may represent present values discounted to the date of the root decision or cost.

Analyzing the Decision Tree

Literature on decision trees often looks at minimizing the risk by selecting decision alternatives that offer maximum net present value or minimum cost. The abundance of such examples prompted us to take a different approach. In particular, we look at the general process of analyzing decision trees on an example related to minimizing schedule duration. This is in response to risks faced by a large group of today's projects that must execute to the fastest possible schedule as an unconditional priority.

Using the decision tree technique requires input from a number of sources. The risk management plan specifies how to use the decision tree to support decisions involving risk. Information contained in the risk assessment matrix about individual risks and their response strategies is also crucial. In our example from Figure 14.11, this information will be funneled into calculating durations for each of the outcomes in the decision tree. Since this involves the project timeline, we also need schedule information for the decision alternatives.

Describe the Decision under Risk

Common sense dictates that in order to make the best decision, we first need to understand the decision context and related risks a project is facing. A convenient way for this is to describe the decision. Here is an example.

Consider a company that competes on time-to-market capabilities. As they are entering the design phase for its new product development project, their goal is to finish it as soon as possible and hit the market before their competitors. While literally each project day is considered to be of extreme priority, the development cost is given lower importance. In such a situation, the project team is attempting to decide on the appropriate design approach to use for this product. A major uncertainty involves the amount of time it will take to design the central module in the product. Three major alternatives are being considered, each one identified with a single word:

- *Green (G)*. Incorporate the routing rules early in the central module design.
- *Yellow (Y)*. Predict the routing rules early and modify them at the end of the central module design.
- *Red (R)*. Incorporate the routing rules at the end of the central module design.

A second uncertainty is related to an off-the-shelf part that goes into the central module, whichever the alternative. Two different vendors produce the part. It is well known that both companies are in a race to release the newest upgrade of the part, and they have announced the same release date. To represent this decision description and enable its analysis, the team must first structure the model.

Structure the Model

The model is drawn from left to right (for a better understanding refer again to the "Five Components of a Typical Decision Tree"). Therefore, draw a decision node (the square marked 1), then add to its right-hand side three branches, for three available design alternatives—the green, yellow, and red (Figure 14.11). Put a chance node (the circles marked 2, 3, and 4) at the end of each branch, followed by two branches, each one for outcomes of probabilistic events—vendor 1 hitting the market first and vendor 2 hitting the market first. As monotonous as structuring the model appears, it is precious in sharpening our understanding. Complex decisions require a level of structure to help simplify the decision options.

Assess the Probability of the Possible Outcomes

The company's new product development project team cannot wait for the actual release of the first-to-market part by vendor 1 or vendor 2 before beginning their own design process. That would significantly extend the module design schedule, jeopardizing the project end date. Therefore, the project team decides to assess the probability of who—vendor 1 or vendor 2—will release the product first. Their research and past performance of the vendors led them to assess that there is 60 percent probability that vendor 1's part will first reach the market. The probability for the part from vendor 2 is 40 percent. Everybody on the team is clear that these are subjective probabilities influenced by their perceptions, beliefs, and historical events.

Determine the Possible Outcomes

The project team has developed initial network schedules for each of the design options—green, yellow, and red—as if the vendor part is already available. The sequence of design activities involved in each option is different, as well as some other activities. Also, although both vendors' parts can be used for the central module, the process of their incorporation into the design is different, causing the duration of each outcome to differ. Because the team's expectation is that the first-to-market vendor part will be released sometime midway through the module design, they need to evaluate how such a release is going to change the initial network schedules. The product of their evaluation is a set of possible outcomes values, also called payoffs. These schedule durations of the outcomes expressed in days are added at the end of each branch (Figure 14.11).

There are two conceptually different parts to a decision tree analysis. Included in the first part are structuring the model, assessing the probabilities of possible outcomes, and their payoffs. This is a particularly unstructured task, requiring a significantly greater proportion of effort. The second part—evaluate alternatives and select the strategy—is the easy part of the model and the heart of the decision analysis under risk. We address it next.

Evaluate the Alternatives

Our objective is to evaluate possible outcomes and select one with the shortest possible schedule. To accomplish this, we need to solve the tree (see "Two Simple Steps for Solving a Tree"). When these steps are applied:

Step 1. Chance Node 2 Expected Value: $(0.60 \times 60 \text{ days}) + (0.40 \times 50 \text{ days}) = 56 \text{ days}$
Chance Node 3 Expected Value: $(0.60 \times 50 \text{ days}) + (0.40 \times 55 \text{ days}) = 52 \text{ days}$
Chance Node 4 Expected Value: $(0.60 \times 55 \text{ days}) + (0.40 \times 55 \text{ days}) = 55 \text{ days}$

Step 2. The best alternative is the alternative with lowest expected value, the shortest schedule—52 days. That means the team goes with the yellow option.

Decision tree analysis enables more than just identifying the best alternatives. Sensitivity analysis, as well as a tornado chart, can also be developed to better understand the decision under risk.[36]

Two Simple Steps for Solving a Tree

The procedure for solving a tree is called "rolling back" or "folding back" the tree. Simply, by starting at the far right of the tree and working back to the left, we solve for the value of each node, annotating it with its expected value. The expected value can be represented as cost (measured in currency) or schedule (measured in time units). Two simple steps for solving a tree are as follows:

1. At each chance node, calculate the expected as the sum of each branch's outcome value (payoff) multiplied by probability. This is the value of the node and the branch leading to it.
2. At each decision node, we find the best expected value alternative. This is the alternative with the highest value (when dealing with present values) and the lowest alternative for expected value (schedule or cost).

When the folding process is completed, the alternative with the best outcome value for the leftmost decision nodes becomes the best alternative.

Using Decision Trees

Theoretically, we can use a decision tree to evaluate *any* decision under risk, regardless of its complexity, as long as the decision and probabilities of the possible outcomes are specified.[37] Practically, this is not the case. Rather, practicing project managers see the tree as a method to address daily problems, requiring a straightforward and quick selection of the best alternative. Why is this? The issue here is that complex decision situations lead to a "combinatorial explosion" and the time associated with it. As we add decision and chance nodes, the trees tend to grow exponentially.[38] For example, multiple alternatives with multiple uncertainties can explode decision trees into hundreds of paths, which is where Monte Carlo is a better analysis approach. Constructing and solving for such trees may take hundreds of hours. This situation normally creates a major constraint. Practicing project managers go to decision trees primarily when they need to swiftly evaluate simple alternatives, pick the best alternative, and go on with their daily routine. Still, the majority of such situations occur in larger projects, although we have seen project managers using two-alternative models with four to six paths very informally with minimum time consumption.

Two extremes can help fathom the time requirements of decision trees. Spending 10 to 15 minutes to construct and evaluate a two-alternative decision tree with four paths seems realistic to many project managers. On the other hand, tens of hours may go into the construction of a decision tree with hundreds of paths. The assumption here is that information necessary to estimate probabilities and outcome values is already available. The analysis of large trees may be a matter of minutes, given the power of professional software necessary to use large trees.

Benefits

Two major benefits seem to motivate project managers to use decision trees. First, decision trees reduce an evaluation and a comparison of all decision alternatives under risk to a single value metric. Simply, this metric indicates the degree of support of project goals. In our example, that metric was time, gauging progress toward the quest for time-to-market speed. In other cases, an additional convenience stems from the fact that most of the time this single value metric is expressed in monetary terms, a universal language of business and projects. Then, this single monetary value combines cost, schedule, and performance criteria.

The second benefit lies in the belief of many project managers that the real value of decision trees is not in the numerical results but rather in their ability to help us gain insights into decision problems. With or without the numerical results, users should understand that decision trees do not provide an entirely objective analysis. In the absence of sufficient empirical data necessary for a complete analysis, many facets of the analysis are rooted in personal judgment—structuring the model and assessing probabilities or payoffs, for example.

Benefit also comes in the knowledge gained about the uncertainty being analyzed through the evaluation and conversation between project team members and subject matter experts. Many times a decision can be made without a decision tree being fully developed or analyzed as a result of having the right conversations between the right people.

THE RISK DASHBOARD

Due to the dynamic nature of project risks, risk monitoring must continue through the life of a project. New risks will be identified, many anticipated risks will disappear, some risks will be mitigated, and some will change in severity due to a change in probability of occurring or a change in potential impact to a project.

Regularly held risk reviews force consistent risk monitoring and enable repeated risk identification, assessment, analysis, and response planning as a project progresses through the life cycle. To facilitate the risk reviews, a number of tools are needed to help project managers efficiently move through the monitoring process. Efficiency comes in the ability maintain discussions at a high enough level that the team does not get mired in risk detail. The best tools to use are the risk identification matrix, the risk assessment matrix, the risk map (all covered in previous sections), and the risk dashboard.

Developing the Risk Dashboard

The risk dashboard is a business intelligence tool that provides the important risk statistics pertaining to a project. The dashboard helps the project manager and his or her top management assess the health of a project as well as possible issues facing the project. An example risk dashboard is presented in Figure 14.12.

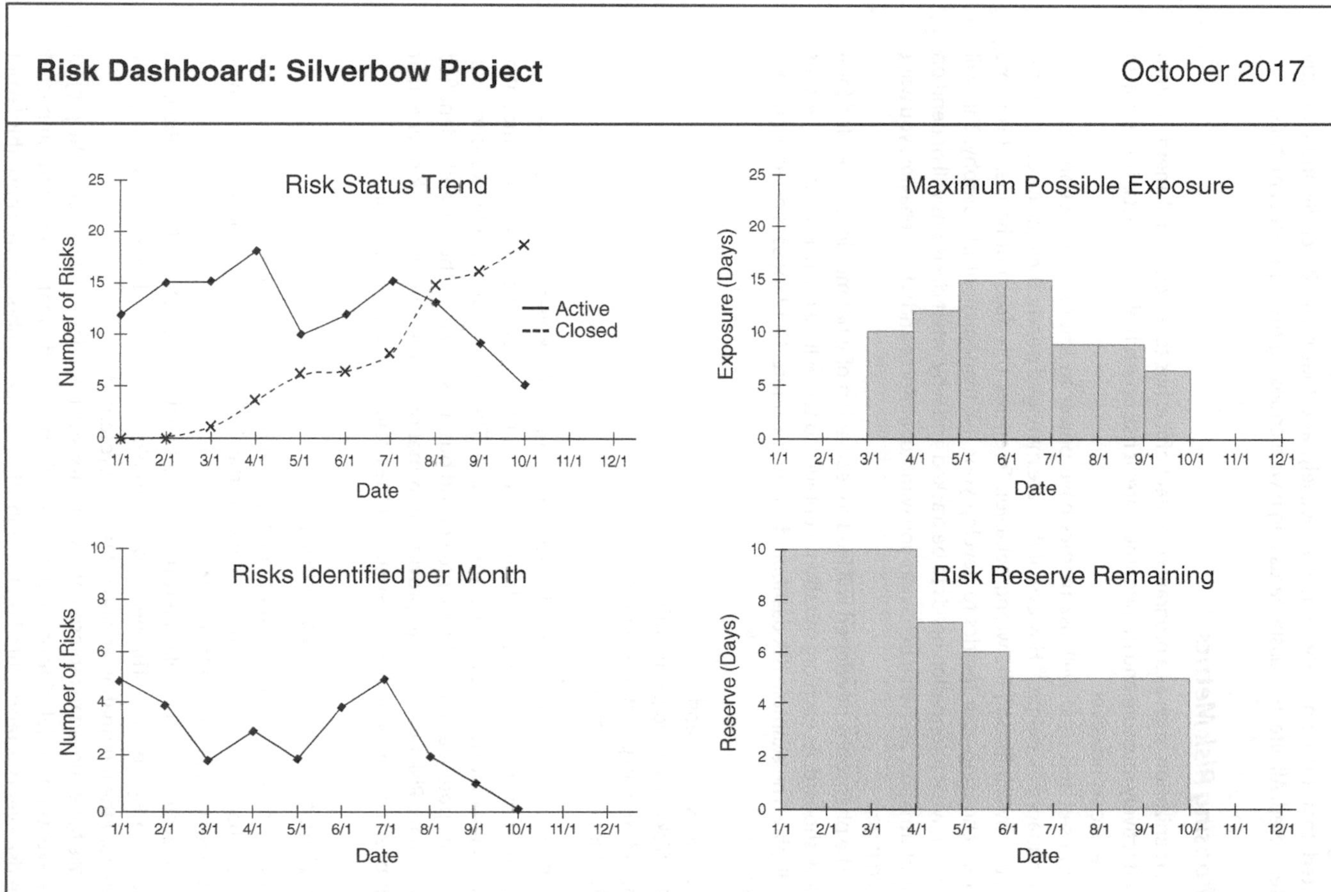

Figure 14.12 Example Risk Dashboard

The risk dashboard needs to be carefully designed. Since it is a project and business intelligence tool, you want to ensure that the most important intelligence is being collected, that it is being represented effectively, and that it can be easily and correctly interpreted. All effective dashboards begin with choosing the right set of metrics.

Choosing Risk Metrics

An organization, such as a program or project office, has to develop a risk metric system to effectively measure and monitor the state of risk on each of its projects. This is not an easy exercise, however.

Chosen and used well, risk metrics have power to improve project and business performance over time.[39] However, if they are managed poorly, they can be counterproductive by reinforcing the wrong behavior. To ensure this does not happen, take time before choosing your metrics to develop your risk management philosophy. This will focus on what information needs to be tracked and why, what decisions will depend on the information, and what behaviors you want to reinforce and what behaviors you want to change.

The philosophy developed will lead to realization of what metrics are needed for a risk dashboard. As each organization is unique, so too will be the risk metrics it chooses to display. The example risk dashboard shown in Figure 14.12 focuses on four primary metrics:

1. Risk status trend
2. Risks identified per month
3. Maximum possible exposure
4. Risk reserve status

Risk status provides a view of the overall number of risk events open and closed over the life of the project to date. Risks identified per month tells the story of the amount of uncertainty facing the project team over time. The maximum possible exposure indicates the potential impact to the project and business based on the currently identified risk events. Risk reserve status provides information on the amount of risk reserve that has been consumed and what remains at the project manager's disposal.

Outline the Dashboard Layout

With the metrics that are going to be represented in the risk dashboard defined, you know what information should be shown in the dashboard. Now you need to determine *how* you want to present that information.

To accomplish this, take a few minutes to sketch the structure of the dashboard as shown in Figure 14.13. This need not be an eloquent sketch, the intent is to logically structure the information and how it will be presented.

We have chosen to represent the summary and trend information about the project risk events on the left side of the dashboard. Summary information about the impact of the risk events is represented on the right side of the dashboard. We chose this layout to represent cause and effect moving from left to right on the risk dashboard.

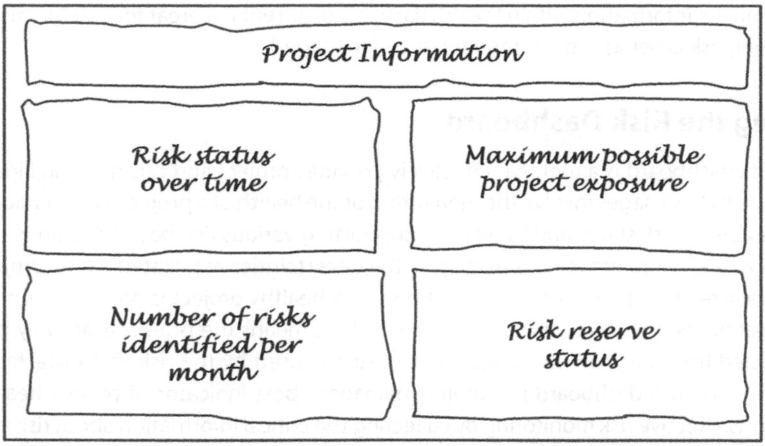

Figure 14.13: Example Risk Dashboard Layout Sketch

The goal is to design the dashboard so that it is comprehensive in content, appealing to the eye of the recipient, and meaningful in its presentation of the information and messages it conveys.

Create the Dashboard Graphics

The job of a dashboard is to display a large amount of critical information about a project in a very concise manner, and in a way that decisions can be made. The risk dashboard should therefore be simple and quick to interpret, with only a few key indicators much like the dashboard in an automobile. The best way to accomplish this is through graphical representation of the data.

Once the key measures and metrics have been chosen, you now have to think about how those measures will be graphically displayed. Trend information is best represented in a line graph, while comparison of information (such as number of risks that are active, inactive, or closed) is best represented in either bar charts or pie charts.

We recommend using common graph styles and keeping them simple. Remember, the primary consumer of the risk dashboard will be your top management team. You will want to spend your time with them talking about the status of risk on your project, not helping them interpret your graphs.

Populate the Dashboard

This final step in development of the dashboard involves collecting, synthesizing, and graphically representing current risk information about your project. While all previous steps are normally performed on a one-time occurrence at the beginning of a project, this step is continuous throughout the life of the project.

It is well known that the best risk management practice is that which is performed on a continuous, or at least periodic, basis due to the dynamic nature of project risk.

Keeping the information within the risk dashboard current is a great forcing function for ensuring risk is being actively monitored and managed.

Using the Risk Dashboard

The risk dashboard is a tool that effectively provides project and business intelligence. As such, its key usages involve the monitoring of the health of a project, communication of the current risk status and trends, and supporting various risk-based decisions.

How well a project team is managing the uncertainties associated with a project is generally referred to in terms of project health. A healthy project is one that is on track to meeting its objectives and business goals. That means the project is actively being protected from the potential negative exposure caused by the risk that uncertainties present. The risk dashboard is a project manager's best indicator of project health. It supports effective risk monitoring by collecting the critical information about risk status and presenting it in a fashion that conveys project health.

Since the dashboard contains critical information about the project health, it also becomes an important communication device. All project stakeholders have a vested interest in the outcome of a project for various reasons, so the stakeholders also have a vested interest in knowing the health of the project. The risk dashboard presents risk status and trend information to project stakeholders in a synthesized and summarized manner. If designed correctly, project stakeholders will be able to determine the health of a project within a few minutes.

The risk dashboard is also a good decision support tool. Good project decisions are those made from information concerning the current status of a project. Better decisions are those that also include information about the current and future risk associated with a project. The dashboard contains the critical information needed to support critical risk-based decisions.

During project planning, the risk dashboard can be used to establish the amount of risk reserve to include for a project. Risk reserve is the amount of time and funding needed above the estimate to reduce the risk of overruns. Risk events *will* occur and not all risk events will be predicted. Additionally, not all identified risks can be avoided or mitigated; therefore, risk reserve is a necessary aspect of every project plan.

As a project progresses through its life cycle, the dashboard is used to determine when risk reserve is needed to complete an aspect of a project (such as completing a milestone) and to track the amount of reserve remaining.

Benefits

All projects (small or large, simple or complex) will benefit from the use of the risk dashboard. The dashboard is an effective tool for the execution of the project that focuses on the state of project health. As such, it provides the rationale for risk-based project decisions that have to be made by the project manager as well as the organization's top managers.

The risk dashboard also provides value by serving as a forward-viewing and predictive project and business intelligence tool. This is especially true if information about

Table 14.5: Situational Use of Risk Management Tools

Situation	Risk Mgmt Plan	Risk ID Checklist	Risk Register	Risk Assess Matrix	Monte Carlo	Decision Tree	Risk Dashboard
Establishes risk management methodology for a project	✓						
Establishes risk response options	✓		✓				
Supports the identification of project risks		✓	✓				
Establishes risk categories		✓					
Establishes a project risk repository			✓				
Used to determine risk severity and prioritization				✓	✓		
Provides focus on the most critical risks			✓	✓	✓		✓
Explores many project scenarios and what-if analysis					✓		
Used to help determine if action and resources are required				✓	✓	✓	✓
Used to determine risk reserve				✓	✓		
Provides information on risk timing			✓				✓
Monitors risk status			✓				✓
Supports qualitative risk analysis				✓			
Supports quantitative risk analysis					✓		

potential risk exposure, risk severity trend, and risk reserve consumption is included in the dashboard.

Finally, the risk dashboard facilitates ongoing risk monitoring and management. It is an unfortunate and too common practice that project risk managagement begins and ends with risk identification and assessment activities. The real value comes from the ongoing monitoring and continuous management of risk as new uncertainties arise during the course of a project.

CHOOSING YOUR RISK MANAGEMENT TOOLS

The tools presented in this chapter are designed for various project risk management situations. Matching the tools to their most appropriate usage is sometimes a bit confusing. To help in this effort, the following table lists various risk management situations and identifies which tools are geared for each situation. Consider Table 14.5 as a starting point, and create your own custom risk management tools of choice to fit your particular project management style.

References

1. Martinelli, Russ, and Jim Waddell. "Managing Program Risk." *Project Management World Today*, September–October 2004.
2. Bostrom, A., Steven P. French, and Sara J. Gotlieb. *Risk Assessment Modeling and Decision Support* (Berlin, Germany: Springer Publishing, 2008).
3. Gilb, T. *Competitive Engineering: A Handbook for Systems Engineering, Requirements Engineering and Software Engineering* (Oxford, England: Butterworth-Heinmann, 2005).
4. Martinelli, Russ, James Waddell, and Tim Rahschulte. *Program Management for Improved Business Results*, 2nd ed. (Hoboken, NJ: John Wiley & Sons, 2014).
5. Project Management Institute. *A Guide to the Project Management Body of Knowledge*, 5th ed. (Drexell Hill, PA: Project Management Institute, 2013).
6. Wideman, M. *Project and Program Risk Management* (Newton Square, PA: Project Management Institute, 1992).
7. Couillard, J. "The Role of Project Risk in Determining Project Management Approach." *Project Management Journal* 26 (4): 3–15, 1995.
8. Sunstein, C. R., and Reid Mastie. *Wiser: Getting Beyond Group Think to Make Groups Smarter* (Boston, MA: Harvard Business Review Press, 2014).
9. Couillard, 1995.
10. Duckert, G. H. *Practicing Enterprise Risk Management: A Business Process Approach* (Hoboken, NJ: John Wiley & Sons, 2010).
11. Wideman, 1992.

12. Graves, R. "Open and Closed: The Monte Carlo Model." *PM Network* 15 (2): 48–52, 2001.
13. Graves, R. "Qualitative Risk Assessment." *PM Network* 14 (10): 61–66, 2000.
14. Project Management Institute, 2013.
15. Person, Ron. *Balanced Scorecards and Operational Dashboards with Microsoft Excel* (Hoboken, NJ: John Wiley & Sons, 2009).
16. Project Management Institute, 2013.
17. Ibid.
18. Wideman, 1992.
19. Project Management Institute, 2013.
20. Smith, P. G., and G. M. Merritt. *Proactive Risk Management* (New York, NY: Productivity Press, 2002).
21. Project Management Institute. *Practice Standard for Project Risk Management* (Drexell Hill, PA: Project Management Institute, 2009).
22. TSO. *Managing Successful Projects with PRINCE2.* (London, England: TSO, 2012).
23. Martinelli and Waddell, 2004.
24. Milosevic, Dragan Z., and P. Patanakul. "Standardization May Help Development Projects," *International Journal of Project Management* 23 (2), 2005, 62–66.
25. Smith and Merritt, 2002.
26. Meredith, Jack R., and Samuel J. Mantel Jr. *Project Management: A Managerial Approach*, 8th ed. (Hoboken, NJ: John Wiley & Sons, 2011).
27. Project Management Institute, 2013.
28. Vose, D. *Risk Analysis: A Quantitative Guide*, 2nd ed. (New York, NY: John Wiley & Sons, 2000).
29. Ibid.
30. Ibid.
31. Schuyler, J. *Risk and Decision Analysis in Projects*, 2nd ed. (Newton Square, PA: Project Management Institute, 2001).
32. Ibid.
33. Crouhy, M., Dan Galai, and Robert Mark. *The Essentials of Risk Management*, 2nd ed. (New York, NY: McGraw-Hill, 2014).
34. Cleland, D., and Lewis Ireland, . *Project Management: Strategic Design and Implementation*, 5th ed. (New York, NY: McGraw-Hill, 2006).
35. Eppen, G. D., C. P. Schmidt, Jeffrey H. Moore, et al. *Introductory Management Science*, 5th ed. (Upper Saddle, NJ: Prentice Hall, 1998).
36. Winston, W. L., and S. Christian Albright. *Practical Management Science*, 5th ed. (Independence, KY: Cengage Learning, 2011).
37. Eppen et al., 1998.
38. Schuyler, 2001.
39. Smith and Merritt, 2002.

15

INFLUENCING PROJECT STAKEHOLDERS

A stakeholder is commonly defined as anyone who has a vested interest in the outcome of a project. More importantly, for a project manager, a stakeholder is anyone who can influence, either positively or negatively, the outcome of their project. This includes people and groups of people both inside and outside the organization. Stakeholder management is a process with which a project manager can increase his or her acumen in managing the political, communication, and conflict resolution aspects of his or her project to ensure a positive outcome.[1]

Quite often, the accountability for success relies more and more on the abilities of the project manager—a person who has limited positional power within the organization, yet still owns the responsibility for project success. Empowerment must come from the building of strong relationships and successfully influencing key stakeholders.[2]

Stakeholders are many and varied on a project, and come to the table with a variety of expectations, opinions, perceptions, priorities, fears, and personal agendas that many times are in conflict with one another.[3] The challenge, therefore, is to find a way to *efficiently* manage this cast of characters in a way that does not become all consuming. Fundamental to efficiency is being able to identify and separate the highly influential stakeholders and then create and execute a stakeholder strategy that can strike a balance between their expectations and the realities of the project.[4]

The tools presented in this chapter are designed to help project managers develop an effective stakeholder strategy to positively influence the attitudes and behaviors of the stakeholders associated with their project. These tools are commonly found in the PM Toolboxes of practicing project managers, and in our view provide the broadest application to project types and sizes. To begin, successful stakeholder influence should always begin with a stakeholder management plan.

THE STAKEHOLDER MANAGEMENT PLAN

The stakeholder management plan is used to establish the framework and methodology the project team will use to identify, categorize, and analyze key project stakeholders in order to develop and implement an effective stakeholder strategy.

Developing a stakeholder management plan in the early stages of a project can help to align your key stakeholders to the goals of the project and to turn key stakeholders into advocates for project success. Having a stakeholder plan in one's PM Toolbox gives project managers an opportunity to proactively influence their stakeholders so when issues arise, the most influential stakeholders are positioned to assist instead of hinder.

Developing a Stakeholder Management Plan

Project managers spend a significant amount of time interfacing and working with the various stakeholders of their project. A stakeholder management plan helps to ensure that this significant effort is not happening in a haphazard manner. Rather, that their efforts are focused on the right set of stakeholders and that they are working toward achieving the desired outcomes.

A good stakeholder management plan is an effective tool to assist a project manager in creation of a focused stakeholder strategy by identifying the correct set of stakeholders, aligning the stakeholders to the goals of the project, and building good personal and professional relationships with key stakeholders.

Stakeholder Management Approach

The stakeholder management plan is a document that is developed as part of the broader project plan and establishes the framework for how a project manager will perform his or her stakeholder management duties. Included in this plan is a general description of the approach used to identify, analyze, and influence the project stakeholders.[5] It should include information about the following:

- *Stakeholder management process.* The plan should identify and describe how management of stakeholders will be performed on the project, including a description of the overall process.
- *Roles and responsibilities.* The overall responsibility for stakeholder management lies with the project manager, but he or she must leverage members of the project team and assign responsibility to them as appropriate. The plan should identify members of the team who have responsibility for managing stakeholders as well as which stakeholders they are responsible for establishing and maintaining a relationship with.
- *Tools.* The plan should describe the various tools that will be used to identify stakeholders and to analyze the stakeholders' interest, attitudes, power, and influence in order to develop and implement the stakeholder strategy.

Identifying Stakeholders

Effective stakeholder management begins with identification of all stakeholders associated with a project. The stakeholder management plan should discuss how the project stakeholders will be identified. It is important that the list of stakeholders should be comprehensive in order to cast a wide net over all players who may have a vested interest in the outcome of the project (see "Even a Salad Vendor May Be a Stakeholder").

Even a Salad Vendor May Be a Stakeholder

This is a bizarre example of what can happen if a comprehensive stakeholder analysis is not conducted on a project. The project involved a fast-track transfer of manufacturing technology from Europe to a Middle East country. Just after the beginning of the project, the project manager got a call from a vendor claiming that all roads leading to his office were blocked by groups of violent people. As a result, some important computer equipment couldn't be delivered. A quick check proved this call correct. The people were local butterhead salad farmers who were unhappy that the foreign contractor was not buying salad from them, rather, importing it from Europe.

The siege went on for days and the delivery delays were impacting the project schedule. Finally, the project manager figured out his mistake—one cannot ignore the relationship with local communities that have a big stake in the project. The salad farmers, in this case, were part of the local community and, therefore, a stakeholder in the project. Once the oversight was discovered, the contractor began buying local salad, and project deliveries proceeded without delays.

Tools such as a stakeholder map are common, and can be effective in helping a project manager identify the various stakeholders. The stakeholder list should include internal stakeholders such as top managers, project governance board members, department or functional managers, support personnel (accounting, quality, human resources), and the project team. External stakeholders should also be listed and may include contractors, vendors, regulatory bodies, service providers, and others. The objective of stakeholder identification is to include anyone who *might* have an influence on the outcome of the project.

Stakeholder idenfication also involves the categorization of stakeholders into the logical groups that they belong to. Such categories may include senior sponsors, executive decision makers, team members, and resource providers to name a few. It is important to realize that some stakeholders may belong to multiple groups. The intent of stakeholder categorization is to bring structure to the stakeholder list based on common interests in the project.

Analyzing Stakeholders

The stakeholder management plan should describe how the project team will analyze its list of identified stakeholders. Stakeholder analysis activities involve determining what type of influence each stakeholder has on the project, such as decision power, control of resources, or possession of critical knowledge, and their level of allegiance to the project. In other words, would the stakeholder prefer the project to succeed, not to succeed, or is he or she indifferent about the outcome of the project?

The stakeholder analysis process is geared toward identifying the subset of stakeholders who should be deemed as *primary stakeholders* along with the reasoning for why they are deemed as key to project success. Primary stakeholders are often those who have the most influence over the outcome of a project (positive or negative) or those who may be the most affected by the project.

Being able to determine the primary stakeholders is critical to the project manager as he or she will not have time to engage with all stakeholders. To effectively build relationships with the right stakeholders, the project manager must flush out who the primary stakeholders are, understand what level and type of influence they have, and determine how they feel about the project. These stakeholders will therefore require more communication and engagement to solicit and maintain their support.

Creating a Stakeholder Strategy

Stakeholder analysis is about sense making. This means understanding the significance of the information gained about the various project stakeholders. Significance of the information is then used to develop a strategy for engaging and managing the *right* set of stakeholders. The right stakeholders are those who have power and influence to affect the outcome of the project.

Most of the literature on stakeholder management immediately classifies the primary stakeholders as those with both high power and strong allegiance to the project. This is significantly inaccurate because it leaves out the most potentially dangerous stakeholders—those with high power and negative allegiance to the project. These people also need to be considered primary stakeholders. Figure 15.1 helps to illustrate

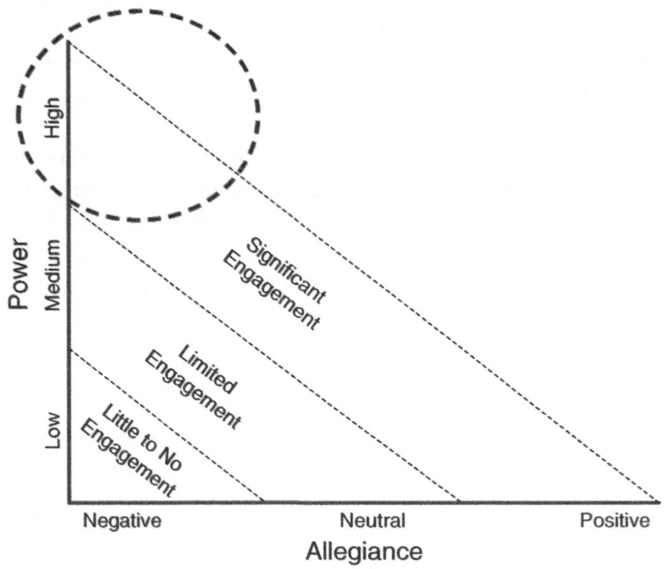

Figure 15.1: The Power/Allegiance Grid

why. Stakeholders with high power and negative allegiance fall in the category of significant engagement required.

If a project manager uses the power/allegiance grid to map their stakeholders, a core stakeholder strategy will begin to emerge. The strategy should consist of a communication and action plan for each of the primary stakeholders. It should also keep the project advocates engaged, describe how they can be used to influence others, and plan how to win over or neutralize the stakeholders who are not current advocates.

Using the Stakeholder Management Plan

Every project manager, whether they are managing a small, simple project or a large and complex project, will benefit from using a stakeholder management plan. All projects involve stakeholders who rarely come to the table completely aligned and fully supportive of a project.

Because of this, a stakeholder strategy needs to be developed and followed to ensure that the right stakeholders are engaged and effectively managed in a manner that helps to set the project up for success. The stakeholder management plan is a project manager's most critical tool for developing an effective stakeholder strategy.

The stakeholder management plan is normally created during the earliest stages of a project, either during project initiation or project planning. Many times, it is incorporated into a larger project plan that covers all aspects of planning a project effectively.

Although the use of the stakeholder management plan should become institutionalized on a project, development of the plan can vary greatly. For smaller projects, only a few hours may be required to conduct a planning session and develop a plan. This time proportionately rises as projects get bigger and more stakeholders become involved. Tens of hours may be necessary to devise a quality stakeholder management plan for a team in charge of a large and complex project.

Even though it is created early, the plan should be viewed and treated as a dynamic tool. This is due to the fact that project stakeholders are often transient, entering and exiting throughout the project cycle. As turnover in stakeholders occurs, it is likely that the stakeholder strategy will need to be modified. Therefore, the stakeholder management plan will need to be modified.

Benefits

Even though every project manager can benefit from using a stakeholder management plan, it brings the most value to project situations that have a larger number of stakeholders and where their opinions and viewpoints about the project are not completely aligned. In these project situations, the plan is a wonderful tool to assist project managers in the creation of structure out chaos.

The greatest benefit derived from the use of a stakeholder management plan is ability to align the various stakeholders to the goals of the project and bring competing agendas and opinions to the surface for debate and discussion.

THE STAKEHOLDER MAP

The stakeholder map is a powerful tool that visually displays all stakeholders who either influence a project or are affected by the project. It takes the outcome of the stakeholder identification brainstorming activities and begins to create order and structure around the individuals and groups associated with a project.

Developing a Stakeholder Map

Stakeholder mapping begins with identification of all possible stakeholders. The emphasis here is on *all* possible stakeholders. The intent is not to create a magic list of stakeholders, but to ensure that a stakeholder who may be later deemed influencially important is not forgotten. Stakeholder identification should be performed without screening, identifying everyone who may have an interest in your project today and in the future. The screening of stakeholders takes place in a following step. Table 15.1 identifies a number of stakeholder sources to consider.

Choose a Stakeholder Map Format

Stakeholder maps come in many different formats. An Internet search will yield many map formats for you to choose from or use as a reference design. However, some forethought is needed ahead of time to ensure you choose the right format for your needs. The key to choosing or designing a format which is right for you comes from

Table 15.1: Sample Stakeholder Sources

Internal Stakeholder Sources	
☑ Project team	☑ Project sponsor
☑ Organizational top managers	☑ Functional (department) managers
☑ Business strategists	☑ Governance bodies
☑ Program or portfolio managers	☑ Information technology group
☑ Internal clients	☑ Customer service group
☑ Human resources group	☑ Finance group
☑ Legal group	☑ Manufacturing group
☑ Procurement group	☑ Quality assurance group
☑ Sales and marketing group	☑ Operations group
External Stakeholder Sources	
☑ Clients or customers	☑ Competitors
☑ Community organizations	☑ Vendors
☑ Regulators	☑ Trade unions
☑ Media	☑ Lobbyists
☑ Users	☑ Venture capitalists

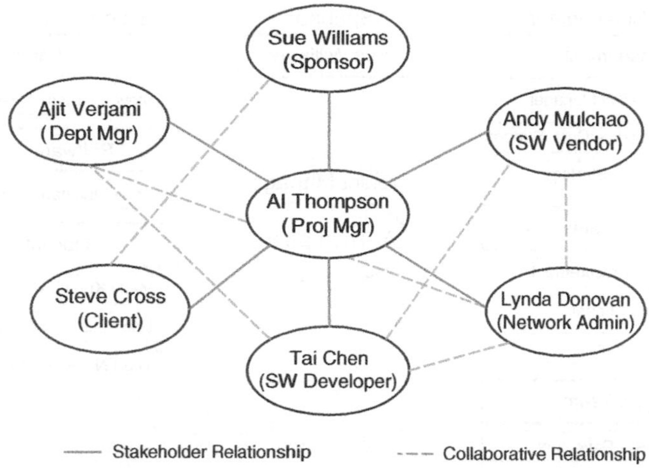

Figure 15.2: Example Stakeholder Map for Simple Projects

understanding how you want to structure your stakeholders, and what information you want to include in your map.

For small and simple projects, try not to over design or complicate what is needed. Figure 15.2 illustrates a simple stakeholder map format that is sufficient for these types of projects.

Here, a relatively small number of stakeholders are involved due to the simplicity of the project. Therefore, not a lot of structure is needed.

For projects that are larger and more complex, additional structure is needed to categorize the various stakeholders. Figure 15.3 shows an example of a stakeholder map that is quite useful for this level of stakeholder organization and structure.

For projects that involve a larger number of stakeholders, particularly external stakeholders, an even more structured stakeholder map is required. Using a table format (Figure 15.4) is effective for this type of project.

This format allows for a great deal of structure, as well as additional information that may be valuable to a project manager during stakeholder analysis activities. This might include information on who the project sponsor is, who the champions and supporters are, and who might be a disrupter.

Populating the Stakeholder Map

Once a map format has been chosen based upon the project type and information needs, the stakeholder map can now be populated with the details extracted from the stakeholder identification activities. If categories or groupings are used, begin by placing the stakeholder names into the map under the appropriate category. We suggest using names (for instance, Lynda Bratz) instead of titles (Director of Data Management). Stakeholder management is about building relationships with people, not titles or job responsibilities.

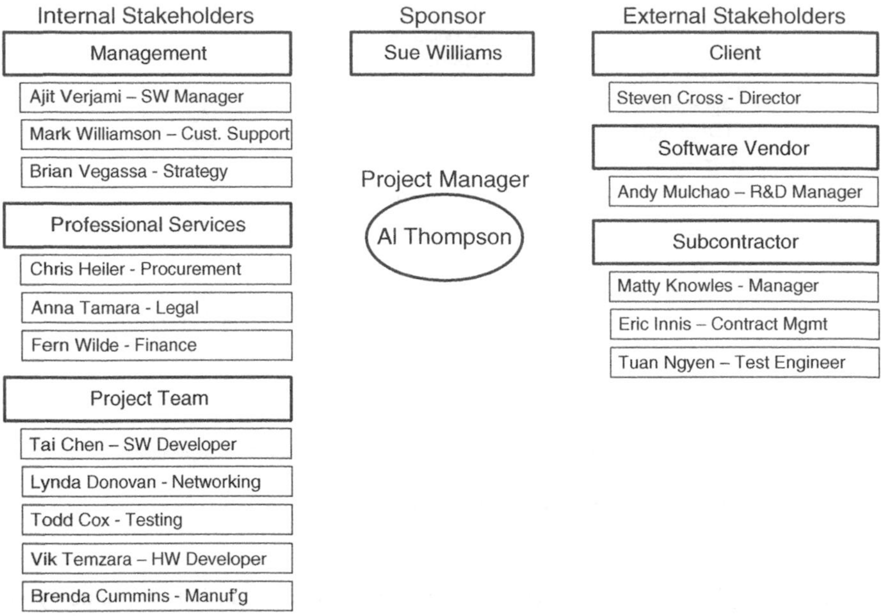

Figure 15.3: Example Stakeholder Map for Complex Projects

Category	Name	Function	Role
Management	Sue Williams	Product Line Manager	Sponsor
	Ajit Verjami	Software Dept Mgr	Resource provider
	Mark Williamson	Customer Support Mgr	Subject matter expert
	Brian Vigassa	Strategist	Champion / Consultant
Professional Services	Chris Heidler	Procurement	Subject matter expert
	Anna Tamara	Legal	Subject matter expert
	Fern Wilde	Contracting	Subject matter expert
Client	Steven Cross	Director	Client
Vendors	Andy Mulchao	Software Vendor	R&D Manager
	Ian Langly	Management Consultant	Subject matter expert
Subcontractor	Matty Knowles	Test Contractor	Manager
	Eric Innis	Test Contractor	Contract Manager
	Tuan Ngyen	Test Contractor	Test Engineer

Figure 15.4: Example Stakeholder Map in a Table Format

Once the stakeholder map is populated, it is a good idea to have someone else review the contents for missing stakeholders. Better to find the missing stakeholders early in the project than to have a senior manager remind you toward the end of a project that you have forgotten to consult a key stakeholder.

Using the Stakeholder Map

The stakeholder map should be created at the earliest stages of a project: ideally, during project initiation when the first stakeholders become aware of an idea that is intended to be converted to an executable project. Waiting too long to create, populate, and use the stakeholder map can have negative affects on a project, particularly if detractors are not identified and champions are not engaged.

The primary use of the stakeholder map is to help the project manager structure the various stakeholders into some form of logical order. This is particularly useful for projects with a large number of stakeholders. Earlier, we described the need to take a systematic approach to stakeholder engagement instead of an ad-hoc approach. The stakeholder map is used to create a systematic view of the various stakeholders.

The map can also be used to perform first level stakeholder analysis if information about stakeholder expectations or how they may be impacted by the project is included. The stakeholder map can be used to describe *why* a particular individual or group has a stake in a project.

The stakeholder map is a very dynamic tool, constantly changing, and for that reason it must be reviewed and updated on a periodic basis. Project stakeholders transition in and out, as do some of their viewpoints and opinions about the project. The stakeholder map is a project manager's best intelligence gathering tool for understanding the changes associated with the stakeholders surrounding their project.

Benefits

Even though every project manager can benefit from using a stakeholder map, it is most valuable in project situations that have a larger number of stakeholders and where their opinions and viewpoints about the project are not well aligned. In these project situations, the map is a wonderful tool to assist project managers in the creation of structure out of chaos.

The stakeholder map brings benefit when used as an intelligence gathering tool as well; particularly when it is used as a transition device from stakeholder identification to stakeholder analysis by documenting why each stakeholder has an interest in the outcome of a project. This first order of intelligence gathering helps a project manager determine if each of his or her stakeholders is in support of the project, is in opposition, or is indifferent about the outcome.

THE STAKEHOLDER ANALYSIS TABLE

Project stakeholders can be many, dispersed, and varied in their viewpoints and characteristics. It is important to do a good job in identifying your stakeholders, but stakeholder identification by itself has limited value for project managers. Developing a deeper understanding of the project stakeholder's interests, opinions, and viewpoints is the necessary next step in the stakeholder management process. This step is commonly referred to as stakeholder analysis.

The purpose of stakeholder analysis is to enable the project manager to identify the individuals and groups that must be interacted with in order to accomplish the project goals. Effective stakeholder interaction is supported by thorough stakeholder analysis activities that allow a project manager to develop a strategy to accomplish the following:

- Identify strategic interests that the various stakeholders have in the project to negotiate a *common* interest.
- Develop plans and tactics to effectively negotiate competing goals and interests between stakeholders.
- Secure active support from project champions (see "What Is a Project Champion?").
- Devise activities to either neutralize or prevent the negative actions of non-supporters.
- Allocate personal and expanded resources to engage with the key stakeholders.

What Is a Project Champion?

A project champion is an informal but important role on a project. Also called the project advocate, a champion is a key stakeholder who is a strong supporter of the project and continuously communicates the benefits and value of the project to the other stakeholders. The project champion often serves as a liaison to top management and is effective in communicating at all levels of the organization. He or she also helps the project manager navigate the political landscape and serves to remove obstacles that are or may prevent progress.[6]

Due to the champion's position and influence within an organization, he or she normally understands the concerns, goals, viewpoints, and opinions of the other stakeholders. For this reason, they are an important individual for project managers to work with during stakeholder analysis activities.

When looking for a project champion, search for individuals who have the following characteristics:

1. *Respected*. A project champion should be someone who is trusted and whose viewpoints and opinions are sought out and considered.
2. *Influential*. A project champion is someone who can prompt action on the part of other stakeholders without relying on his or her positional power.
3. *Strong communicator*. In order to effectively promote the value of the project and influence action, a project stakeholder has to be a strong communicator.
4. *Politically savvy*. A project champion should be knowledgeable about the formal and informal political landscape and effective at using their political capital to realign personal agendas to the goals of the project.

Many of the stakeholder analysis activities are about prioritizing the project stakeholders. A small subset of your stakeholders, commonly called your key or primary stakeholders, possess a significant amount of organizational influence to either advance your project or block its progress. Either way, these stakeholders have to be identified through a filtering and prioritization process.

Effective stakeholder analysis also helps a project manager identify key stakeholder relationships such as power and interest or influence and level of engagement.

There are a number of very effective tools that project managers use to analyze their stakeholders. Some of these tools are detailed in the sections that follow, beginning with the stakeholder analysis table.

Developing a Stakeholder Analysis Table

The stakeholder analysis table is a tool that is used for a number of stakeholder analysis purposes, the first of which is capturing information about stakeholder expectations and reservations. The table documents and communicates this information in a concise and simple manner as demonstrated in the example stakeholder analysis table (Table 15.2).

As stated previously, a primary outcome of stakeholder analysis is gaining alignment of common stakeholder interests and negotiating competing interests. Introductory meetings should be used to understand each stakeholder's understanding, interpretation, and opinion of the project for which they are a stakeholder (see "Key Questions to Ask Your Stakeholders"). Additionally, the meeting can be used to collect information on what each stakeholder believes their role is on the project, what decisions they

Table 15.2: Example Stakeholder Analysis Table

Name	Assumed Role	Expectations	Reservations	Provides to Project	Decision Control
Sue Williams	Sponsor	Project meets all business and execution goals	Firm's abiility to develop the new capability	Direction and decisions	Gate approvals
Ajit Verjami	Department manager	No expectations for this project	Believe another project provides a better solution	Resources	Resource allocation decisions
Steven Cross	Client	Project completed under budget	Timeline is very aggressive	Funding	Gate approvals
Danielle Carvalho	Subject expert	Project will stay on schedule and complete on time	Already committed to two other projects	Time and expertise	None
....	....	....	....	....	....

believe they are charged with making, and what they are investing in the project (e.g., resources, budget, or time). This information should then be entered into the stakeholder analysis table for future reference and use.

Key Questions to Ask Your Stakeholders

Understanding the needs, expectations, and potential issues of project stakeholders is crucial to the success of any project. The work you do to familiarize yourself with your stakeholders and learn as much as possible about them may make or break the outcome of your project.

The following questions can serve as a starting point for collecting the necessary information you will need to fully understand and analyze your stakeholders:

1. Who will receive the output of your project?
2. What direct benefit will the recipient gain from your project output?
3. Who will be working with you to execute your project?
4. Who are the subject matter experts that you should consult?
5. Who is the client or customer?
6. What resources can I use for the project?
7. Who are the primary detractors of the project?
8. Who controls the project budget?
9. How does the stakeholder define project success?
10. What are the business goals driving the need for the project?
11. How does the stakeholder see their role in the project?
12. What worries the stakeholder about the project?
13. Do any stakeholders have conflicts of interest?
14. What information does the stakeholder need about the project?
15. Is there an alternative project that would be a better investment?

Using the Stakeholder Analysis Table

The stakeholder analysis table, like all stakeholder management tools, should first be created and implemented during the earliest stage of a project, normally project initiation. First, information about who the project stakeholders are is directly transferred from the stakeholder map to the stakeholder analysis table prior to conducting your first introductory discovery meetings with identified stakeholders.

The table then becomes the single source of information about the project stakeholders from which further stakeholder analysis activities are performed. To begin the analysis, you should use the information within the stakeholder analysis table to focus on four key pieces of information:

1. Determination of who the key stakeholders of the project are.
2. Assessment of stakeholder alignment to the project goals, scope of work, and project outcomes.

3. Identification of potential conflicts of opinion between stakeholders.
4. Identification of the project advocates and nonsupporters.

It is far better to understand if your stakeholders agree with the project goals that you have identified and documented early in the project cycle instead of waiting until it is too late to make adjustments to the goals or to work with stakeholders to get alignment. The project goals in turn affect the scope of work to be performed and the outcomes of that work. Stakeholder redirection is a primary cause of scope changes, so the earlier the changes are executed, the less impact there will be to the project (see Chapter 8).

Unless you have a very small number of project stakeholders, it is unrealistic to believe that there will not be conflicting opinions and interests between the stakeholders. This is common, and for this reason it needs to be uncovered and understood in order to develop a strategy and tactics to broker this conflict of interest. Left unchecked, conflicting interests and opinions are a leading cause for project failures.[7]

The stakeholder analysis table is also used to provide additional structure to your list of stakeholders and to pinpoint the most important stakeholders who you will need to work with to insure project success. To do this, critical information has to be collected and analyzed to understand which stakeholders will be most involved, what their roles will be, and how they will contribute to the project.

Project managers cannot afford to be naïve in thinking that all stakeholders want their project to succeed. Unfortunately, this is not always the case. Because of this, the stakeholder analysis table can be used to begin determining who your supporters and nonsupporters are. Specifically, pay close attention to the information that is contained in the "reservations" portion of the table. Typically, nonsupporters will bring their reservations to light during the discovery conversations with the project manager. Normally, an additional level of analysis is needed to determine if stakeholders who are not advocates may in fact become blockers to project success. However, at this stage project managers should be able to separate their advocates from the nonsupporters.

Benefits

In order to develop a strategy to influence and manage the stakeholders on a project, you must first develop an understanding of the individuals and groups you consider your stakeholders. The stakeholder analysis table creates value by providing a means to get beyond stakeholder identification and helps project managers learn about stakeholder expectations, concerns, issues, and ideas.

In the process, the table can also help to verify whether alignment among stakeholders exists concerning the project goals. If not, it enables early adjustment of the goals and associated scope of work necessary to accomplish the goals.

It is not uncommon for conflicting expectations and viewpoints to emerge between stakeholders. By using the stakeholder analysis table to document the expectations, reservations, and desired outcomes from the project for each stakeholder, the conflicting opinions rise to the surface where they can be addressed and compromises brokered. Also through this process of listening to the project stakeholders, potential issues and risks will emerge. Again, bringing them to the surface so they can be managed.

The stakeholder analysis table helps project managers pinpoint the key project stakeholders and helps them focus their stakeholder engagement strategy and tactics on the individuals and groups that have the most influence on a project. This is beneficial because without a focused strategy, a project manager can spend an inordinate amount to time and energy managing project stakeholders to the detriment of the other aspects of their project management duties.

Finally, one of the greatest benefits of the stakeholder analysis table is an intangible benefit. The use of the table becomes a means for establishing and building a relationship with the project stakeholders. To populate the table with sufficiently adequate information requires the project manager to interface with and listen to the individuals involved. The act of listening is a powerful tool for building relationships.

THE STAKEHOLDER EVALUATION MATRIX

The stakeholder evaluation matrix, also commonly called the stakeholder grid, is used to perform a deeper level of stakeholder analysis by assessing important relationships between stakeholder characteristics, the most common relationship being a stakeholder's organizational influence and his or her allegiance to a project. This relationship is demonstrated in Figure 15.5.

The primary purpose of the stakeholder evaluation matrix is to assess the political landscape surrounding the project which is commonly created by the various stakeholders involved (see "I Never Wanted to Be a Politician"). The matrix helps to create political awareness. This awareness is critically important as a project manager goes about creating a stakeholder strategy. Specifically, it allows he or she to assess whether they have the political clout to influence the power players, or if they have to enlist the help of other stakeholders.

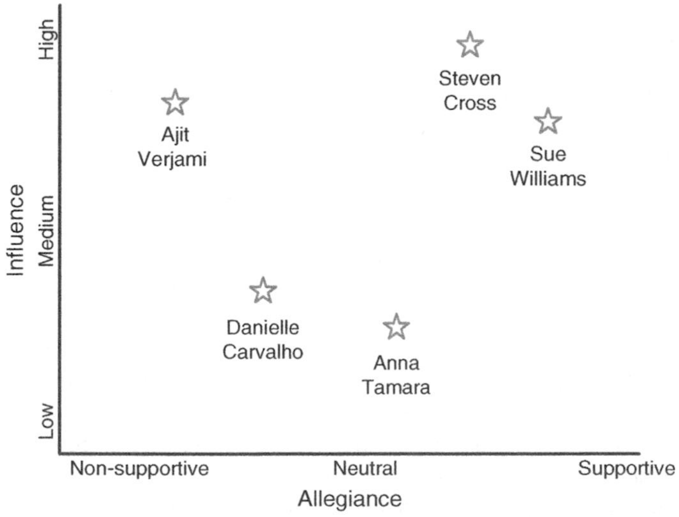

Figure 15.5: Example Stakeholder Evaluation Matrix

I Never Wanted to Be a Politician

Joseph Algere is a bit perplexed and more than a little frustrated. Having just returned to his office from a meeting with the senior manager of his division, he wonders why several key project team decisions concerning the features of the new patient scheduling system have been overturned by senior management—especially since the decisions were approved just two weeks ago during the project plan approval meeting. Additionally, Joseph found out that his senior system architect had been reassigned to a research project last week by the functional manager of the architecture group. Now that the new system has fewer advanced features, the schedule has been extended, and the development cost has increased, he has to revise the project business case to determine if the project is still viable from a business perspective.

What Joseph is not aware of is that the senior leaders of the firm have not fully empowered their project managers, a situation that is not uncommon. This leaves the project managers in a tenuous position—fully responsible for the business success of the project, but limited authority over the resources and decisions to make it happen. As a consequence, project managers can become immersed in the organization's political posturing, where personal and functional agendas and priorities take precedence over a project's business results. Due to the nature of organizational behavior, a project manager must accept that they are at times in the political hot seat, and learn the nuances of managing their stakeholders accordingly.

Organizational politics originate when individuals drive their personal agendas and priorities at the expense of a cohesive corporate agenda. The basis of organizational politics is really twofold: one's desire to advance within the firm, and one's quest for power (usually in the form of controlling decisions and resources). In a silo-structured organization, politics originate when the functional managers develop objectives that support the specific long-term goals of their department. But what happens if these functional objectives do not support, or worse yet, are in direct conflict with the strategic goals of the company?

Unfortunately, no organization structure or business model will eliminate political behavior within a firm, as it's a natural part of the dynamics involved when people work together. Therefore, both senior managers and project managers must take action to prevent an organization's projects from becoming negatively impacted by corporate politics.

As Patrick Lencioni states in this book titled *Silos, Politics, and Turf Wars,* "If there is a place where the blame for silos and politics belongs, it is at the top of the organization. Every departmental silo can ultimately be traced back to the leaders of those departments who have failed to understand the interdependencies that exist among [the departments]".[8] For project management to be effective, senior leaders must make the tough decision to shift the balance of power within the organization from the functional managers to the project managers, and continue to ensure that the functional organizations are fully supporting the projects.

The project manager must also actively manage the politics surrounding his or her project. It is important that the project manager possess both a

(*continued*)

keen understanding of the organization, and the political savvy to build strong relationships to effectively leverage and influence the power base of the company. Company politics are a natural part of any organization, and the project manager should understand that politics is a behavioral aspect of project management that he or she must contend with in order to succeed. The most effective method for playing the political game is to leverage the project stakeholders and powerful members of one's network who can help achieve the project objectives.

Developing a Stakeholder Evaluation Matrix

An effective evaluation matrix can only be created once the stakeholders have been identified and the key stakeholders pinpointed. This requires critical input from the information contained in the stakeholder map and the stakeholder analysis table.

Document the Primary Stakeholders

At this stage of stakeholder analysis, project managers should be focused on those stakeholders who have the most impact on a project, particularly those who can impact critical decisions or resource assignments: a reminder again to also include stakeholders who have the ability to negatively impact the project.

The number of primary stakeholders will typically relate to the size of a project. Larger projects normally require broader organizational and partnership involvement; and, subsequently, more stakeholders will be involved.

The primary stakeholders should be documented in the stakeholder analysis table and carried over to the stakeholder evaluation matrix.

Evaluate the Level of Stakeholder Influence

For each primary stakeholder, evaluate the level of influence you believe they have on the outcome of the project. Focus on those individuals who can affect critical project decisions (such as phase gate approvals), those who are providing resources to the project, those who control all or portions of the project budget, and those who have expressed reservations about the project.

Plot the stakeholders with repect to their level of influence on the stakeholder evaluation matrix by assessing whether you believe a stakeholder has a high, medium, or low level of influence on the project. Do this by asking the question, "To what degree can the stakeholder exert influence to change the course of the project?"

Evaluate the Level of Allegiance

The second vector of evaluation focuses on each stakeholder's attitude toward a project, and how their attitude will affect their level of support. Your evaluation is concentrated on trying to determine if a stakeholder is supportive, nonsupportive, or essentially indifferent toward your project, and then developing a further determination on the *level* of support or nonsupport.

Plot the Stakeholders on the Matrix

Each stakeholder can now be given a location on the matrix based upon your assessed level of influence and allegiance to the project. When completed, project managers have a graphical representation of their stakeholders from which they can then analyze the political landscape of the project.

Using the Stakeholder Evaluation Matrix

The primary use of the evaluation matrix is to analyze the current landscape of primary stakeholder's influence and attitude toward a project, and to begin formulating a strategy for how you can move the stakeholders to a more favorable position on the matrix if needed.

Begin the evaluation process by adding a grid to the matrix as shown in Figure 15.6. We recommend beginning with a simple 2 × 2 grid that will divide the matrix into four quadrants. If additional resolution is needed, a 3 × 3 matrix can be used.

The stakeholders in quadrant A possess high influence and low, or indifferent, allegiance to the project. These stakeholders constitute the highest risk to the project, and will require significant focus and attention.

The stakeholders in quadrant B possess high influence and positive allegiance to the project. These individuals or groups are potential project champions. Work will be required to ensure the continued support of these stakeholders. These individuals or groups may be instrumental in helping to move other stakeholders, especially those in quadrant A, to more favorable positions on the matrix.

The stakeholders in quadrant C are generally nonsupportive of a project but possess little influence within the organization to affect the project outcome. These stakeholders

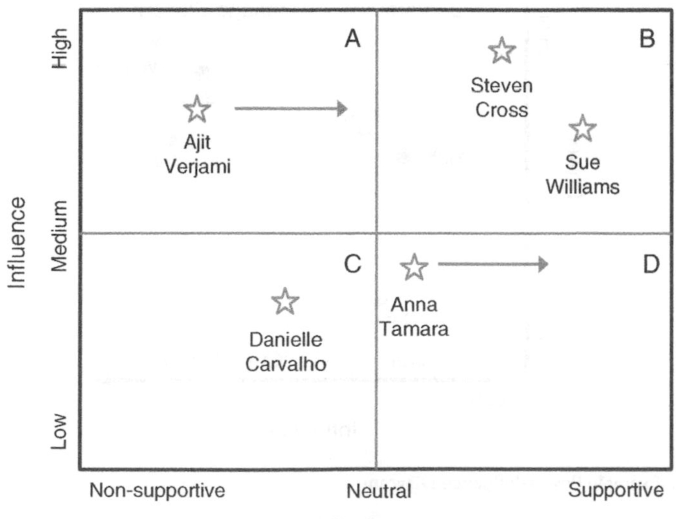

Figure 15.6: Example Stakeholder Evaluation Matrix with Grid

cannot be ignored, however, as organizational and political landscapes are dynamic. It is not uncommon for a stakeholder originally placed in quadrant C to gain influence (e.g., through a promotion to a position of power) throughout the course of a project. If they remain nonsupportive, once they gain additional influence, an additional risk to the project is introduced.

The stakeholders in quadrant D are supportive of the project but again possess little influence over the project outcomes at the current time. Use these stakeholders as part of your project coalition of support by keeping them informed and engaged. Often, these stakeholders possess specific knowledge and viewpoints that can be leveraged to support specific decisions or actions you are trying to influence.

Plot Stakeholder Engagement Movement

Using the information gained by segmenting the evaluation matrix into a grid, physically draw the various stakeholder movements that will be necessary to secure positive support for the project (Figure 15.7). This provides a visual representation of the *future state* of the political landscape. This future state provides clarity and focus for the development of your stakeholder strategy and engagement actions.

Be realistic when plotting your planned stakeholder movement. For instance, it is not realistic to believe you can move a highly influential stakeholder with strong aversion to your project to a position of strong support. In reality, the best scenario may be that you will be able to move this stakeholder to a neutral position. This in itself is a wonderful risk mitigation strategy.

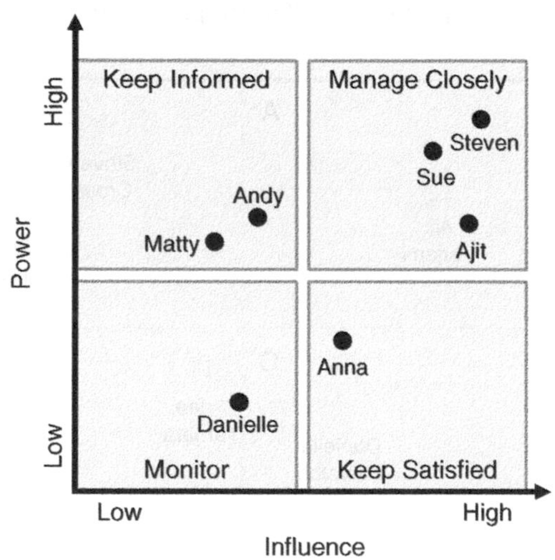

Figure 15.7: Example Power-Influence Diagram

Variations

The power-influence diagram is a variation of the stakeholder evaluation matrix that focuses directly on each key stakeholder's power within the organization and the level of influence they have on a project's outcome (see "The Difference between Power and Influence"). Figure 15.7 illustrates an example of a typical power-influence diagram.

Like the stakeholder evaluation matrix, the power-influence diagram is divided into a 2 × 2 matrix, which creates four distinct quadrants that represent stakeholder engagement strategies. Determining a stakeholder's level of power and influence is a subjective exercise, so trying to determine *exactly* where stakeholders are located on the matrix is not a value-add activity. Making a determination of which quadrant they reside in does add value, however, as each quadrant defines the type of engagement action to pursue.

The quadrant defined as high power (positional and decision authority) and high influence (ability to change the project outcome) requires the project manager to manage these stakeholders closely (the engagement action).

The quadrant defined as high power and low influence requires engagement actions that focus on keeping these stakeholders informed of the project progress.

The quadrant defined as low power and high influence includes stakeholders who do not necessarily possess a high degree of organizational power, but can influence a project's outcome. This requires stakeholder actions that ensure these stakeholders remain satisfied that their interests are being served by the project.

Finally, the quadrant defined by low power and low influence requires little to no stakeholder engagement action. Good risk management practices require that these stakeholders should periodically be monitored, however. Power and influence within an organization can change over the course of a project which may require more in-depth engagement if a stakeholder gains additional organizational power or influence during the life of a project.

The Difference between Power and Influence

If you are aiming to plan and execute a successful project, you will need to be successful at stakeholder management. Beyond stakeholder identification and ranking, successful project managers know that they must discern the power and influence each stakeholder has among others interested in the outcome of the project. Often, however, even among some of the most seasoned managers, power and influence gets confused. The following characterizes and distinguishes the two.

(continued)

Power	Influence
Positional	Personal
Formal	Informal
Assumed	Earned
Control	Consent
Management	Leadership
Command	Request
Dictate	Dialogue

Stakeholder management is needed because the fact is, some people have more power than others and some have more influence than others, and the project manager must understand and then be able to leverage each stakeholder's power and influence for project success. A stakeholder can be an enabler or constrainer to any project's success.

It is important to realize that neither power nor influence is bad, although either can certainly be abused. Power is an ability someone has to direct the behaviors and actions of others. Similar but different is influence, which is a personal capacity that one has to affect the behavior of someone.

To manage both stakeholder power and influence to increase the probability of project success, project managers must connect people, share information, build commitment, and make decisions. To do so, determine the power or influence each stakeholder has with others and match the power or influence as a lever to those engaged and impacted by the project.

Another variation of both the stakeholder evaluation matrix and the power-influence diagram is a tool called the powergram (see Figure 15.8). The powergram focuses exclusively on the power dynamics between project stakeholders.[9] It is especially useful in project situations where achieving alignment between stakeholders is a challenge caused primarily by differences in opinion or historical tensions between individuals.

The powergram displays the power structure between your key stakeholders, which can provide insights into where stakeholder tension exists and how other stakeholders can be used to positively influence and change the power dynamics.

Developing a set of power relationship rules is necessary to construct an effective powergram. We suggest the use of the following rules:

1. The size of a circle denotes the amount of organizational power a stakeholder possesses. The larger the circle, the more power they possess.
2. A double line represents a *positive* relationship between two stakeholders.
3. A single line with a strike through it represents a *negative* relationship between two stakeholders.
4. The shorter the line connecting two stakeholders, the stronger the relationship between them.

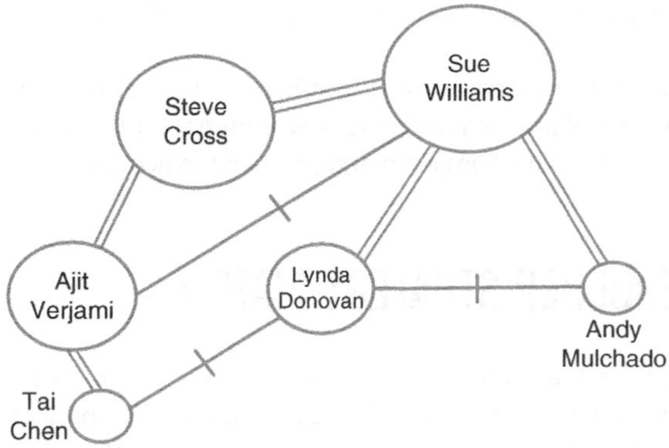

Figure 15.8: Example Powergram

If you have created a graphical stakeholder map (such as the one shown in Figure 15.2), the map can be used as the basis of the powergram relationship structure. This is the approach we used to create the powergram in Figure 15.8.

The following power dynamics can be derived from the powergram. Sue Williams (project sponsor) and Steven Cross (client) are the most powerful stakeholders, with Ajit Verjami (department manager) relatively powerful as well. The powergram also shows that there is a strained relationship between Williams and Verjami, but the positive relationship between Williams and Cross, and between Cross and Verjami, can be leveraged if needed. Finally, Lynda Donovan (network administrator) has negative relationships with both Chen and Mulchao. As a result, the positive relationships between Williams and both Mulchao and Donovan may be needed to gain alignment between these three project stakeholders.

The powergram is a very simple tool to create and does not require a high degree of precision and accuracy to be useful in helping project managers to make sense of the power dynamics surrounding their project.

Benefits

The value the stakeholder evaluation matrix and its variations provide to a project manager first and foremost is in the enablement of a deeper understanding of the project's key stakeholders: specifically, an understanding of the stakeholder's attitudes toward the project evaluated against their organizational power and ability to influence critical decisions and project outcomes. This creates important knowledge concerning which stakeholders create risk to the project and which stakeholders can be leveraged to advance the project. This knowledge is of course crucial for formulating an effective stakeholder strategy.

The simple and graphical format of the tool also offers benefits by providing a visual representation of the stakeholder evaluation results. This visual representation makes it immediately clear which stakeholders will require the most focus and attention. When stakeholder future state annotations are included in the graphs, the project manager gains a quick understanding of a number of engagement actions needed.

If used to evaluate positional power and influence, the matrix also serves to increase a project manager's political acumen. It is now understood that political acumen is a necessary competency that all project managers need to acquire and hone over time.

THE STAKEHOLDER STRATEGY MATRIX

The primary goal of stakeholder engagement activities is to establish stakeholder alignment to the strategic goals, intended business benefits, project objectives, and success criteria of a project. It involves putting action to the stakeholder strategy through the building of professional relationships to influence for project advocacy and to monitor stakeholder actions, words, and decisions. The project manager should maintain focus on the primary stakeholders to prevent the job of stakeholder engagement from being all-consuming at the detriment of other critical aspects of managing a project.

Stakeholder engagement activities will test the courage of project managers who need to be brave and bold when faced with building relationships with stakeholders who are not a fan of the project, or may be professionally threatened by the outcome of the project. Do not follow the human tendency to avoid these stakeholders. Rather, seek them out and, most importantly, listen to what they have to say. Only by listening can a project manager begin to find middle ground to use as a means to positively influence.

The stakeholder strategy matrix is a valuable tool to ensure that you are focused on the right stakeholders and that you thoughtfully develop a stakeholder strategy to gain maximum stakeholder alignment and to effectively maneuver the political landscape surrounding your project.

Developing a Stakeholder Strategy Matrix

All the tools in this chapter to this point have been focused on identifying the project stakeholders and analyzing their interests from multiple perspectives. This has all been focused on collecting information and developing knowledge about the stakeholders in order to create an effective strategy for engagement and communication. The stakeholder strategy matrix (shown in Figure 15.9) is used as a repository that explicitly illustrates the information and knowledge gained from a project manager's stakeholder identification and analysis activities. The information contained in the tool will become a strong basis for development of a stakeholder strategy.

Construct the Matrix

The first step in developing the stakeholder strategy matrix is creating a structure that demonstrates the information you choose to evaluate in the creation of your stakeholder strategy.

Stakeholder Name	Relationship to Project	Level of Influence	Allegiance to Project	Resources					
				People	Money	Material	Facilities	Knowledge	Decisions
Sue Williams	Sponsor	H	⇧		X			X	X
Ajit Verjami	Functional Manager	H	⇩	X		X			X
Steven Cross	Software Vendor	L	▭	X			X	X	
Lynda Donovan	Core Team Member	M	⇧					X	

Influence Key: L = Little or no influence
M = Some influence
H = Considerable influence

Allegiance Key: ⇧ Positive allegiance
▭ Neutral allegiance
⇩ Negative allegiance

Figure 15.9: Example Stakeholder Strategy Matrix

Using the matrix in Figure 15.9 as an example, we have chosen to structure the matrix around the following key information we chose to develop our stakeholder strategy:

- Names of the primary stakeholders.
- Stakeholder's relationship to the project.
- Stakeholder's level of influence or organizational power.
- Our perception of their allegiance to the project.
- What project resources they control.

Keep in mind that this is only an example. It is up to each project manager to determine what information they need in order to develop the stakeholder strategy for their project, and structure their stakeholder strategy matrix accordingly.

Populate the Matrix

Now is the time when a project manager gets to use the information and knowledge they have gained from the stakeholder analysis activities by populating the matrix with the information desired. In our example, this information includes the names of each primary stakeholder, a description of their relationship to the project, a determination of whether they possess a high, medium, or low level of organizational influence, and if we believe they are an advocate for our project, a potential detractor, or are basically indifferent. The last piece of information needed in our example is an evaluation of whether each stakeholder has control over project budget, people, decisions, and so on.

When the stakeholder strategy matrix is fully populated, a project manager is in possession of highly valuable information about their set of primary stakeholders with which they can use to create a stakeholder strategy.

Using the Stakeholder Strategy Matrix

Now, let's be honest. Widely sharing the information shown in Figure 15.9 may be a career-limiting move. Detailed stakeholder analysis information should be considered confidential and tightly controlled by the project manager. Why do we say this? Because true and honest stakeholder analysis brings to the surface the realities of corporate politics and unique biases that have to be managed properly; usually that means actively, but delicately.

Many project managers have told us that they do the analysis in their head, which is good. However, when they hold the information in their heads, it tends not to be used consistently or correctly, which is to sort the primary stakeholders from the remainder of the stakeholders and to develop a concise stakeholder strategy. Performing the additional step of documenting the stakeholder analysis information, either physically or electronically, helps to maintain history and clarity in developing the stakeholder strategy.

As stated earlier, the strategy should focus on keeping the project's advocates engaged, describe how they can be used to influence others, and plan how to win over or neutralize the stakeholders who are not current advocates.

The stakeholder strategy should consider the following aspects:

- What is wanted or needed from each stakeholder?
- What is the message that needs to be delivered to each stakeholder?
- What is the best method and frequency of engagement and communication with each stakeholder?
- Does the strategy reflect the interests and concerns of each stakeholder?

See "Being a PM Would Be Easy if It Weren't for People" for an opportunity to practice developing a stakeholder strategy. Keep in mind, there is no right right strategy to employ. However, some will be more effective than others.

Being a PM Would Be Easy if It Weren't for People

The Department of Social Services for a large government agency has initiated a project to improve community response to its "Illnesses Awareness Services" campaign. As the project manager who was recently hired to lead the project, and who is new to the agency, Soo Lee has decided to complete a stakeholder analysis and strategy for the project.

From personal discussions with her manager, the department manager, and others within the department, this is what she currently knows about some of the key people associated with her project:

- The sponsor, Sue Williams, is a strong advocate for the project and seems to be quite influential within the agency.

- The functional managers, Steve Sainz, Diane Best, and Ajit Verjami, all supply resources to the project. They are also on the project's management review committee. All but Ajit Verjami are aligned to your project; Ajit believes an alternative solution and project should have been selected and funded.
- Lee's project core team consists of three members: Tai Chen, Jose Hernandez, and Lynda Donovan. She knows little about her core team at this point, but she has some concern about Lynda Donovan. She has learned that Lynda had applied for the project manager position for this project but was the alternate choice at the end of the selection process.
- The agency has hired an outside software development company to create a new web site and discussion portal for the Department of Social Services. Initial development and go-live of the web site will be managed as part of the project. Multiple attempts have been made to contact the project manager for the software vendor, but he has not responded to Lee's messages.

Using the information above as well as the stakeholder analysis information contained in Figure 15.10, can you develop a four-point stakeholder engagement strategy?

Benefits

The stakeholder strategy matrix provides a number of important benefits for a project manager. First, the tool helps a project manager pinpoint the right stakeholders to engage—those that can have the greatest affect (either positive or negative) on the project outcome.

The matrix also helps a project manager focus on the right actions that will provide the most results: for example, using a project champion to influence another stakeholder who may feel negatively or indifferent toward the project.

By using the tool to make sense of the organization's influence and power structure, a project manager can develop strategies to leverage that structure to the benefits of the project.

Finally, the stakeholder strategy matrix can help project managers establish a strong coalition of support that can be called on when their projects run into difficulties and challenges.

CHOOSING YOUR STAKEHOLDER MANAGEMENT TOOLS

The tools presented in this chapter are designed for various project stakeholder management activities and situations. Matching the tools to their most appropriate usage is sometimes a bit confusing. To help in this effort, Table 15.3 lists various stakeholder management situations and identifies which tools are geared for each situation. Consider this table as a starting point, and create your own custom project situation analysis and tools of choice to fit your particular project management style.

Table 15.3: Situational Use of Stakeholder Influencing Tools

Situation	Stakeholder Mgmt Plan	Stakeholder Map	Stakeholder Analysis Table	Stakeholder Evaluation Matrix	Stakeholder Strategy Matrix
Establishes stakeholder management methodology for a project	✓				
Identify project stakeholders		✓			
Creates stakeholder structure		✓			
Tracking stakeholder changes		✓			
Determine stakeholders interest in the project			✓		
Determine how stakeholders will be impacted by the project			✓		
Identify primary stakeholders			✓	✓	✓
Identify advocates and detractors			✓	✓	✓
Assess stakeholder alignment to project goals			✓	✓	
Determine opposing stakeholder opinion			✓		
Used to devise activities to prevent negative action of nonsupporters			✓		✓
Assess stakeholder characteristics				✓	✓
Used to develop a stakeholder management strategy			✓	✓	✓

References

1. Pinto, Jeffrey K. *Power and Politics in Project Management* (Newtown Square, PA: Project Management Institute, 1998).
2. Roeder, Tres. *Managing Project Stakeholders: Building a Foundation to Achieve Project Goals* (Hoboken, NJ: John Wiley & Sons, 2013).
3. Bourne, Lynda. *Stakeholder Relationship Management* (Burlington, VT: Gower, 2009).
4. Pinto, 1998.
5. Singh, Harjit. *Mastering Project Human Resource Management: Effectively Organize and Communicate with All Project Stakeholders* (New York, NY: Pearson FT Press, 2014).
6. Eskerod, Pernille. *Project Stakeholder Management: Fundamentals of Project Management* (Burlington, VT: Gower, 2013).
7. *IBM Systems Magazine*. www.ibmsystemsmag.com/power/Systems-Management/Workload-Management/project_pitfalls/?page=2. Accessed April 2015.
8. Lencioni, Patrick. *Silos, Politics and Turf Wars, a Leadership Fable about Destroying the Barriers that Turn Colleagues into Competitors*. (San Francisco, CA: Jossey-Bass, 2006).
9. Andler, Nicolai. *Tools for Project Management, Workshops and Consulting: A Must-Have Compendium of Essential Tools and Techniques* (Erlangen, Germany: Publicis Publishing, 2011).

FINAL THOUGHTS ON THE PM TOOLBOX

Like all tools, project management tools are not meant to be a panacea that will automatically make someone a better project manager. Becoming a better project manager is accomplished through experience and continuous building of skills and competencies. Rather, tools are designed to help a project manager become more effective and efficient in performing the key project management practices and in creating the various project outcomes and deliverables required of project-based work.

For example, if we need to define the project scope, we can use the scope statement, statement of work, and work breakdown structure (WBS) or program work breakdown structure (PWBS) for this purpose. Similarly, if we are faced with having to accelerate our project schedule, we can enlist the help of the schedule crashing technique or perhaps the critical chain schedule tool. There are tools available for nearly every project situation known.

The primary point we have put forward in this book is that there are two main approaches to choosing and using your project management tools: (1) select your tools one at a time when a situation and need arise, or (2) create a PM Toolbox systematically that provides a higher level of utility.

Creating a PM Toolbox requires a higher level of thinking and design. The toolbox approach, if implemented to its full value, aims to establish and maintain alignment between business strategy and project execution, between strategic goals and project deliverables, and between the work of the top managers and the project managers of the company.

The demands of the modern business world have changed and are continuing to change as companies manage more and more projects for all aspects of their business. Such changes necessitate a new best practice, which this book was created to enable—creating and using a PM Toolbox approach to enable a higher level of project management practice and performance.

INDEX

D

Printed and bound by CPI Group (UK) Ltd, Croydon, CR0 4YY
27/10/2024
14580315-0004